PLANT BREEDING INSTITUTE
CAMBRIDGE
2 8 SEP 1966

Received ....
Origin ....
Classmark .... EA 36 ....
Current No ....

The Library
John Innes Centre
Colney Lane
Norwich NR4 7UH

Class No: 58.032 SLA

Standard No: A6078

D1807137
978 94 010 3595 5

003972

CZECHOSLOVAK ACADEMY OF SCIENCES

Scientific Editor
Dr. Bohdan Slavík

Scientific Adviser
Prof. Dr. Miroslav Penka

Foreign Language Editor
Dr. A. E. Ridesová

Graphic Design by Josef Prchal

CZECHOSLOVAK ACADEMY OF SCIENCES

WATER STRESS IN PLANTS

Proceedings of a Symposium
held in Prague,
September 30 — October 4, 1963

Edited by
Bohdan Slavík

Dr. W. JUNK PUBLISHERS — THE HAGUE — 1965

© Nakladatelství Československé akademie věd 1965

Printed in Czechoslovakia

Symposium
on Water Stress in Plants

CONTENTS

The study of water stress is one of the most interesting subjects in the investigation of water relations in plants. From the theoretical point of view it is concerned with investigating the mechanisms of the distribution and movement of water in the plant organism and the way in which physiological processes are influenced by water deficiency. From the practical point of view, water deficiency is a major factor limiting plant production.

It has been progressively shown that water deficiency is not by far only a factor in plant life in dry climates, that obvious wilting is not the first warning sign of water deficiency and that moderate water stress, caused by temporary negative water balance during the day, affects physiological activity and decreases production in the ecological conditions of the temperate zone. In addition, even general water deficiency is not today confined to arid or semi-arid zones and to the absolutely dry season of the year. The tremendous consumption of water in our civilization has become today, even in the temperate zone, an important competitor with the plant cover. The study of water relations from the aspect of water stress is, therefore, important both theoretically and practically.

I assume, therefore, that it was useful, important and interesting to meet in a symposium on water stress in plants and to discuss, as far as possible, in detail problems which are obviously among the main, whose solution would help plant physiology in increasing and improving plant production. Since "Multum non multa" is still valid, I think that the discussion of one theme from many angles, which formed the agenda of our symposium, was useful and contributed to increasing knowledge in this branch of science.

I believe, therefore, that these "Proceedings", containing papers contributed to the symposium and authorized recordings of the discussions, gives a valuable survey of the problems of water deficiency and of the results of its study. It shows what we know of the problem today and where there are white spots on the map of our present knowledge. Attempts to synthesize our present conceptions will be found as well as endeavours to outline the best directions for further work.

I am very glad that the first symposium on water stress was held in Prague, and if this established a new tradition, if new friendships were made and old acquaintances renewed, then, in addition to its scientific purpose, the human aims of our symposium will also have been fulfilled.

Allow me to thank all who made our symposium possible, in the first place the Academy of Sciences and Academician C. Blattný, Director of the Institute of Experimental Botany, Academician B. Němec, the founder of Czechoslovak plant physiology, and all who by their presence contributed to the work and results of the symposium, and finally all who by their untiring work created the conditions for the symposium to be held. I would further like to express my sincere thanks to Dr. J. Čatský, C.Sc. and Dr. Z. Šesták, C.Sc. for their very willing help in the preparation of these Proceedings, Dr. A. E. Ridesová for linguistic co-operation and also many others for further help.

Bohdan Slavík

Backa, P.	Výzkumný ústav závlahového hospodárstva, Karloveská cesta 9, Bratislava, Czechoslovakia.
Bezuidenhout, S. J. P. K.	University of Stellenbosch, S. Africa; present address: Botanisches Inst. d. Techn. Hochschule, Rossdörferstr. 140, Darmstadt, Federal Republic Germany.
Blattný, C.	Inst. exp. Botany, Czechoslovak Acad. Sci., Na Karlovce 1, Praha 6-Dejvice, Czechoslovakia.
Bozhenko, V. P.	Bot. Inst., Acad. Sci. U.S.S.R., ul. Popova 2, Leningrad P-22, U.S.S.R.
Čatský, J.	Inst. exp. Bot., Czechoslovak Acad. Sci., Flemingovo 2, Praha 6-Dejvice, Czechoslovakia.
Duffek, J.	Vysoká škola zemědělská, katedra zahradnictví, Technická 3, Praha 6-Dejvice, Czechoslovakia.
Gloser, J.	Přírodovědecká fakulta University J. E. Purkyně, katedra fysiologie a genetiky rostlin, Kotlářská 2, Brno, Czechoslovakia.
Gusev, N. A.	Biol. Inst., Kazan State Univ., ul. Lobachevskogo 2/31, Kazan, U.S.S.R.
Hodaňová, D.	Inst. exp. Bot., Czechoslovak Acad. Sci., Flemingovo 2, Praha 6-Dejvice, Czechoslovakia.
Hygen, G.	Bot. Inst., Vollebekk, Norway.
Jarvis, M. S.	Inst. physiol. Bot., Univ. Uppsala, Sweden.
Jarvis, P. G.	Inst. physiol. Bot., Univ. Uppsala, Sweden.
Ješko, T.	Chem. Inst., Slovak Acad. Sci., Bratislava, Czechoslovakia.
Jiroušek, J. M.	Komise pro vodní hospodářství ČSAV, Valdštejnská 14, Praha 1-Malá Strana, Czechoslovakia.
Kousalová, I.	Výzkumný ústav obilnářský, Koperníkova 1, Kroměříž, Czechoslovakia.
Kozinka, V.	Inst. Bot., Slovak Acad. Sci., Dept. Plant. Physiol., Dúbravská 26, Bratislava IX, Czechoslovakia.
Kraslová, J. A.	Vysoká škola zemědělská, Technická 3, Praha 6-Dejvice, Czechoslovakia.
Krejcarová, M.	Inst. exp. Bot., Czechoslovak Acad. Sci., Flemingovo 2, Praha 6-Dejvice, Czechoslovakia.
Larcher, W.	Inst. Bot., Sternwartestr. 15, Innsbruck, Austria.
Lebedev, G. V.	Inst. Plant Physiol., Acad. Sci. U.S.S.R., Leninsky prospekt 33, Moskva V-71, U.S.S.R.
Lerch, G.	Inst. Bot., Pädagogische Hochschule, Potsdam, present address: Inst. f. Kulturpflanzenforschung, Abt. f. ökol. Pflanzenphysiol., Am Drachenberg 1, Potsdam-Sanssouci, GDR.

Majerník, O.   Inst. Bot., Slovak Acad. Sci., Dúbravská 26, Bratislava IX, Czechoslovakia.

Makkink, G. F.   Inst. voor biologisch en scheikundig onderzoek van landbouwgewassen, Bornsesteeg 65/67, Wageningen, Netherlands.

Meinl, G.   Inst. f. Pflanzenzüchtung der DAL, Gross-Lüsewitz/Rostock, German Democratic Republic.

Müller-Stoll, W.R.   Inst. f. Kulturpflanzenforschung, Abt. f. ökol. Pflanzenphysiol., Am Drachenberg 1, Potsdam-Sanssouci, German Democratic Republic.

Nátr, L.   Výzkumný ústav obilnářský, Koperníkova 1, Kroměříž, Czechoslovakia.

Nečas, J.   Výzkumný ústav bramborářský, Valečov, pošta Okrouhlice u Havl. Brodu, Czechoslovakia, present address: Laboratory for exp. Algology, Inst. Microbiol., Czechoslovak Acad. Sci., Třeboň.

Němec, B.   Na Václavce 7, Praha-Smíchov, Czechoslovakia.

Penka, M.   Vysoká škola zemědělská, lesnická fak., Lesnická 37, Brno, Czechoslovakia.

Polster, H.   Inst. f. Forstwissenschaften, Graupa bei Pirna/Sa., German Democratic Republic.

Pospíšilová-Šanderová, J.   Inst. exp. Bot., Czechoslovak Acad. Sci., Flemingovo 2, Praha 6-Dejvice, Czechoslovakia.

Prát, S.   Inst. Plant. Physiol. and Soil Biol., Charles Univ., Viničná 5, Praha 2-Nové Město, Czechoslovakia.

Rufelt, H.   Inst. physiol. Bot., Univ. Uppsala, Sweden.

Rychnovská, M.   Bot. Inst., Czechoslovak Acad. Sci., Stará 18, Brno, Czechoslovakia.

Rypáček, V.   Katedra fysiologie rostlin, Přírod. fak. University J. E. Purkyně, Brno, Czechoslovakia.

Scheumann, W.   Inst. f. Forstwissenschaften, Graupa bei Pirna/Sa., German Democratic Republic.

Segeťa, V.   Ústřední výzkumný ústav rostlinné výroby, Praha 6-Ruzyně 507, Czechoslovakia.

Sekerka, V.   Inst. Plant. Physiol., Fac. Sci., J. A. Komenský Univ., Odborárske 12, Bratislava, Czechoslovakia.

Šesták, Z.   Inst. exp. Bot., Czechoslovak Acad. Sci., Flemingovo 2, Praha 6-Dejvice, Czechoslovakia.

Slavík, B.   Inst. exp. Botany, Czechoslovak Acad. Sci., Flemingovo 2, Praha 6-Dejvice, Czechoslovakia.

Slavíková, J.   Inst. Bot., Charles Univ., Benátská 2, Praha 2-Nové Město, Czechoslovakia.

Smetánková, M.   Ústř. výzkumný ústav rostlinné výroby, Praha 6-Ruzyně 507, Czechoslovakia.

Solarová, J.   Inst. exp. Bot., Czechoslovak Acad. Sci., Flemingovo 2, Praha 6-Dejvice, Czechoslovakia.

Strebeyko, P.   Inst. Plant Physiol., Univ. Warsaw, Krakowskie Przedmieście 26/28, Warszawa 64, Poland.

Sveshnikova, V. M.   Inst. Bot., Acad. Sci. U.S.S.R., ul. Popova 2, Leningrad P-22, U.S.S.R.

Tichá-Möllerová, I.   Inst. exp. Bot., Czechoslovak Acad. Sci., Flemingovo 2, Praha 6-Dejvice, Czechoslovakia.

Úlehla, J.   Výzkumná stanice základní agrotechniky a hnojení, Pohořelice u Brna, Czechoslovakia.

Václavík, J.	Inst. exp. Bot., Czechoslovak Acad. Sci., Flemingovo 2, Praha 6-Dejvice, Czechoslovakia.
Vartapetyan, B. B.	Inst. Plant Physiol., Acad. Sci. U.S.S.R., Leninský prospekt 33, Moskva V-71, U.S.S.R.
Vicherková, M.	Katedra fysiologie rostlin, Přírod. fak. University J. E. Purkyně, Kotlářská 2, Brno, Czechoslovakia.
Vinš, B.	Výzkumný ústav lesního hospodářství a myslivosti, Strnady 167, Zbraslav n. Vlt. II, Czechoslovakia.
Visser, W.C.	Inst. Land and Water Management Res., P. O. Box 35, Wageningen, Netherlands.
Weatherley, P. E.	Dept. Bot., The Univ., St. Machar Drive, Old Aberdeen, Great Britain.
Zemánek, M.	Výzkumný ústav obilnářský, Koperníkova 1, Kroměříž, Czechoslovakia.
Zwicker, R.	Inst. f. Acker- und Pflanzenbau, Karl-Marx-Univ., Johannisallee 19, Leipzig 05, German Democratic Republic.

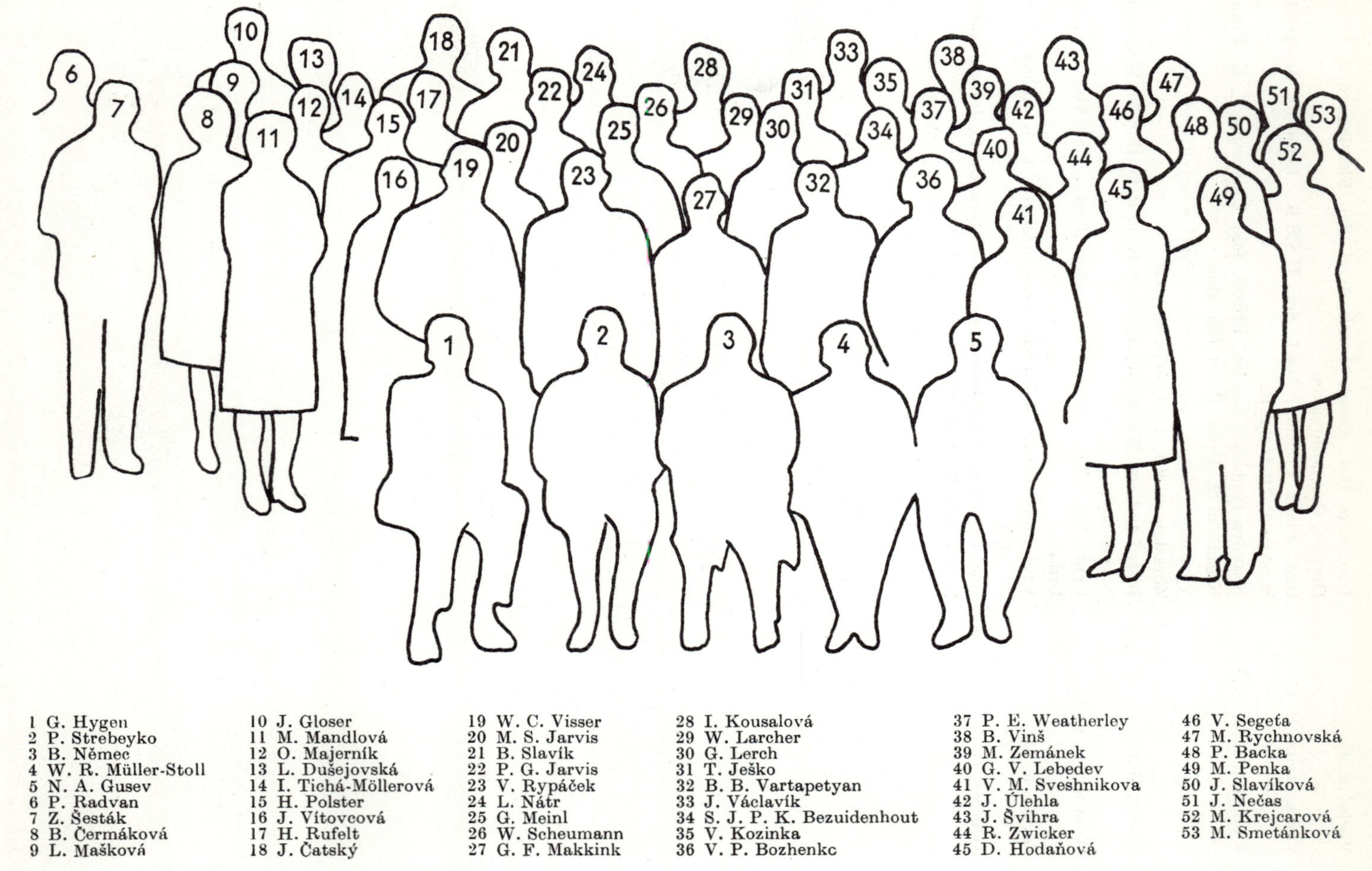

1 G. Hygen	10 J. Gloser	19 W. C. Visser	28 I. Kousalová	37 P. E. Weatherley	46 V. Segeťa
2 P. Strebeyko	11 M. Mandlová	20 M. S. Jarvis	29 W. Larcher	38 B. Vinš	47 M. Rychnovská
3 B. Němec	12 O. Majerník	21 B. Slavík	30 G. Lerch	39 M. Zemánek	48 P. Backa
4 W. R. Müller-Stoll	13 L. Dušejovská	22 P. G. Jarvis	31 T. Ješko	40 G. V. Lebedev	49 M. Penka
5 N. A. Gusev	14 I. Tichá-Möllerová	23 V. Rypáček	32 B. B. Vartapetyan	41 V. M. Sveshnikova	50 J. Slavíková
6 P. Radvan	15 H. Polster	24 L. Nátr	33 J. Václavík	42 J. Úlehla	51 J. Nečas
7 Z. Šesták	16 J. Vítovcová	25 G. Meinl	34 S. J. P. K. Bezuidenhout	43 J. Švihra	52 M. Krejcarová
8 B. Čermáková	17 H. Rufelt	26 W. Scheumann	35 V. Kozinka	44 R. Zwicker	53 M. Smetánková
9 L. Mašková	18 J. Čatský	27 G. F. Makkink	36 V. P. Bozhenkc	45 D. Hodaňová	

OPENING ADDRESS AT THE RECEPTION MEETING

by Academician *Ctibor Blattný*

Ladies and Gentlemen,

As Director of the Institute of Experimental Botany of the Czechoslovak Academy of Sciences, I wish to welcome all of you who have come to take part in this Symposium on Water Stress in Plants. We considered it important to call this meeting, because plants, as the producers of living matter, are the most important factor in the nutrition of mankind. It is therefore necessary in our branch of science more than in any other, to probe more closely into plant physiology using all the modern methods that are at our disposal. We have not at all underestimated the empirical work of the unknown millions who discovered the fundamental facts about the life of crop plants and of course we have never underrated the scientific efforts of workers everywhere in the past hundreds of decades and years. We know that science can only advance on the foundation created by earlier scientists. This holds fully in the plant physiology. Since it is quite clear that the limiting factor in the production of living substance by plants are the water relations, it is necessary to devote primary attention to this problem. That this interest is general, is shown by the presence here of the representatives of so many nations. This also ensures that our proceedings will have a high scientific level, for which aim I wish you every success. I cannot finish, however, without wishing you a very pleasant time in our country, every happiness, friendly cooperation and also good weather.

OPENING ADDRESS OF THE FIRST WORKING SESSION

by Academician *Bohumil Němec*

Ladies and Gentlemen,

I have the privilege of welcoming you to our country and I do it with great pleasure. You came to participate in a symposium on water and plants and the questions you will discuss are extremely important for biology, for the prosperity of plants and, therefore, for the existence and the prosperity of mankind. The great poet Pindaros said "Ἄριστον τὸ ὕδωρ and a Roman poet has written: "Ab aqua principium aquacium omnia plena." The great philosopher Thales taught that water is the principle of everything and that we all revert to water and not to dust and ashes. Alchemists already came to the conclusion that "corpora non agunt nisi liquida" and van Helmont, as a result of his water cultures, came to the conclusion that plants form their bodies from water. Sachs began his work on water cultures here in Prague and he wrote in his Vorlesungen über die Pflanzenphysiologie that water uptake, transpiration and the assimilation of carbon dioxide are the main processes determining the form and structure of green plants. His first lecture at the Congress of German Physicians and Naturalists in 1856 dealt with the uptake, conduction and evaporation of water. In our country Prof. Farský, nearly 100 years ago, measured the lengths of roots of plants growing in water cultures and was one of the first botanists who tried to solve the question of the significance of mineral elements in plants by water cultures. I will mention the great work done by my pupil Prof. Iljin on stomatal movements and the conditions of this process. Meanwhile physiological research found how important are water relations for plant growth, respiration and the photosynthetic process. Because production of plant substance and especially of organic substances is so important for the life of all heterotrophic organisms and for mankind especially, the study of the relation between water and green plants has become very important and is one of the first problems of experimental botany. I hope your symposium will have good results and I wish to your work the best success.

I open the symposium.

WATER STRESS AND WATER UPTAKE

Thursday, October 1

Chairmen: *B. Němec* and *N. A. Gusev*
Secretary: *Z. Šesták*

THE PROBLEM OF WATER OUTFLOW FROM ROOTS

W. R. MÜLLER-STOLL

Institut für Kulturpflanzenforschung, Abteilung für Ökologische Pflanzenphysiologie (Potsdam),
Deutsche Akademie der Wissenschaften zu Berlin, Deutsche Demokratische Republik

In general, it is assumed that plant roots are merely organs of water absorption and that water is conducted from root to shoot. It is not taken into consideration that roots may be able to exude water and then water streams in the reverse direction from the leaves to the roots. Recently, it has been stated, though not without opposition, that roots of spermatophytes should be able to effect active water secretion excellently, provided the leaves absorb water from a moist atmosphere. This would be a process analogous to guttation, but in the opposite direction. In several papers Breazeale et al. (1950, 1951, 1953) described experiments made at the Arizona Agricultural Experimental Station which in their opinion demonstrated active water secretion from the roots. They assert to have demonstrated that liquid water vigorously leaks out from roots if the shoots are kept in an artificial fog atmosphere, and even, though less strongly, if the shoots are located in air of high water vapour saturation. They claim to have observed this water outflow from root system hanging freely in an empty glass flask as well as from roots dipping into liquid, and also, from plants rooted in soil. In the latter case soil moisture in the rhizosphere is said to have increased due to active water secretion up to field capacity and even above it (Breazeale et al. 1950). If plants with their roots were sealed in flasks, the exuded water gathered within the containers, or increased the volume of the solution into which the roots were dipping. The authors believed they had proved that the pressure induced within the plants owing to water uptake into leaves was greater than that after water absorption by roots. The leaf was said to be a much more efficient organ for water absorption from the liquid as well as from the vapour phase. It was reported that tomato shoots absorbed water from the vapour phase down to 75—85% relative humidity (Breazeale et al. 1955). The pressure supposedly exerted by tomato roots secreting water was measured by means of manometer tubes inserted into the bottom of flasks in which plants were sealed with roots dipping into solution. After 3 to 4 days the water column in the manometers had reached 100 cm., after 14 days 140 cm. (Breazeale and McGeorge 1953).

Several authors objected to Breazeale's notions. Haines (1952) repeated

Breazeale's experiments with tomato plants inset in empty glass flasks. Breazeale's results could be reproduced, but, according to Haines, should be explained in a different way. Water accumulation within the flasks occurred to a greater extent only under significant temperature fluctuations. At widely constant temperature only minimum moisture was exuded from the roots. Air pressure differences between the sealed interior of the flasks and the environment might also have caused water exudation by way of the inter-cellular spaces. If those sources of mistake were excluded Haines could not find any essential water accumulation within the containers.

Wiersma and Veihmeyer (1954) checked Breazeale's experiments with root systems of tomatoes dipping into liquid water. They found that the rise in pressure within the flasks measurable by manometer tube was not due to active water secretion but to the metabolic activity of microorganisms (fermentation processes). Pressure would increase to the same extent if detopped root systems were sealed into flasks, water absorption from above being thus impossible.

Finally Höhn (1954) tested Breazeale's results, and, also, could not find any signs of active water secretion from roots. Höhn worked only in vapour-saturated atmosphere (without fogging) and left it undecided whether Breazeale's results might be true with shoots in direct contact with liquid water. Höhn failed to get any considerable water uptake by leaves from the vapour phase nor water exudation from roots comparable to Breazeale's statements. He had to lower the relative humidity within the flasks to 85% (by means of sulphuric acid) in order to force as great an evaporation from the roots as indicated by Breazeale's papers. Plants treated in that way died after a few days, since their shoots were unable to supply the roots deprived of water with sufficient moisture from the surrounding atmosphere. Gessner (1956) compiling the literature on this problem came to a strange opinion. He compared Breazeale's notions with those of Duvdevani* (Israel) on absorption of dew by leaves which process is also said to cause a lively water exudation from roots (experiments with empty glass containers and moistening the above-ground plant organs by artificial dew), and came to the following conclusion (l.c., p. 233): "The corresponding results of Breazeale and Duvdevani do not leave any doubt that the plant is able not only to cover its own water deficit by absorbing dew water but also to secrete water into the soil of the root horizon. Thus the plant is building up a water supply outside its body from which its water need is covered when, on hot days, transpiration loss will increase again".

Although Gessner confined the problem of efficient water absorption by leaves to liquid water, he clung to the conception of active water secretion

---

* Literature cited by Gessner (1956, p. 244) as to Duvdevani's paper (Amer. J. Bot. 1955) is wrong; we could not find where this paper mentioned by Gessner has been published.

from roots. This positive review to Breazeale's opinion by Gessner induced us to make some further investigations on this opinion.

Methods

Before the experiments tomato and maize plants were grown in a greenhouse in soil or in nutrient solution up to the required size.

1. According to Breazeale's method plants grown in nutrient solution were sealed with their root systems in empty Erlenmeyer flasks by rubber stoppers and a paraffin-vaseline mixture. A curved capillary tube inserted into the stopper secured air pressure balance. Before charging the flasks were empty and dry or contained water levels of various heights so that root tips could dip in or not.

2. The plants were cultivated in water proof plastic pots filled with different soil material (pure sand, sandy garden mould, sandy loam). Before the experiments the cultures were dried up to a different degree, e.g. nearly to the wilting point, and soil moisture was measured by taking soil samples. After that the soil surface in the pots was sealed over with a paraffin-vaseline mixture; capillary tubes for air pressure balance were also inserted. During the actual treatment the experimental sets according to (1) or (2) were kept under a plastic screen or a big thick-walled glass cover (infection globe) and sprayed with dust of tap water produced by a bouncing-nozzle (system Tegtmeier). The constant spray under the screen secured that temperature could change merely within a very small scale. In the experiments with plants rooted in soil no temperature fluctuation within the pots was observed. We took strict care that no spray water could penetrate the seal on the top of the containers.

Results

Roots never exuded liquid water actively. As far as moisture accumulated in the empty flasks it was condensation water which evaporated from the root surface and condensed on the glass walls. If root tips dipped into water some moisture was lost from the flasks inspite of spraying. This confirms that plants with sufficient water supply in the root horizon transpire water even in a moisture-saturated atmosphere. If the flasks contained some water, without the root tips dipping in, water dust treatment did not cause any essential change of weight. If plants were placed with their roots into dried flasks a different amount of condensed water accumulated on the glass walls. We could confirm that this water condensation increased with increasing temperature fluctuations and remained very small under widely constant temperature (Haines 1952). After a 24 hours' water dust treatment of tomato plants sealed in Erlenmayer flasks according to Breazeale's method, the containers (without plants) showed the following gain or loss of weight:

Arrangement of test	Mean value	Range
Root tips dipping into water	—0·79 g.	—0·24 ... —1·39 g.
Shallow water layer, roots not dipping in	+0·03 g.	0·00 ... +0·08 g.
Root systems placed into dry flasks	+0·70 g.	+0·36 ... +1·08 g.

Further experiments were carried out with young tomato plants grown in nutrient solution but exhibiting water stress after drying up to a different degree. The plants inserted into dry, empty glass flasks showed the following water content of their shoots and root system after a 48 hours' spraying or water uptake from a vapour-saturated atmosphere (water content in % of fresh weight, mean values of 5 tests each):

| Treatment for 48 hours | Plants at beginning of experiment | | | |
| | not yet wilted | | wilted | |
	shoot	root	shoot	root
Plants treated with water dust	90·6	85·8	89·9	84·8
Plants kept in vapour-saturated atmosphere	89·7	81·7	86·2	80·5
Control plants (in normal laboratory air)	87·8	78·0	82·0	71·5

It is well known that plants suffering from water deficit will be saturated by aerial water sources. The table above shows that the saturation process does not restrict itself to the shoots, and water deficit in the roots is compensated, too. If plants are not fully water-saturated, in fact an inverse water flow may occur, by water uptake into the leaves. It will be efficient, however, merely up to saturation of roots, at best up to a compensation of evaporation loss from the roots. In all experiments considered small amounts of droplets condensed on the glass walls independent of the kind of treatment, even from the roots of the more or less wilted control plants heavily affected by water stress. Water uptake from vapour-saturated atmosphere was significantly less than under spraying which restored full water saturation of shoots

ivity in the vessel elements prevents greater differences of water saturation in root tissues. Moisture gradients existing between different ranges of the rhizosphere, therefore, cannot be reflected adequately by the saturation deficit in different parts of a root system, since under usual conditions slow water movement in soil is contrasted by the much more efficient water transport within the roots. It must be supposed that by this means water might even move from moister to drier parts of the rhizosphere. Water absorbed from moister parts of the rhizosphere may, after having been transported within the roots, be withdrawn again by drier parts of the soil. In this way a rapid though mostly not very extensive water translocation in soil is possible by means of the conducting capacity of root tissues as several authors could demonstrate (literature cited by Bormann 1957).

Thorough studies with *Fraxinus* roots recently carried out by Slavíková (1963) ascertained that the suction force of the soil and the suction force of the roots embedded in it are not always in a dynamic equilibrium as hitherto mostly has been assumed; under certain conditions the suction pressure in different parts of a root system may be nearly equal notwithstanding that the moisture content in the surrounding soil varies in a very wide range. Under favourable moisture conditions the suction force in the roots increases normally in centripetal direction whereas in soil strongly dried up Slavíková (1964) observed an inverted suction pressure gradient turned to the peripheral parts of the root system. Under such circumstances a centrifugal water translocation within the roots takes place (Hunter and Kelley 1946), and dry soil may even suck out water from the roots.

There has been much discussion on the ecological importance of water uptake by above-ground organs from dew, fog etc. Comparison between the amount of dew water absorbed by aerial tissues and the amount of water necessary for maintaining the whole water balance showed that dew water uptake cannot be of great ecological importance, except in specific cases. On the other hand, it could be proved repeatedly, e.g. by Steubing (1955), that even under Central European conditions dew favours growth and productivity of plants. Our evidence that water absorbed by aerial organs can reach even the soil of the rhizosphere may perhaps be of some value in explaining the obviously greater efficiency of dew water absorption as hitherto computed. At slight nightly rainfall, too, which does not wet a dry soil down to the root region, moisture could be transported by a reverse water flow into the rhizosphere.

Water absorption by aerial organs is said to be of particular efficiency in more or less arid regions (e.g. Went 1953, Duvdevani 1953). In areas with big diurnal temperature fluctuations, expecially when dew point is frequently crossed, liquid water will condense upon plants in a large scale. Then aerial water supply might be of decisive importance for plant life, because it can be expected that moisture is transported in an efficient amount from shoots through roots into soil. Such conditions may exist, in particular, in semiarid to arid highlands with steppe or semidesert vegetation or in mountainous districts in continental climates (Sveshnikova 1956). For ecological judgment, however, the mere fact of a certain process is less important than its quantity. For the present we are far from any knowledge on quantitative relations necessary for an ecological understanding of the questions mentioned here.

Acknowledgement is expressed to Mrs. Helga Thust who carried out most of experiments reported here. Furthermore, the author is indebted to his coworker Dr. G. Lerch for the final experiments and preliminary arrangement of the present paper.

Summary

1. The statement of Breazeale et al. (1950, 1951, 1953) that plant roots are able to affect an excellently active guttation-like water secretion has been criticized by several authors. A positive review of Breazeale's ideas by Gessner (1956) stimulated us into further investigations in this problem. We found, however, no signs of any active water exudation from root systems of tomato and maize plants sealed with their roots into glass flasks, the shoots being exposed to water dust or to vapour-saturated atmosphere. The results of Breazeale et al. must be explained by evaporation and condensation processes in the empty glass containers used by them. Certainly very small temperature fluctuations may cause noticeable water condensation on the glass walls.

2. Water deficits in root systems are compensated rather quickly by water supply to the shoots alone, but this water absorption and the downward flow of water to the roots, thereby provoked, comes to an end when saturation maximum has been restored in the whole plant. It could be confirmed that above-ground water uptake from a vapour-saturated atmosphere is much less efficient than from the liquid phase (water dust).

3. In plants rooting in dry soil a reverse water stream and a water outflow into soil takes palce, provided the shoots are treated with liquid water. In contrast with Breazeale's opinion on no account do roots partake in the outflow. A soil of high diffusion pressure deficit sucks water from roots supplied with moisture by a downward flow of water absorbed by above-ground organs from the aerial source. This water outflow from roots into soil is a merely physical process caused by an inverted water saturation gradient between roots and soil. The possible ecological importance of such a moisture accumulation in the soil of the rhizosphere by a reverse water flow from shoots through roots is discussed.

References

Bormann, F. H.: Moisture transfer between plants through intertwined root systems. — Plant Physiol. *32* : 48—55, 1957.

Breazeale, E. L., McGeorge, W. T.: Exudation pressure in roots of tomato plants under humid conditions. — Soil Sci. *75* : 293—298, 1953.

—, —, Breazeale, J. F.: Moisture absorption by plants from an atmosphere of high humidity. — Plant Physiol. *25* : 413—419, 1950.

—, —, —: Water absorption and transpiration by leaves. — Soil Sci. *72* : 239—244, 1951.

Duvdevani, S.: Absorption of moisture by plants from dew and fog. — Current Sci. *22* : 6, 1953.

Gessner, F.: Die Wasseraufnahme durch Blätter und Samen. — Encyclopedia of Plant Physiology, edited by W. Ruhland, *3* (Water Relations of Plants): 215—246. Berlin-Göttingen-Heidelberg 1956.

Haines, F. M.: The absorption of water by leaves in an atmosphere of high humidity. — J. exper. Bot. *3* : 95—98, 1952.

Höhn, K.: Untersuchungen über das Wasserdampfaufnahme- und Wasserdampfabgabe-Vermögen höherer Pflanzen. — Beitr. Biol. Pflanzen *30* : 159—178, 1954.

Hunter, A. S., Kelley, O. J.: The extension of plant roots into dry soil. — Plant Physiol. *21* : 445—451, 1946.

Slavíková, J.: A critical evaluation of the determination of the root suction force. — Acta Univ. carol., Biol. (Prague) *1963* (3) : 245—254, 1963.

—, Horizontaler Gradient der Saugkraft eines Wurzelastes und sein Zusammenhang mit dem Sassertransport in der Wurzel. — Acta Horti bot. pragensis *1963* : 73—79, 1964.

Steubing, L.: Studien über den Taufall als Vegetationsfaktor. — Ber. dtsch. bot. Ges. *68* : 55—70, 1955.

Sveshnikova, V. M.: [Studies of water relations in plants of Eastern Pamir.] (Russ.) — Bot. Žurn. *41* : 1137—1144, 1956.

Went, F. W.: Desert grown plants absorb dew on leaves. — Crop and Soils *5* : 9, 1953.

Wiersma, D., Veihmeyer, F. J.: Absence of water exudation from roots of plants grown in an atmosphere of high humidity. — Soil Sci. *78* : 33—36, 1954.

THE RELATIONSHIP BETWEEN THE SAP EXUDATION RATE AND THE EXUDATION LAG

J. ÚLEHLA

Central Research Institute of Plant Production, Praha, Research Station of Soil Management and Fertilization, Pohořelice near Brno, Czechoslovakia

Various aspects of the sap exudation from the stumps of detopped plants have been studied (Filippov 1954, 1956; Kramer 1956).

Sap exudation has usually been measured by means of graduated capillaries or tubes, fastened to the detopped stumps and filled with water. In some cases temporary negative exudation, i.e., absorption of water from the capillary by the stump has been observed.

In the previous paper on sap exudation from our laboratory the use of measuring strips was described (Úlehla 1963). Results obtained by this method led us to the idea that it could be of some interest to study the relationship between the exudation rate and the exudation lag, i.e., the time which elapses between the detopping of the plant and the beginning of exudation.

Material and Methods

The sap exudation was studied in plants of maize, variety Čejčský hybrid C, spaced 10 by 60 cm., and in plants of a spring wheat, variety Přerovská PK.

The maize plants were cultivated under field conditions of an experiment with different sowing rates, conducted by Jelínek on the experimental fields of the Research Station at Pohořelice (Southern Moravia). The maize was sown on May 7, it emerged on May 21, headed from July 29 to August 2 and was harvested on October 10.

The wheat was cultivated under the conditions of a field experiment conducted by Straňák to find out the effects of combinations of different cultivation treatments and different ways and rates of mineral fertilization in the above-mentioned locality. The wheat was sown on April 6, it emerged on April 20, was in full heading on June 24, and was harvested on August 20. In the plots that had received no rolling, the emergence, heading and harvest were delayed. The rolled plots gave the mean total yields of 118·7 and 116·1 metric centners per hectare; the unrolled ones, 109·2 and 105·5 metric centners per hectare. The first values stand for the yields following treatment with a normal plough, the second values, for those following treatment with a rotary plough.

The sap exudation was studied using measuring strips. Narrow filter paper

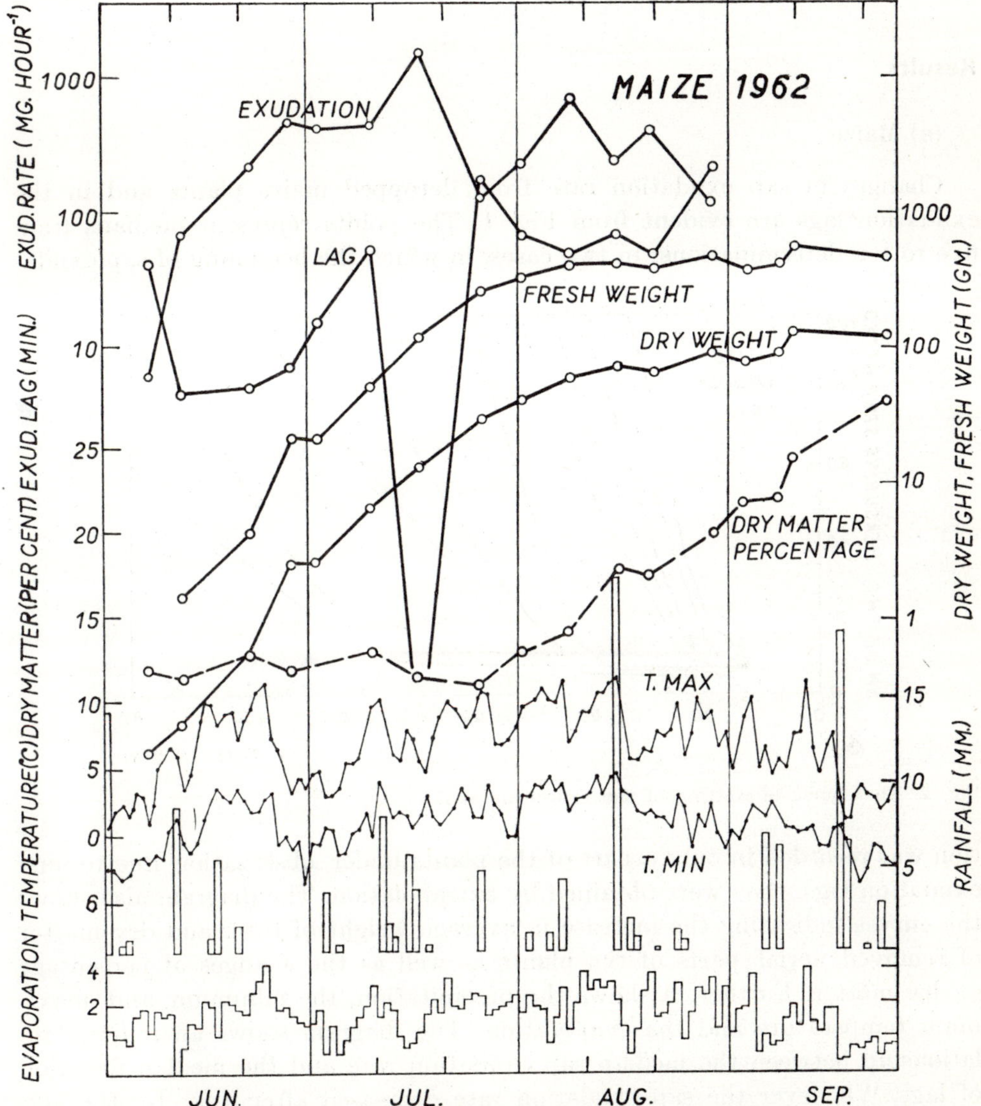

Fig. 1. Changes in exudation rate, exudation lag, fresh and dry weight and dry matter percentage in maize plants during the growing season 1962.

strips (e.g., 10 × 10 mm.), previously impregnated with water soluble stain, are fastened to the detopped stumps and protected against evaporation by

means of test tubes or polyethylene cover bags. The stain, eluted by the sap
absorbed into the strip, visualizes the front of advancing liquid. The experiments usually began at 9 a.m.

Results

(a) Maize

Changes in sap exudation rate from detopped maize plants and in the
exudation lags are evident from Fig. 1. The points represent medians from
five to ten determinations; in two cases, in which the beginning of sap exuda-

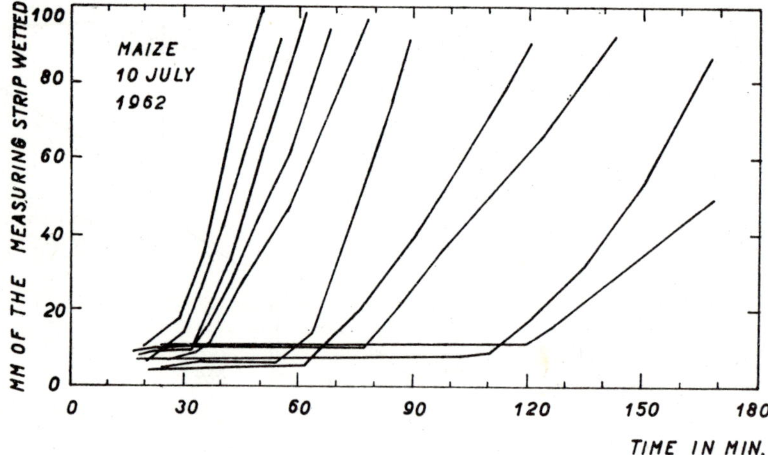

Fig. 2. The course of wetting of the measuring strips.

tion was recorded in only a part of the plants under observation due to long
exudation lags, they were obtained by extrapolation. The diagram also shows
the curves indicating the increase in average weight of fresh and dry matter
of removed aerial parts of the plants as well as the changes of percentage
of dry matter. Further, it shows the precipitation, the minimum and maximum temperature and the evaporation. The diagram shows an indirect relationship between the median sap exudation rate and the median duration
of lags. Whenever the sap exudation rate decreased after July 10, the lags
increased in length and vice versa. At the same time, substantial differences
could be observed among individual plants both in the rate and in the lags
of sap exudation, as seen in Fig. 2, showing the course of wetting of the
measuring strips on July 10.

Fig. 3 shows the relationship between the logs of sap exudation rate and the logs of duration of the lags in individual plants. In the diagram, the values obtained on June 28, July 2 and July 10 are plotted separately as on those days, the beginning of sap exudation and thence the duration of the lags were determined in all or most of the plants under observation. In the period following July 10, the duration of the lags increased and the beginning of sap exudation was recorded in only a minor part of the plants. For this reason, the values of sap exudation rate and exudation lags obtained on July 26 and August 1 as well as on August 8 till August 27 were plotted as joint groups. Calculated regression lines were fitted to the plotted values. Minor differences in the slope of the regression lines may be due to the dispersal of the plotted values. The constants b and a for the individual regression lines of the type u = by + a

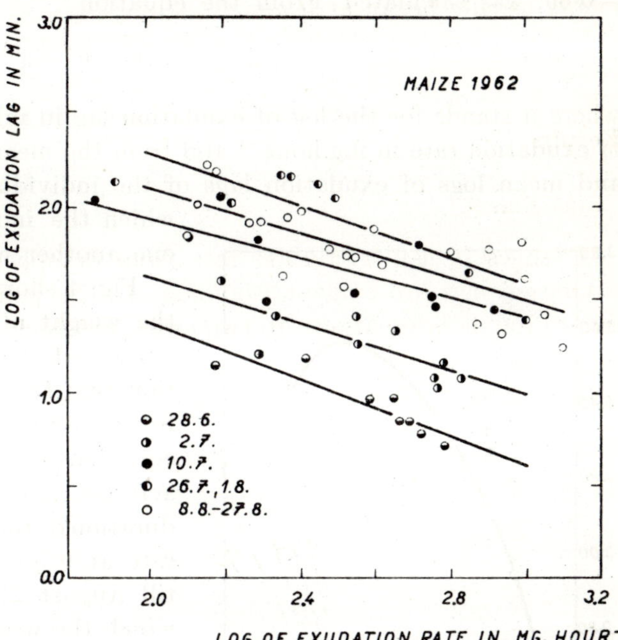

Fig. 3. The relation between the log of exudation rate and the log of exudation lag in maize plants at different times during the growing season 1962.

Table 1.

Data for regression lines of the relation between logs of sap exudation rate and logs of exudation lags. For symbols see text.

Date 1962	n	v	u	b	a	B	A
June 28	8	2·59	0·92	—0·78	2·96	—0·65	2·61
July 2	10	2·57	1·25	—0·64	2·91	—0·65	2·93
July 10	10	2·45	1·69	—0·54	3·02	—0·65	3·28
July 28 to August 1	8	2·45	1·95	—0·71	3·71	—0·65	3·55
August 8 to August 27	23	2·58	1·74	—0·63	3·36	—0·65	3·42

u = B.v + a

u = mean log of exudation lag in min.

v = mean log of exudation rate in mg.hour⁻¹

are given in the table 1. Using all the b values of the individual regression equations, the weighted mean value of the regression coefficient B, equalling —0·65, was calculated. From the equation

$$u = -\,0{\cdot}65\,.\,v + a$$

where u stands for the log of exudation lag in min. and v stands for the log of exudation rate in mg.hour^{-1} and from the mean logs of sap exudation rate and mean logs of exudation lags of the individual groups, the A values in which the individual groups differ from one another were calculated.

Fig. 4 shows the relationship between the weight of dry matter of the aerial parts and the A values. On the assumption that the points represent a coherent function this diagram and ther egression equation given above enable us to characterize the relationship between the duration of the lags and the sap exudation rate at 9 a.m. on any day from June 28 till August 27 except for those days on which the weather conditions limited the transpiration to such an extent that the exudation lag did not appear at all. This, for instance, was the case on July 17.

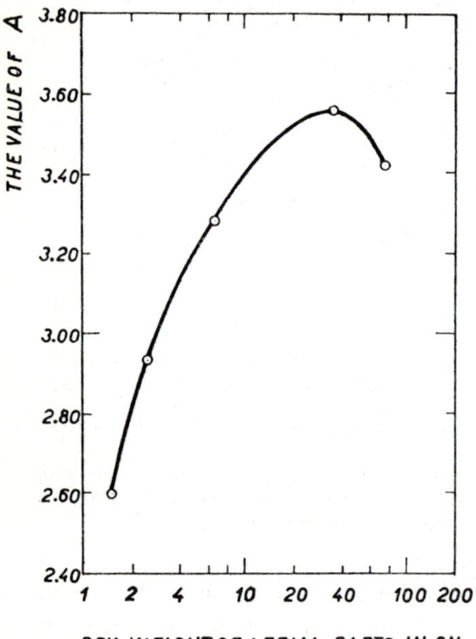

Fig. 4. The relation between the A values and the mean fresh weight for maize plants of different age. A values calculated according to $\bar{u} = 0{\cdot}65\,\bar{v} = A$, where $\bar{u}$ stands for the log of the median exudation lag and $\bar{v}$ for the log of the median exudation rate.

(b) Wheat

The sap exudation from the stumps of the main stems of spring wheat was recorded seven times during the growing season on a total of 328 plants. The relationship between the logs of sap exudation rate and the logs of exudation rate and the logs of exudation lags in exuding plants is shown in the diagram on Fig. 5. The regression of the log of the exudation lags on the log of the sap exudation rate is insignificant for the values obtained on June 8 when the sap exudation was rather high and the lags rather short in all plants under observation. On June 20, the differences among the individual plants increased. This was mainly due to limited soil water supply in certain observation sites, resulting in decreased sap exudation, as appeared from a comparison of the records for individual sites on sap exudation of wheat and on the degree of wilting of sugar beet plants cultivated as the subsequent crop as stated in a separate paper (Úlehla in press). The soil water supply was different because layers of gravel sand, inhibiting the growth of the roots, occur

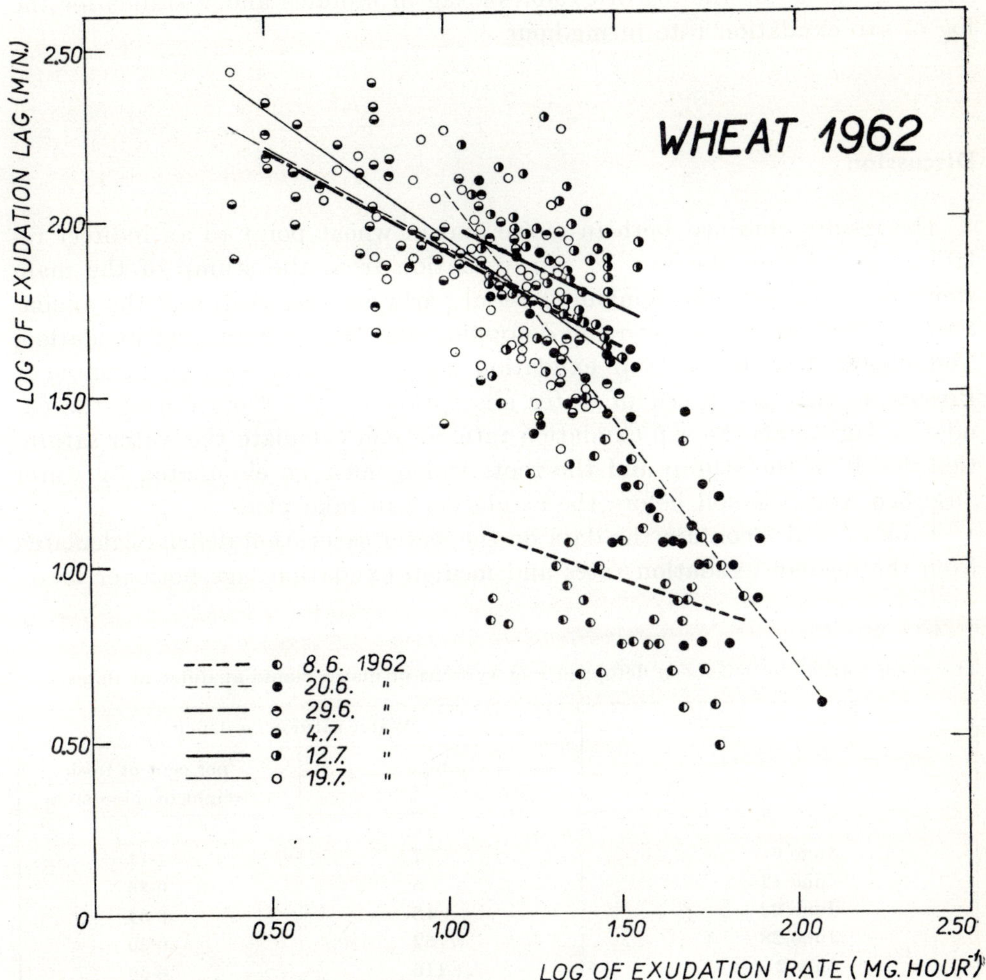

Fig. 5. The relation between the log of exudation rate and the log of exudation lag in wheat plants at different times of the growing season 1962.

at various depths of the experimental field. The regression line for June 20 shows a slope significantly greater than 1 (P < 0·01).

Calculated regression lines fitted to the groups of points standing for results of subsequent observations show a slope smaller than 1 and do not differ from one another appreciably either in their slope and their mutual

3*

position. Thus, the relationship between the sap exudation rate and the exudation lags may be expressed by a single equation

$$u = -0.60v + 2.56$$

where u stands for the log of exudation lag in minutes and v stands for the log of sap exudation rate in mg.hour^{-1}.

Discussion

The results obtained both in maize and in wheat point to an indirect relationship between the rate of sap exudation from the stump of the main stem (provided that the remaining aerial parts are removed) and the exudation lag that elapse between the detopping and the beginning of exudation. The plants with higher sap exudation rate show shorter exudation lags. Presuming that the speed of water absorption by the roots during the exudation lag equals the sap exudation rate, we can calculate the water saturation deficit in the stump and the roots which must be eliminated by water absorbed from the soil before the exudation can take place.

Tables 2 and 3 contain the data on the water saturation deficits calculated from the median exudation rates and median exudation lags both for maize

T a b l e 2.

The water saturation deficits in detopped root systems of maize plants at different dates

Date 1962	Water saturation deficit	
	mg.	per cent of fresh weight of main stem
June 6	7	1·11
June 12	5	0·36
June 22	18	0·34
June 28	62	0·30
July 2	116	0·56
July 7	380	0·75
July 17	0	0·00
July 28	320	0·12
August 1	247	0·08
August 8	552	0·13
August 17	273	0·07
August 20	316	0·08
August 25	417	0·09
September 3	not estimated	

and wheat. In the absence of data on the volumes or weights of root systems of the investigated plants the saturation deficits were related to the fresh weights of the removed above ground parts. The resulting values are given in the last columns of the tables. It is seen from the tables that in maize the

Table 3.

The water saturation deficits in detopped root systems of wheat plants at different dates

Date 1962	Water saturation deficit	
	mg.	per cent of fresh weight of the main stem
June 8	5	0·19
June 20	15	0·21
June 29	13	0·16
July 4	12	0·13
July 12	26	0·31
July 19	20	0·24
July 25	not estimated	

mean saturation deficit amounted to some 0·5% of the fresh weight of aerial growth in the first part of the growing season, varying around 0·1% later on. In wheat the mean deficit amounted to 0·2% of fresh weight of the main stem.

As the values of the regression coefficients b in the regression equations do not equal 1, we obtain different values of the saturation deficits for plants with different sap exudation rates. According to the calculation, maize plants with a tenfold sap exudation rate show about a twofold water saturation deficit. However, these data should be accepted with reserve as the deficit is determined by extrapolation.

The data on water saturation deficits show that under the conditions of soil moisture permitting the detopped root systems to start exudation in a reasonable period of time as was the case during the months of June, July and August in the present study, the saturation deficit is quite independent of the exudation rate and the duration of the exudation lags. It depends above all on the size of the experimental plants.

We may assume that the water saturation deficits reflect the elastic deformation of the plant body, caused by tension in the conducting system. Emmert (1961) considers the elastic nature of the xylem wall as a possible source of error in the estimation of the volume of xylem conduits in the stem and petioles of young bean plants. He states that. "The relaxation of tension as the stem is severed could lead to a springing back of conduit walls and a resultant increase in the respecitve (measured) volume".

Elastic deformations may be assumed to accompany the diurnal variations in water absorption by the stumps of detopped sunflower plants rooting in dry soil, observed by Hagan (1949). This author found that the behaviour of the root system indicated variations on hydration of the protoplasm amounting to 0·5—1·0% of the total root volume. The data on the water saturation deficits in the root systems of maize and wheat plants show that the respective volume changes are of the same order as those given by Hagan. In both cases the volume change may be limited by the elasticity of the roots.

Summary

The relationship between the rate of sap exudation from the detopped stump of the main stem and the duration of the exudation lag (the time which elapsed between the detopping of the plant and the start of exudation) was studied.

Plants of spring wheat and of maize (stand density 10 × 60 cm.), grown under field conditions, were used. Sap exudation was determined by means of measuring strips, the use of which was described in a previous paper.

Individual plants showed considerable differences in the rate of bleeding and in the length of the lag phase. The plants with the higher rate of bleeding generally showed shorter exudation lags.

The relationship between the exudation rate and the duration of the lag changed during the growth period. The young plants, at the same rate of bleeding, showed shorter lags than the older plants.

The sap deficit, defined as the product of the exudation rate and the length of the lag, corresponded in the case of wheat to about 0·2% of the fresh weight of the main stem and in the case of maize to about 0·5% during the period before the plants reached the height of 70 cm. and to about 0·1% afterwards.

Note added at proof: Assuming in our maize plants the same shoot/root ratios as those derived from data published by Foth (Agron. J. *54* : 49—52, 1962) the respective values of sap deficit in maize amount to 0.4 and 0.9 % of the fresh weight of roots, pointing thus to the relative constancy of this value.

References

Emmert, F. H.: Volume determination of xylem conduits in stem and petioles of *Phaseolus vulgaris* using radiophosphorus. — Physiol. Plant. *14* : 470—477, 1961.
Filippov, L. A.: [Physiological determination of cotton plant irrigation.] In Russian. — Izv. Timirjazev. s.-ch. Akad. *6* : 129—136, 1954.

—: [On the problem of the critical soil moisture for plants in connection with diurnal changes of cotton plant bleedings.] In Russian. — Dokl. Akad. Nauk SSSR *106* : 145—147, 1956.

Hagan, R. M.: Autonomic diurnal cycles in the water relations of nonexuding detopped root system. — Plant Physiol. *24* : 441—454, 1949.

Kramer, P. J.: Physical and physiological aspects of water absorption. — *In*: Ruhland, W. (ed.): Encyclopaedia of Plant Physiology, Vol. 3., Berlin, Springer Verlag, pp. 124—159, 1956.

Úlehla, J.: The effect of decreasing soil moisture on the water relations of sunflower seedlings. — Biol. Plant. *3* : 312—319, 1961.

—: Changes in sap exudation of maize and occurence of lags in exudation during the growing season. — Biol. Plant. *5* : 190—197, 1963.

Discussion

P. E. Weatherley: Is the extrapolation Dr. Úlehla has shown us really justified ? Would not the water potential depression in the plant at the instant of severance be greater than during subsequent steady exudation ? Thus the immediate uptake would be faster than the exudation rate and the extrapolation would give a lower value than the actual uptake occurring during tha lag.

J. Úlehla: The remark of Prof. Weatherley is fully justified and the data have to be taken with reserve. Nevertheless, it seems useful to make the calculations because they give at least a first approximation to the values of water saturation deficit of the roots caused by tension in the conducting elements depending on transpiration.

B. Slavík: With regard to the remark of Prof. Weatherley, one can assume that the deficit values in roots obtained by extrapolation of exudation curves are minimum values. They might be greater.

J. Úlehla: This has to be shown because some other interpretations of exudation lags seem possible, e.g. there may be a water deficit in the soil layer immediately in contact with the absorbing surface of the root. Then the extrapolation would probably give higher values of water saturation deficit than the true values.

THE CELL SAP CONCENTRATION IN MAIZE PLANTS AT CONTROLLED SOIL MOISTURE STRESS

J. VÁCLAVÍK*

Central Research Institute of Plant Production, Prague, Czechoslovakia

The osmotic pressure of the cell sap reacts sensitively to changes in external and internal conditions and is therefore suitably used to characterize the actual state of water balance of the plant.

The absolute height of the osmotic pressure of the cell sap of a given tissue is determined partly by hereditary fixed conditions of phylogenesis of the plant species and partly by individual adaptation to external conditions during ontogenesis which delimits a certain range within which these values vary. It is, therefore, not possible to agree with the generalization of certain values to characterize the water supply conditions for a number of plants, as done by several authors (Lobov 1949, 1957, Babushkin 1959, etc.).

Changes within the given range are brought about partly by physiological factors which are not directly connected with those influencing the cell sap osmotic pressure (such as changes in the quantitative composition of the cell sap due to photosynthesis, respiration, etc.) and partly by the conditions of the external environment (particularly soil moisture, temperature and relative air humidity) which determine the state of the dynamic balance between water uptake and water loss of plants.

Below, the results are given of experiments investigating changes in the cell sap concentration and growth of plants under conditions of controlled soil moisture. The experiments were planned to resolve the question of when, i.e. at what soil moisture, does temporary water deficit change into permanent deficit in plants.

Material and Methods

Maize, variety Stupická raná, in the phase of 4—6 leaves was used as material. The plants were cultivated in Mitscherlich vegetation pots with clay soil (from the humus horizon of eluated chernozem), under different conditions of soil moisture.

* Present Address: Institute of Experimental Botany, Czechoslovak Academy of Sciences, Prague-Dejvice, Czechoslovakia.

Soil moisture was determined gravimetrically by weighing the entire vegetation pot.The increase of soil moisture or its maintainance at a definite level was carried out by watering to a previously calculated weight.

The osmotic pressure of the cell sap was determined using the refractive index (Lobov 1949, Slavík 1959) on the basis of the linear relationship between the refractive index and osmotic pressure which was demonstrated experimentally by cryoscopic determination in the same material (Václavík 1964).

Plant growth was measured by the increase in length of the entire plant and by determining the average number of leaves.

Results and Discussion

On the basis of the present results (Václavík 1964), we reached the conclusion that, in maize in the phase of 4 to 6 leaves, the influence of soil moisture begins to make itself felt in changes in the cell sap concentration only when it falls to below 50% of maximal capillary capacity. This was best demonstrated by the results of an experiment in which changes in cell sap concentration and in the longitudinal growth of the plant were investigated during gradual decrease in soil moisture (Fig. 1). With a fall in soil moisture to below 50% cell sap concentration begins gradually to increase. The growth curve, which is almost linear from the beginning, makes a bend at a soil moisture level of about 50%, gradually becoming horizontal (inhibition of longitudinal growth). Variations or apparent increase in cell sap concentration at a soil moisture from 100—50% can be explained by the transition from heterotrophic type of nutrition at the beginning of the growth phase to later autotrophic nutrition.

With higher values of soil moisture, i.e. above 50% of maximal capillary capacity, the actual cell sap concentration is primarily determined by the diurnal changes in water deficit in plants. This is proved by the results of the two subsequent experiments.

In the first experiment there were three variants with soil moisture maintained at 90, 70 and 50% maximal capillary capacity. In the phase of about 3 leaves, further watering was stopped in one half of each of the variants and on the subsequent days changes in cell sap concentration were investigated in the morning and at mid-day (Fig. 2). It is seen that plants cultivated at constant soil moisture, particularly the 90 and 70% variants, have marked diurnal maxima and minima, varying within approximately the same range. In the non-irrigated group of each variant the curve of cell sap concentration is similar to that of the continuously irrigated group of the variant, with the same maxima and minima, up to a fall in soil moisture to 50%. With a soil moisture of 50—25% diurnal maxima and minima are still apparent in the presence of a continuous increase in absolute values. The factor of soil moisture begins to come into effect and diurnal water deficit changes into a permanent water deficit. When soil moisture falls below 25% a permanent

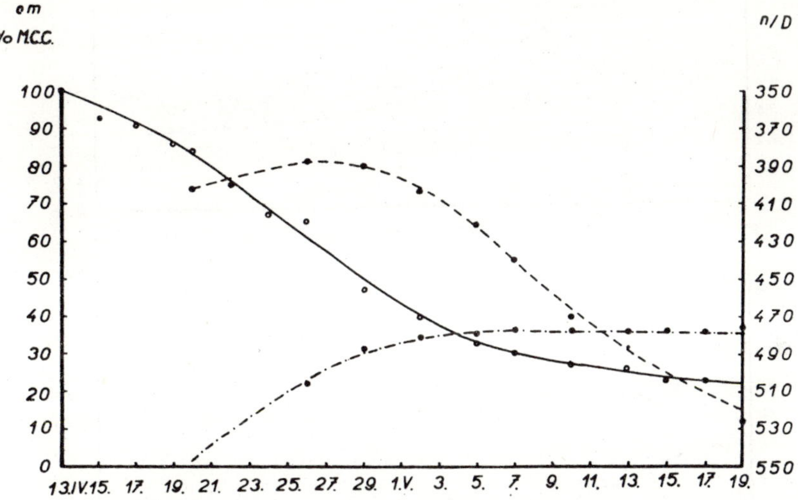

Fig. 1. Cell sap concentration and plant growth with gradual fall in soil moisture
————— soil moisture (in % max. capillary capacity)
- - - - - - - - refractive index
— · —— · — maximal length of plant (in cm.)

water deficit develops and the effect of atmospheric factors on cell sap concentration is already quite obscured by the limiting factor of soil moisture. The curve of cell sap concentration values becomes irregular when compared with the expected curve of diurnal water deficit in plants — secondary maxima and minima appear (Slavík 1951).

This is best seen in the next figure (Fig. 3) which shows the diurnal dynamics of cell sap concentration over several days with a gradual fall in soil moisture from 30% maximal capillary capacity. The curve of cell sap concentration in the control variant, continually saturated to 70% soil moisture, is shown for comparison. The temperature and relative air humidity varied to approximately the same extent in the course of the experiment. In the non-irrigated variant, in addition to the continuous almost linear increase in diurnal maxima and minima, after the first steep rise between 7 and 11 a.m. we see a sudden fall, usually to below the level of the morning value, i.e. a secondary minimum, and then a new rise, i.e. a secondary maximum.

In agreement with the previous results, the effect of irrigation is reflected in the diurnal curve of cell sap concentration only after a fall in soil moisture

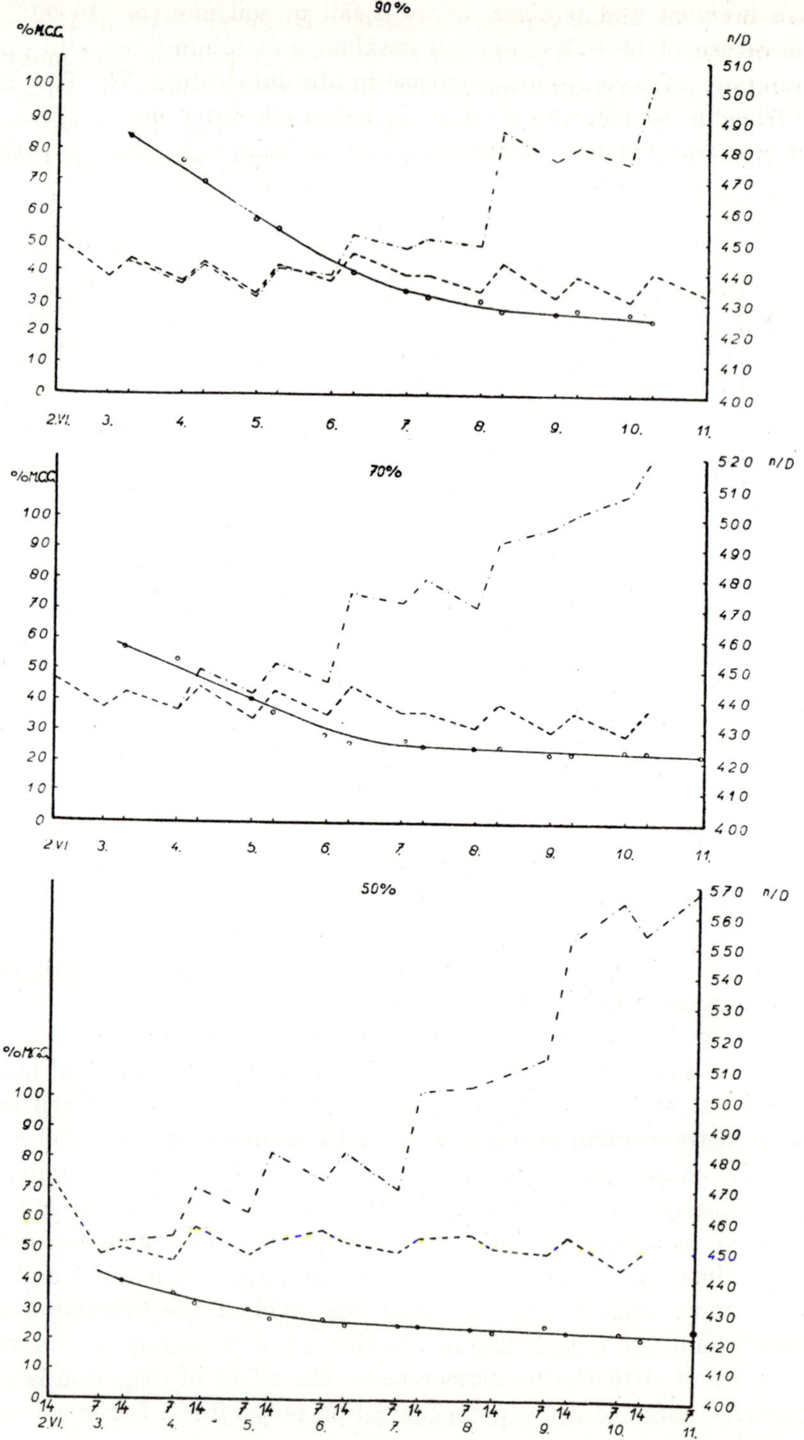
90%
%MCC
n/D
100
510
90
500
80
490
70
480
60
470
50
460
40
450
30
440
20
430
10
420
0
410
400
2.VI. 3. 4. 5. 6. 7. 8. 9. 10. 11.
%MCC
520 n/D
70%
100
510
90
500
80
490
70
480
60
470
50
460
40
450
30
440
20
430
10
420
0
410
400
2.VI 3. 4. 5. 6. 7. 8. 9. 10. 11.
50%
570 n/D
560
550
540
530
%MCC
520
510
100
500
90
490
80
480
70
470
60
460
50
450
40
440
30
430
20
420
10
410
0
400
14 7 14 7 14 7 14 7 14 7 14 7 14 7 14 7 14 7
2.VI. 3. 4. 5. 6. 7. 8. 9. 10. 11.

to below 50% maximal capillary capacity. However, even in this case the reaction of the plant to irrigation does not only depend on soil moisture before irrigation but also on atmospheric factors on the day after irrigation, as follows from the results of a further experiment (Fig. 4). The experiment was divided into 4 irrigation variants with soil moisture maintained at 90, 70, 50

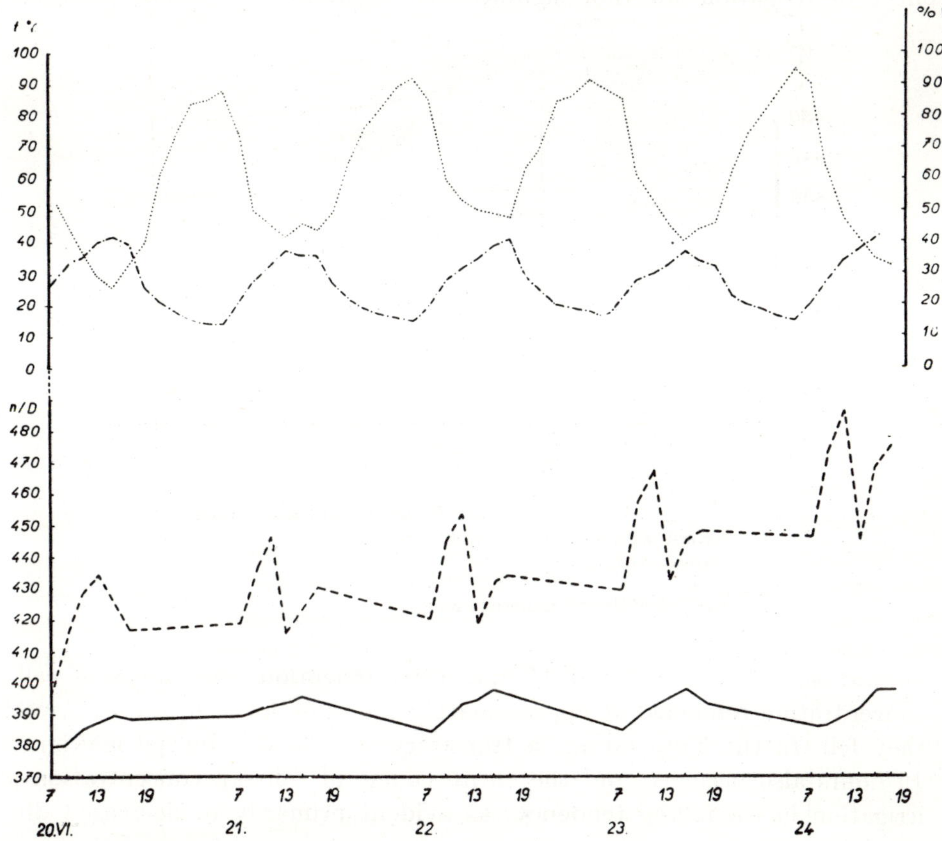

Fig. 3. Diurnal changes in cell sap concentration at 70% soil moisture and with gradual fall in soil moisture from 30%

      ——————— refractive index at 70% soil moisture

      - - - - - - - - refractive index at gradual fall in soil moisture from 30%

      — · — · — air temperature

      · · · · · · · · · relative air humidity

←

Fig. 2. Cell sap concentration at constant soil moisture (a: 90%, b: 70%, c: 50%) and with gradual fall in soil moisture after stopping irrigation

      ——————— soil moisture after stopping irrigation

      - - - - - - - - refractive index at constant soil moisture

      — · — · — refractive index with gradual fall in soil moisture and 30% maximal capillary capacity. In the phase of 4—5 leaves all plants were irrigated at one time in the early morning hours to 100% maximal capillary capacity. The figure shows the course of cell sap concentrations of the given variants on the day before and for two days after irrigation. Investigations were made at 4-hour intervals. In variants 90% and 70%, the effect of irrigation was not significantly apparent on diurnal changes. In

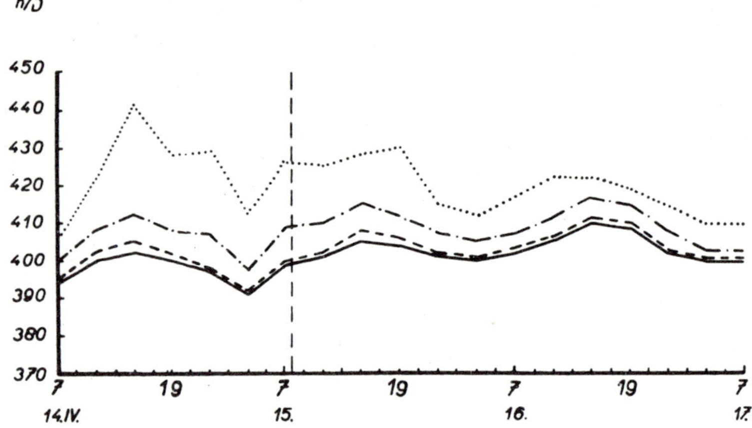

Fig. 4. Diurnal changes in cell sap concentration before and after irrigation
——————— variant with soil moisture 90%
- - - - - - - variant with soil moisture 70%
— · —— · — variant with soil moisture 50%
· · · · · · · · · variant with soil moisture 30%

variant 50% between 7 and 11 a.m. after irrigation, the values of cell sap concentration remained at approximately the same level, in the 30% variant they fell. In the 30% variant a transitory shift of the diurnal maximum to 19 hours also occurred and the entire course of cell sap concentration after irrigation has a falling tendency, as evident primarily in decreased diurnal maxima and minima.

All the observations made hitherto are well demonstrated in the subsequent, final experiment. In this experiment the plants were cultivated at 70% soil moisture up to the phase of 4 leaves. Then in some of the pots further saturation was interrupted and gradually renewed in the course of desiccation. Samples for determining cell sap concentration were collected alternately in the morning and at mid-day (Fig. 5). The effect of renewed saturation begins to become apparent only after a fall in soil moisture to below 50% maximal capillary capacity. At a soil moisture of 45% (II), further increases in cell sap concentration values to mid-day maxima still occur on the day after renewal of irrigation. These increases, however, are more gradual in comparison with the course of diurnal changes in cell sap concentration in

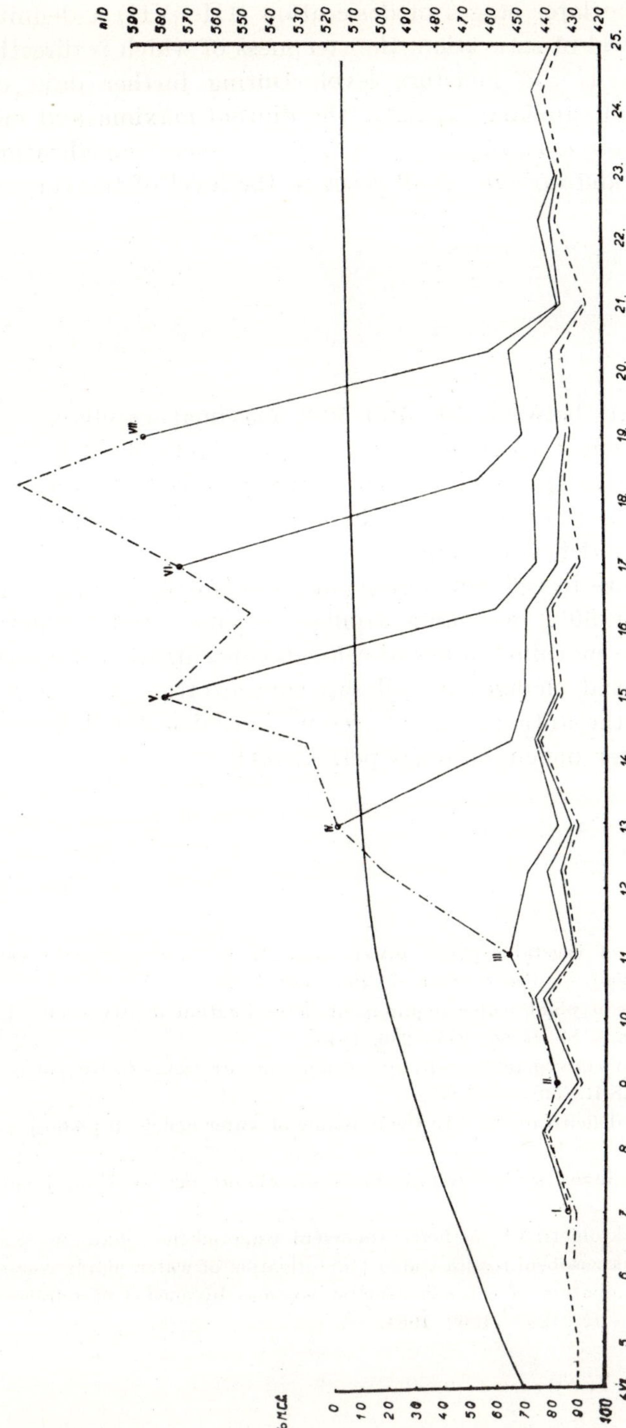

Fig. 5. Cell sap concentration after restarting irrigation at different stages of fall in soil moisture

—— soil moisture in nonirrigated variant
—— refractive index after restarting irrigation
—·—· refractive index in nonirrigated variant
—···— refractive index at constant soil moisture
I—VIII restarting of irrigation
Samples for determining cell sap concentration were collected alternately in the morning and at mid-day.

non-irrigated plants. In the later stages of desiccation (III—VII) a definite fall occurs even after renewal of saturation, the steepness of which is directly proportional to the previous soil moisture level. During further days of saturation to 70% maximal capillary capacity, the diurnal maxima and minima again become more obvious and after six days a general equalization of cell sap concentration values occurs in all cases to the level of the control variants (II—VII).

Summary

Changes in soil moisture between 100 and 50% maximal capillary capacity have no significant effect on the growth of maize plants (4—6 leaves) or on their cell sap concentration. The diurnal changes in cell sap concentration are due to atmospheric factors being similar to the anticipated course of diurnal temporary water deficit in plants.

The effect of soil moisture in cell sap concentration begins to be apparent only when it falls to below 50% maximum capillary capacity and is shown in a gradual increase in the morning values of cell sap concentration. Growth inhibition occurs and diurnal changes in cell sap concentration become irregular as compared with the anticipated changes in diurnal water deficit in the plants (temporary water deficit becomes permanent).

References

Babushkin, L. N.: [Diagnostics of vegetable plants water requirements in irrigation by cell sap concentration.] In Russian. — Fiziol. Rast. *6* : 480—483, 1959.

Lobov, M. F.: [On the methods of plant water requirement determination in irrigation.] In Russian. — Dokl. Akad. Nauk SSSR *66* : 277—280, 1949.

—: [Diagnostics of irrigation terms in vegetable crops by cell sap concentration.] In Russian. — Publ. House Acad. Sci. U.S.S.R., Moscow 1957.

Slavík, B.: K dynamice vodního deficitu rostlin [To the dynamic of water deficit in plants]. — Preslia *27* : 124—153, 1955.

—: The relation of the refractive index of plant cell sap to its osmotic pressure. — Biol. Plant. *1* : 48—53, 1959.

Václavík, J.: Použitelnost refraktometrické methody stanovení osmotického tlaku buněčné šťávy při zjišťování podmínek zásobení rostlin vodou [Investigation of water supply conditions of plants using determination of cell sap osmotic pressure by means of refractometry]. — Rostlinná Výroba *12* : 1255—1260, 1964.

Discussion

P. G. Jarvis: Osmotic values, however they are measured (whether cryoscopically or by refractometry) are possibly not the best thing to measure to understand the physiological effects of the state of water in the cell. The osmotic values give only a crude estimate of one of the potentials contributing to the total water potential, or free energy state, of water in the cell — viz: the osmotic. It ignores hydration components which depend on adsorption, colloidal imbibition and the structure of the protoplasm and cell walls. It is not a good measure of the cell osmotic potential either, because to measure the osmotic value the cell sap has to be extracted and in this process the solutions in the cell and cell walls become mixed. I believe that if we wish to try to relate a physiological process such as photosynthesis or frost resistance to the state of water in the cell, we should try to relate it to the water potential or the free energy of the water, taking into account all the various components, the importance of some of which may well have been underestimated in the past.

J. Václavík: Cell sap obtained from tissue killed at 100° C contains the components of cell water which are freest and most freely movable (vacuolar sap, the water in conducting elements, all solution water). Its relative homogeneity and relatively uniform cytological origin is indicated by the circumstance that when expressing increase in pressure above a certain minimum has almost no effect on the osmotic pressure. The osmotic pressure of cell sap reacts sensitively to changes in water balance in the tissue. Therefore it is often used to characterize these relative changes. On the assumption of a certain state of equilibrium between the individual water fractions in the cell it is possible to use osmotic pressure obtained in this way from cell sap as an indicator of relative changes in the level of hydration. At the same time, I am aware that the osmotic potential of vacuolar cell sap in situ is not always identical with the osmotic pressure of expressed sap.

THE EFFECT OF DECREASE IN SOIL MOISTURE ON THE OSMOTIC VALUES OF CELL SAP IN THE INDIVIDUAL ORGANS OF THE POTATO PLANT

J. NEČAS

Institute for Potato Research, Havlíčkův Brod, Czechoslovakia

Previous work on the influence of decreased soil moisture on the osmotic values of cell sap expressed from the entire overground parts of potato plants (Nečas 1962) did not indicate its effect on the individual organs. The present work was concerned with this problem. Its aim was to reach greater precision in evaluating the regulatory systems of the water regime of potatoes as expressed in changes in the osmotic values of the cell sap from the leaves, petioles and stems, under the influence of changed soil moisture. In order to make the necessary number of analyses in the required time the methods of the experiment were slightly modified as against those used in the previous work.

Material and Methods

M a t e r i a l: in the first stage of the study reported here the variety Kardinal was used. The tubers for planting were obtained in the breeding value V_1 to minimize the danger of virus diseases in the experimental material. The seed tubers were assorted to a weight of 90 ± 5 g.

M e t h o d: The experimental plants were cultivated in pots for potatoes in the greenhouse. The pots were filled with suitable soil with an absolute water capacity of 48% (according to Kopecký). The relative soil moisture was calculated on this value throughout the period of vegetation. The experiment had two control variants (K_1 and K_2) where the soil moisture was maintained at the given level throughout the vegetation period and three experimental variants (V_1, V_2, and V_3) where the level of soil moisture was decreased for only one phase of development of the experimental plants.

S c h e m e   o f   t h e   e x p e r i m e n t:

K_1: 35% relative soil moisture for whole vegetation period.

K_2: 70% relative soil moisture for whole vegetation period.

V_1: 35% relative soil moisture from planting to budding and then 70% relative soil moisture to end of vegetation period.

V_2: 35% relative soil moisture from budding to the end of flowering, 70% relative soil moisture at all other phases of vegetative period.

V_3: 35% relative soil moisture from end of flowering to beginning of physiological ripening, 70% relative soil moisture at all other phases of vegetative period.

T a b l e· 1.

Analysis of variance of total osmotic values measured in the whole research in the glasshouse in one day in the season (June 8).

Variation caused by	N	$S(x - \bar{x})^2$	V	F	s
variants	4	3038·87	759·71	223·44[++]	
organs	2	1110·82	555·41	163·35[++]	
time of day	1	62·21	62·21	18·29[++]	
individual variations	3	27·70	9·23	2·71[++]	
variants × organs	8	165.93	20·74	6·39[++]	
variants × time of day	4	82·33	20·58	6·05[++]	
variants × individual variations	12	171·03	14·25	4·68[++]	
organs × time of day	2	37·62	18·81	5·53[++]	
organs × individual variations	6	11·38	1·89	0·55	
time of day × individual variations	3	92·95	30·98	9·11[++]	
residuum variance	74	252·16	3·40		1·85
Sum	119	5053·00			

The soil moisture in the vegetation pots was determined by weighing and adding water according to the calculation of weight loss with a growth correction.

For each variant there were 38 vegetation pots each with one plant. 32 plants were used for laboratory analysis (8 samples of 4 plants each), 4 plants were left to determine yield and 2 pots were left for the measurement of evaporation from the soil.

Samples were taken at weekly intervals for determining osmotic values. Of the 4 samples, half were taken in the morning and half in the afternoon, in order to compare the effect of rising daily water deficit in the tissues on the osmotic values of their cell sap in the individual overground organs of the potato. The values given are thus the mean of four values. The ratio of individual variability to total variability is indicated by variance analysis and shown in Tab. 1.

M e t h o d   o f   a n a l y s i s: The samples were divided into leaves, petioles and stems and the cell sap was expressed from each type of organ together. It was immediately inactivated (at 100° C) and centrifuged. The supernatant was stored for further analysis at 0° C.

Total osmotic values for sap from individual organs were determined cryoscopically. The contribution of electrolytes to total osmotic values was determined from conductometric measurements. The fraction attributable to reducing substances was determined by Bertrands method without hydrolysing. Preliminary experiments showed that the contribution of substances reducing Fehling's solution after hydrolysis to total osmotic values is very small as compared with the other fractions.

The total osmotic values after the appropriate corrections were expressed according to the lowering of the freezing point in atmospheres using Walter's tables (1936). The electrolyte fraction was also expressed in atmospheres corresponding to the equivalent values of KCl solutions according to conductivity. The fraction of reducing substances was also expressed in atmospheres corresponding to equivalent solutions of glucose according to the reduction of Fehling's solution.

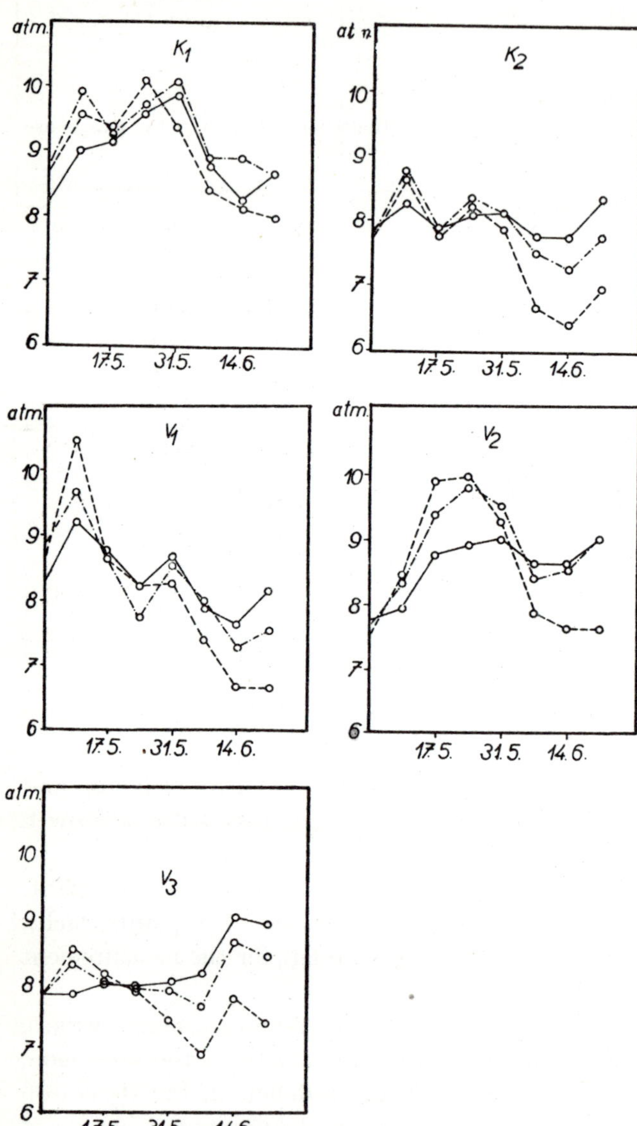

Fig. 1. The course of the total osmotic values in the individual overground organs of the plants (———— leaves, - - - - stems, — · — · — petioles) in all variants of the experiment (K$_1$, K$_2$, V$_1$, V$_2$, V$_3$) during the whole season. Abscissa: date, ordinate: osmotic value in atm.

Results nad Discussion

The effect of decreasing soil moisture on total osmotic values: is shown not only in the position of the range of variation of the measured values but also in its width. The position of this range is evidently determined by permanent water deficit in the plant and its width is probably conditioned by the daily water deficits in the tissues, depending on the weather. With decreased soil moisture the daily water deficits reach high values if the plant does not bring effective adaptive regulatory systems into action. As found in the variety Kardinal in 1961 (Fig. 1) the position of the range of total osmotic values of the sap from leaves of the two control variants during the vegetation period differed by about 1·5 atm. The width of this range in the variant with a higher level of soil moisture was about 0·5 atm., whereas in the variant with a lower relative soil moisture it was about 1·5 atm. This shows that in the variety Kardinal, no particularly

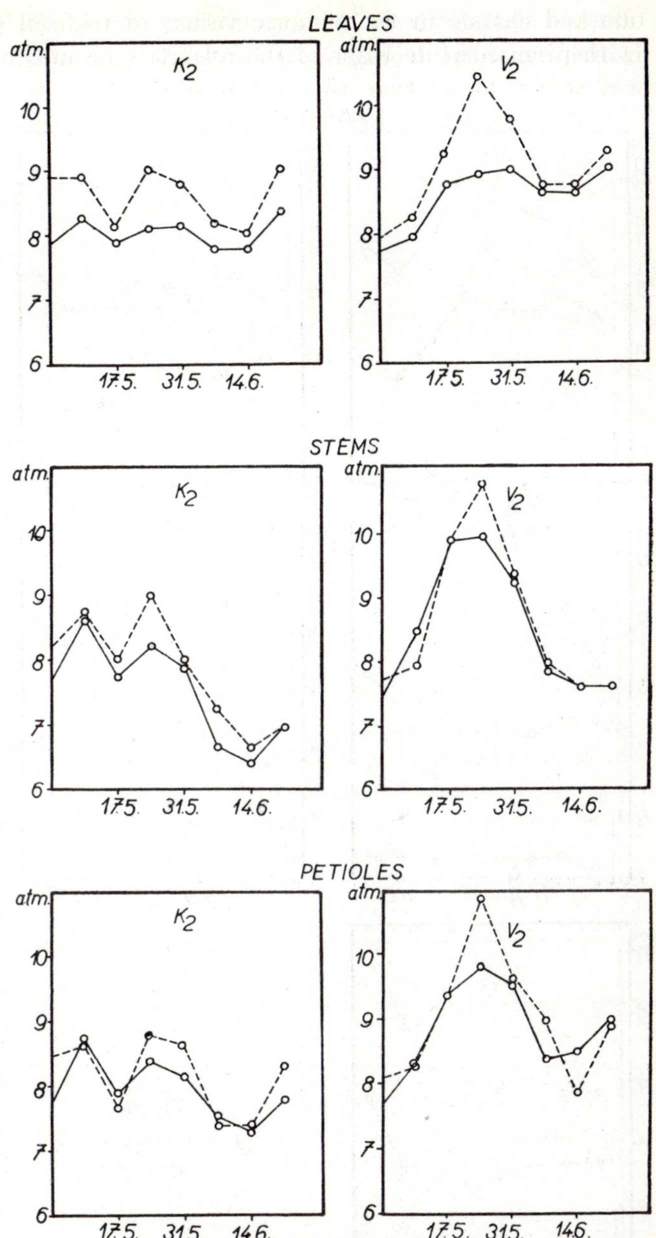

Fig. 2. The influence of the time of day (——— forenoon, - - - - - afternoon) on the course of the total osmotic values in the leaves, stems and petioles during the whole season in two variants of the experiment (K_2 and V_2). Abscissa: date, ordinate: osmotic value in atm.

marked change in the osmotic values of the cell sap of the leaves occurred
with permanent decrease of the relative soil moisture to one half. This means
that osmotic relationships were not changed directly in plants but that other systems of the water regime were brought into action to prevent basic changes in the presence of permanently decreased relative soil humidity. The daily variations in the values during the vegetation period according to the weather suggest that these are not active but passive changes produced by a negative balance in water uptake and water loss. The changes in the main fractions of total osmotic values suggest this, as we shall see later.

Comparison of the effect of a sudden decrease in relative soil moisture at a definite phase of the growth period followed by maintenance of relative soil moisture at the new low level shows that sudden drought does not lead to essentially higher total osmotic values. This supports the correctness of the above suggestion that potato plants tend rather to prevent increasing osmotic values of

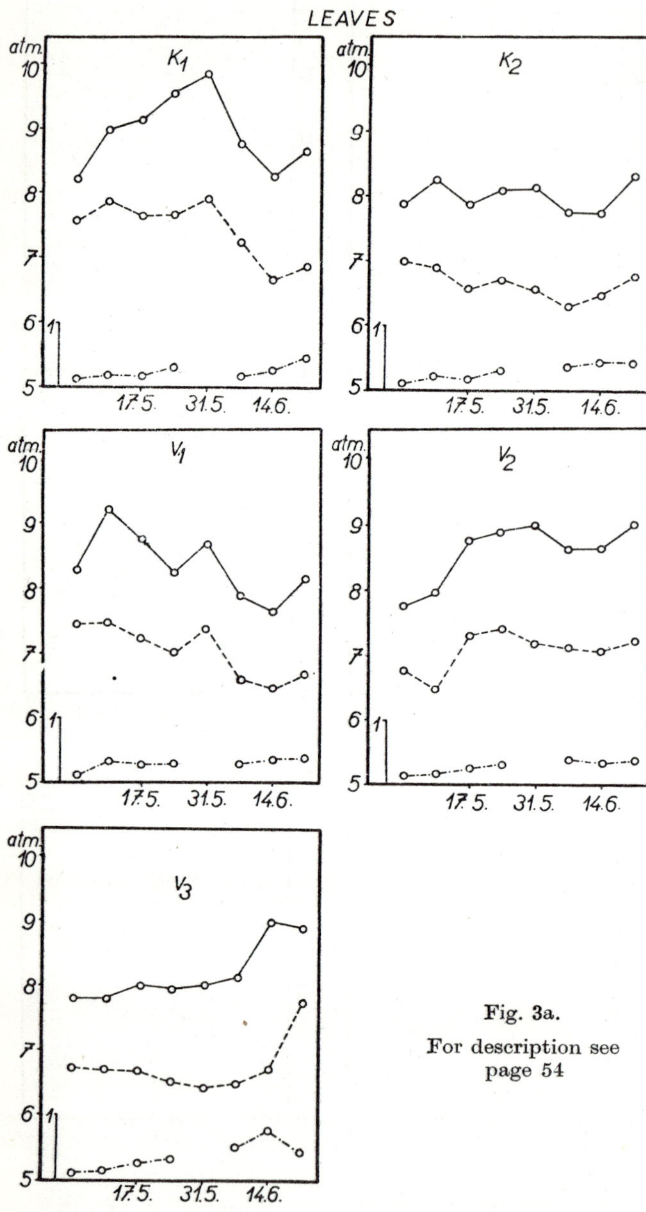

Fig. 3a.

For description see
page 54

the cell sap of their overground organs rather than to regulate the increase
by holding water and thereby preventing loss by transpiration.

The influence of ontogenesis on total osmotic

v a l u e s: Comparison of the curve of total osmotic values in the cell sap of the different organs shows a different seasonal trend. At the beginning of the vegetation period the differences among the single organs are relatively small. However, they increase with the progression of vegetation and towards the end of the vegetation period they are already very marked, especially at higher levels of relative soil moisture. It must be borne in mind that differences of this type between K_1 and K_2 could be produced by differences in physiological age in plants of the same chronological age, since decreased relative soil moisture retards development.

The least significant vegetative trend in osmotic values is found in leaves which also show the smallest fluctuations in values on different days. It follows from this that the leaves have the greatest capacity of buffering osmotic values of cell sap throughout the vegetative period. The petioles behave as adjusting organs between leaves and stems. The greatest osmotic gradients

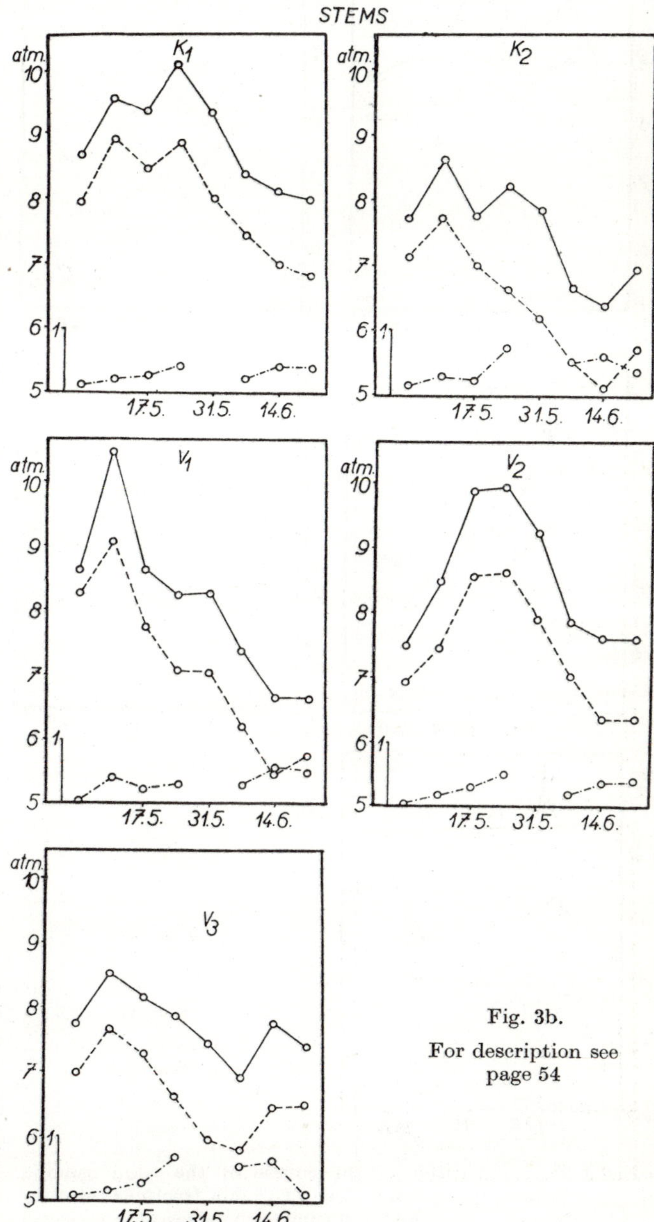

Fig. 3b.

For description see
page 54

are reached, as already stated, towards the end of vegetation. It is evidently associated with a decrease in the activity of the above mentioned equilibrating system.

PETIOLES

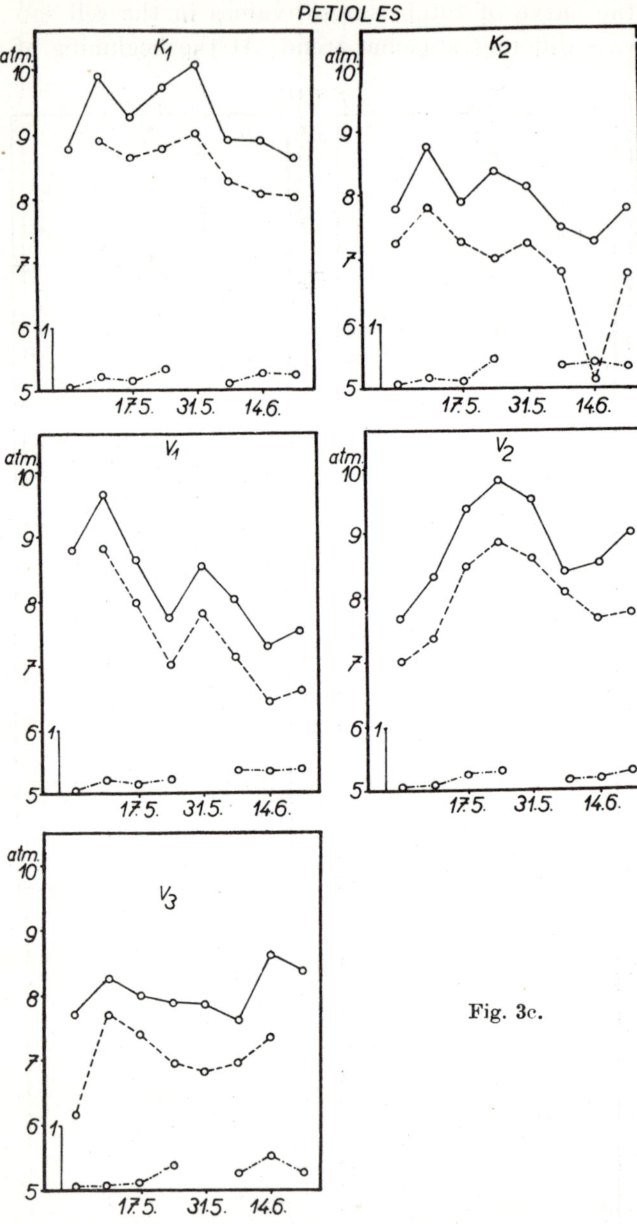

Fig. 3c.

Fig. 3. The comparison of the course of the total osmotic
values (————) and their two main fractions: electrolytes (- - - - -) and reducing substances (— · — · —)
in the individual overground organs of the plants in
all variants of the experiment during the whole
season. The osmotic values belonging to reducing
substances fluctuate in the range of one atmosphere.
Abscissa: date, ordinate: osmotic value in atm.

The effect of
the time of day
on total osmotic
values: The effect of
the daily water deficit in
the single organs of po-
tatoes on total osmotic
values was most marked
in the leaves. The maxim-
al difference between forenoon and afternoon val-
ues in the same variant
in the leaves was 1·5 atm.
This indicates that in
potatoes the effect of the
day can reach the ap-
proximately the same
result as the permanent
or prolonged reduction
of relative soil moisture
to one half. This suggests
that this is actually only
the effect of a negative
water balance in the
plants and not the result
of regulation by the plant.
Micromethods for the de-
termination of osmotic
values would certainly
give more satisfactory
results. In this case it
would be possible to dif-
ferentiate the individual
tissue systems according
to whether a rising water
stream with mineral salts
or a falling stream with
translocation of assimil-
ates predominates, al-
though such a clear differentiation is very diffi-
cult.

The effect of decreased soil moisture on the main fractions of total osmotic values: As already established from previous work (Nečas 1962) the total osmotic values in sap expressed from potato plants is formed mainly by electrolytes and from the other fractions, the most important are reducing substances and among them mainly sugars. If the total osmotic values attributable to both these fractions are compared with the total osmotic values in the single organs it is seen that the greatest difference is found in the leaves. This suggests that greater amounts of further undetermined substances take part in making up the osmotic values in the leaves than in the stems and petioles.

The effect of permanent and temporary decreased soil moisture plays no important role in the ratio of the main fractions of total osmotic values. It would seem that there is no significant regulation of the two main fractions and that the plant is more or less passive in

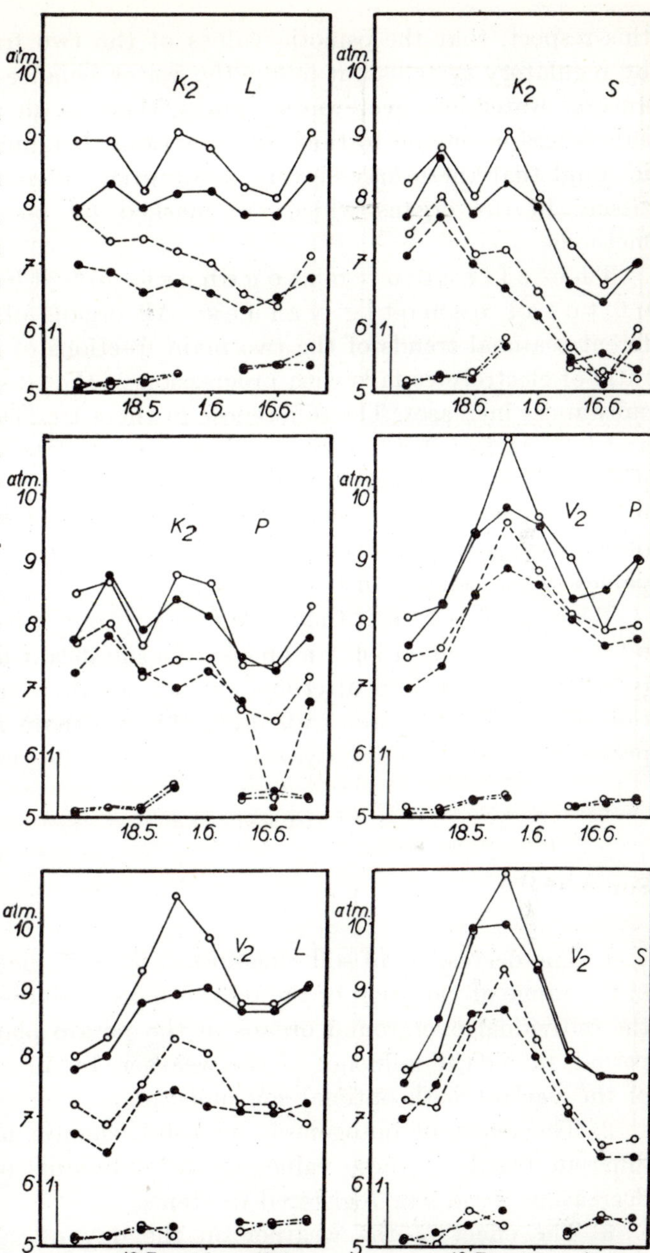

Fig. 4. The influence of the time of day (● : forenoon, ○: afternoon) on the course of the total osmotic values (———) and their two main fractions: electrolytes (- - -) and reducing substances (—·—·—) in the leaves (L), stems (S) and petioles (P) during the whole season in two variants of the experiment (K₂ and V₂). The osmotic values belonging to reducing substances fluctuate in the range of one atmosphere. Abscissa: date, ordinate: osmotic values in atm.

this respect, that the osmotic values of the two fractions are not governed by regulatory systems but follow the water balance. With increasing dominance of water loss over water uptake, there is an increase in the amount of substances belonging to these two fractions. It is necessary, however, to bear in mind that there may be very substantial differences among the individual tissues in the organs which are masked by pooling the sap for macro-methods.

The effect of ontogenesis on the main fractions of total osmotic values: All organs of potato plants show different seasonal trends of the two main fractions of total osmotic values. The ratio of electrolytes falls with progression of the season while that of reducing substances increases. The differences in these trends are greatest in the stems and less marked in the leaves. The necessary data were not obtained for explaining these differences in the two main fractions of total osmotic values in the course of the vegetative season. The differences among the organs obviously arises from their different abilities to buffer changes in osmotic values in the cell sap of their tissues.

The effect of time of day on the main fractions of total osmotic values: No significant differences were registered between forenoon and afternoon values which would suggest the participation of regulatory systems affecting the two main fractions of total osmotic pressure.

Summary

1. The decreasing of soil moisture to 50% of the optimal, led to a shifting in the range of variation of the total osmotic values of cell sap expressed from the individual overground organs of the potato plant, during the vegetative period, under the influence of the weather, by 10 to 15% of absolute values of the variant with optimal soil moisture.

2. The effect of ontogenesis on total osmotic pressure was displayed in different trends of these values in the individual organs. The most marked decreasing trend was registered in stems.

3. The effect of the weather on the day of sampling and the time of day (forenoon or afternoon) also leads to maximal variations of 10 to 15% from the mean value for the variant and is most marked with decreased soil moisture.

4. The ratio of the main fractions of total osmotic pressure (electrolytes and reducing substances) does not change basically with decreased soil moisture. The seasonal trends, however, are different. The electrolyte fraction

decreases towards the end of the vegetation period, while the reducing substances increase. The weather and time of day have no significant effect on these trends.

5. It can be concluded from these results that potato plants do not use regulatory systems to control the osmotic values of the cell sap of individual organs under conditions of decreased soil moisture, but that a negative balance in water loss and water uptake is the true cause of changes in osmotic values.

6. These findings are also important for the prediction of irrigation requirements in potato crops by measuring osmotic or refractometric values.

References

Nečas, J.: Der Einfluss sinkender Bodenfeuchte auf die osmotischen Werte der oberirdischen Teile der Kartoffel. — Fragen der Pflanzenzüchtung und Pflanzenphysiologie (Symposium). Tagungsberichte d. Dtsch. Akad. d. Landwirtschaftwiss. Nr. *48* : 113—124, 1962.

Walter, H.: Tabellen zur Berechnung des osmotischen Wertes von Pflanzensäften Zuckerlösungen und einiger Salzlösungen. — Ber. dtsch. bot. Ges. *54* : 328—399, 1936.

WATER STRESS IN CELL, TISSUE, AND PLANT

Tuesday, October 1, and Wednesday, October 2

Chairmen: *N. A. Gusev* and *P. E. Weatherley*
Secretary: *Z. Šesták*

SOME INVESTIGATIONS ON WATER DEFICIT AND TRANSPIRATION UNDER CONTROLLED CONDITIONS

P. E. WEATHERLEY

Department of Botany, Aberdeen University, Great Britain

The transpirational flux of water from soil to atmosphere is perhaps not in itself of direct physiological importance to the plant, but the water deficits which are a necessary concomitant of the transpirational flux have a profound effect on growth and metabolism. It was for this reason that when, some years ago, I was faced with a study of the water relations of the commercial cotton crop in E. Africa, I concentrated on water deficits rather than transpiration rates. A technique was developed for estimating the average water deficit in the leaves of a group (plot) of plants growing in the field. This consisted of punching leaf disks and expressing their water content as a percentage of their water content when fully turgid (relative turgidity). The disks were rendered turgid by floating on water (Weatherley 1950, Barrs and Weatherley 1962).

Such relative turgidity values were measured at sunrise and in the middle of the day from time to time throughout the growing season of the cotton. It was found (Weatherley 1951) that the leaves usually suffered a considerable water deficit even at sunrise, in other words there was not full recovery during the night. This failure to recover was confirmed by a study of the diurnal pattern of changes in relative turgidity (Fig. 1). It will be seen that there was little rise in relative turgidity after midnight in spite of the transpiration being virtually zero. This steady diurnal maximum appeared to represent some sort of equilibrium.

These sunrise values of relative turgidity were found to vary from week to week, and were related to the evaporation from an atmometer during the previous twenty four hours i.e. days of high evaporation were followed by a low relative turgidity at sunrise (see Fig. 2).

The explanation of the data in Fig. 1 and 2 was thought to be as follows: — During the day, when transpiration is high, the soil tends to become dry round the vigorously absorbing roots. During the night when transpiration and hence root absorption becomes very low, rewetting of these dry zones occurs but this, it is supposed, is partial only, since the water in these zones soon passes into the state represented by a "wetting curve" and thus movement of water into them from the surrounding soil becomes very slow. It is suggested that the relative turgidity attained during the night represents an equilibrium between the plant and these dry zones of soil, the plant so to speak acting as a tensiometer. The more intense the transpiration during a given day the greater the degree of drying round the roots and the lower

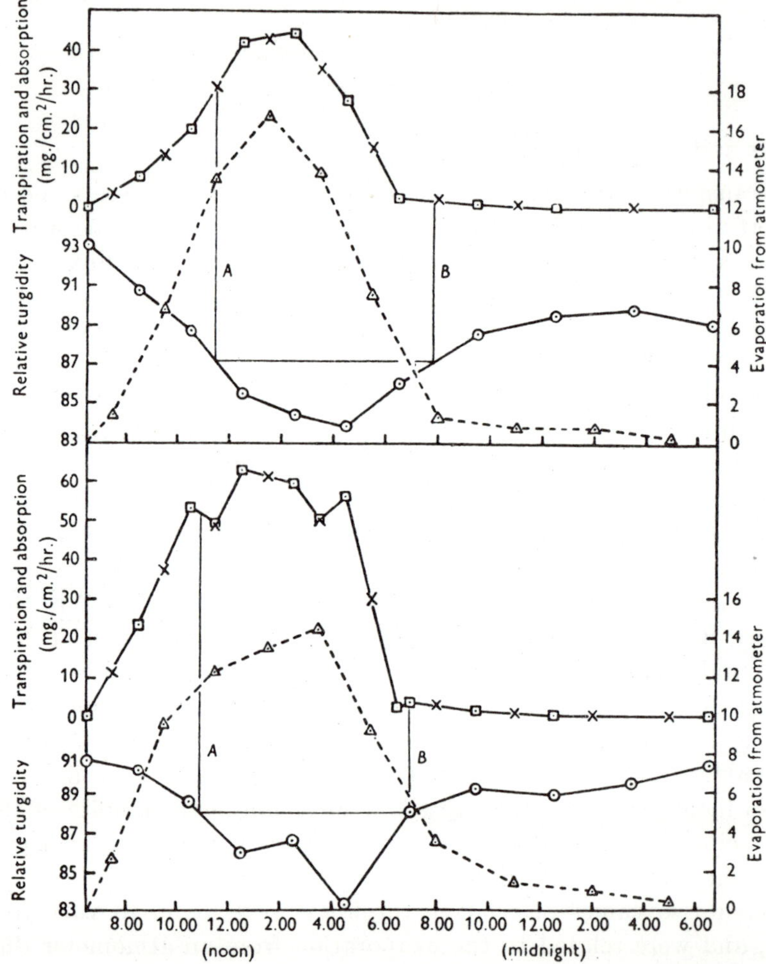

Fig. 1. Curves showing diurnal changes in transpiration (□————□), absorption (×————×), relative turgidity (⊙————⊙) and evaporation (△————△), through two 24 hr. periods (not consecutive). Upper part, first period; lower part, second period. Data for *Gossypium hirsutum*. (From Weatherley 1951.)

the equilibrium relative turgidity the following night. Thus if we assume that evaporation and transpiration are correlated we can see how the correlation in Fig. 2 arises.

The field studies presented in 1951 seemed explicable in terms of this hypothesis, but supporting evidence has only recently been obtained (Macklon

1962) using controlled environment equipment. With this it has been possible to determine the part played by the soil in the development of water deficits in the leaves by comparing the responses of plants growing in water culture and soil to regular patterns of change in transpiration rate induced by changing the atmospheric humidity.

Materials and Methods

Plants of *Ricinus communis*, the castor oil plant, were used. These were grown in standard water culture solutions or in garden compost according to requirements. For the duration of the experiments a number of plants were transferred to a climatological wind tunnel (Weatherley and Barrs 1959) (see Fig. 3) which permitted constant conditions of

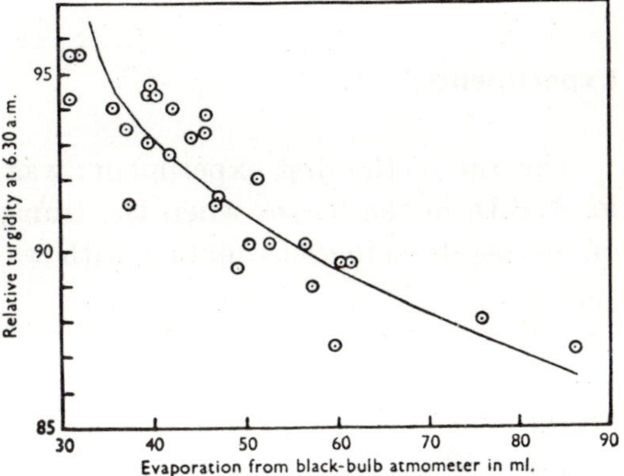

Fig. 2. Curve showing the relationship between relative turgidity at sunrise and evaporation during the previous 24 hr. Data for *Gossypium hirsutum*. (From Weatherley 1951.)

light, temperature, windspeed and humidity to be maintained. All these factors could also be changed rapidly from one level to another. In the experiments to be described all were kept constant except the humidity which was raised or lowered in order to change the rate of transpiration of the plants.

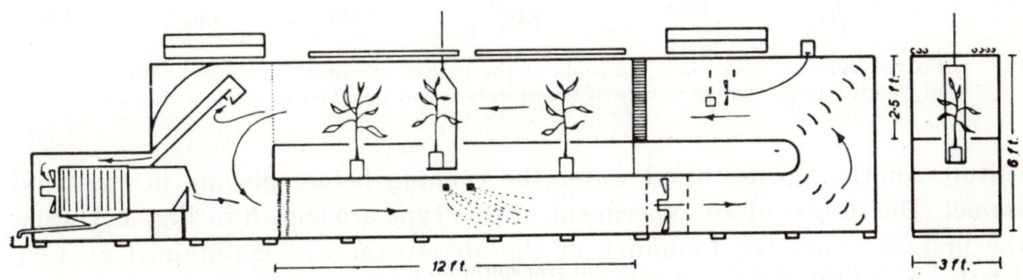

Fig. 3. Elevation diagrams of the climatological wind tunnel. (From Macklon 1962.)

The transpiration rate was followed on one of the plants which was suspended on a self recording balance (see Fig. 3). From time to time disks of leaf tissue were punched and their water potential depression (this will be referred to as W.P.D. which is synonymous with diffusion pressure deficit) was measured using a vapour-pressure device (Macklon and Weatherley). In these studies it was thought that W.P.D. would be more informative than relative turgidity with which it is of course closely related (Weatherley and Slatyer 1957).

Experiments

The aim of the first experiments was to see whether there was an increase in W.P.D. of the leaves when the transpiration rate was increased, the roots of the plants being surrounded with water. Six plants were transferred from

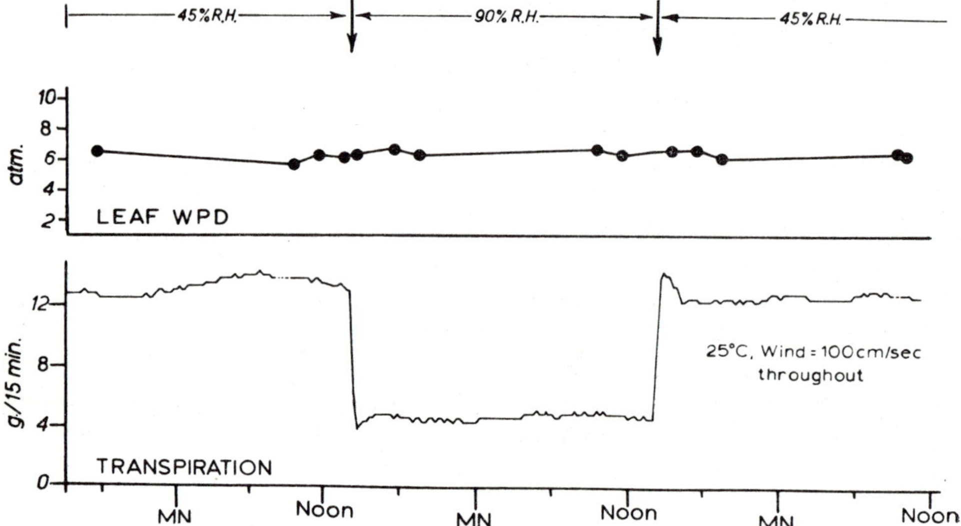

Fig. 4. Curves showing that when the roots of the plants *(Ricinus communis)* were immersed in water, a large change in rate of transpiration had no effect in the leaf water potential.

culture solution to deionised water the evening before placing in the wind tunnel. The results of an experiment of this type are shown in Fig. 4. During the first 24 hours the humidity of the air stream was maintained at 45% relative humidity. Then the humidity was raised to 90% for 24 hours and finally the humidity was returned to 45%. In response to the rise in humidity the rate of transpiration declined to one third of its previous value but the W.P.D. of the leaves remained at about 6 atm. throughout, and seemed quite independent of the rate of transpiration. A similar lack of response in W.P.D. was found when the changes were reversed and when other patterns of transpiration change were induced by changes of humidity.

The results of a similar type of experiment except that the plants were

rooted in boxes of soil are shown in Fig. 5. Before the experiment the soil
in the boxes was brought to field capacity so that at the beginning of the
experiment the soil was somewhat below field capacity and declined throughout. But the bulk of the soil in the boxes was quite moist even at the termination of the experiments. The results are in striking contrast to those just
described. Here the increase in rate of transpiration resulting from lowering
the humidity led to a marked rise in the W.P.D. of the leaves. When the
transpiration rate was made to decline again there was a recovery, the W.P.D.
falling to near its former value.

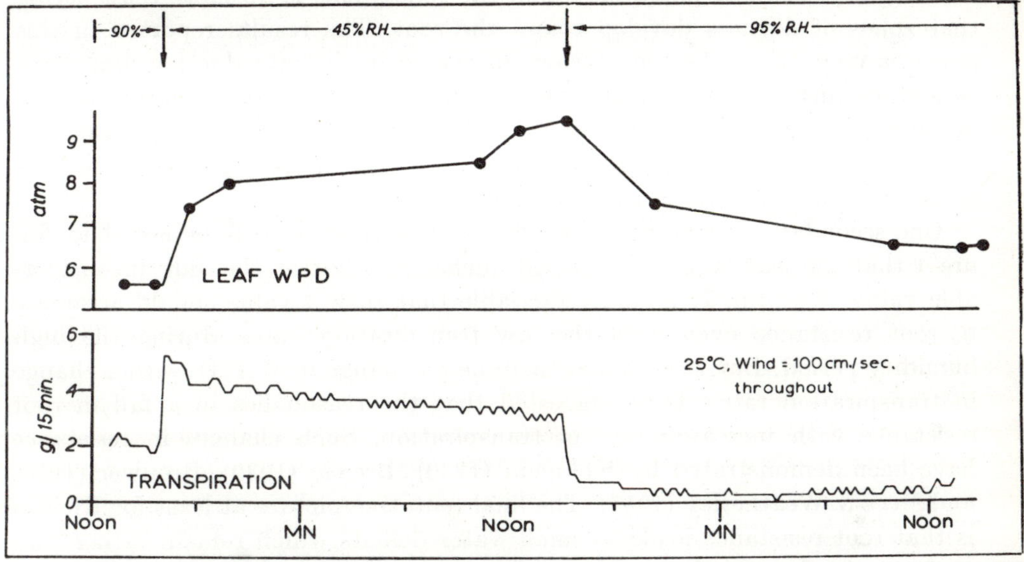

Fig. 5. Curves showing that when the plants *(Ricinus communis)* were rooted in moist soil,
changes in transpiration rate caused marked changes in leaf water potential.

Discussion and Conclusions

The results from the soil rooted plants (Fig. 5) are basically similar to those
which have frequently been obtained under field conditions (e.g. Fig. 1) reflecting diurnal changes in the microclimate. As the humidity falls during
the day, transpiration rises and the water deficit in the plant increases (i.e.
W.P.D. rises). In the evening a reverse sequence of events occurs. The current
interpretation of this sequence would be that the roots offer a resistance to
the transpiration stream so that as the rate increases, the W.P.D. rises pro-
portionately (Ohm-type law. See van den Honert 1948). That this is not so

is demonstrated for Ricinus plants at least, by our first experiments, for when the roots were surrounded by water, no increase in W.P.D. occurred when the transpiration was increased. The cause of the W.P.D. increase in soil rooted plants must lie, therefore, not in root resistance, but in the soil itself. It is suggested that it is the resistance to water movement i n t h e s o i l which leads to water deficits arising in transpiring plants. The roots of transpiring plants abstract water from the soil immediately surrounding them and this sets up a gradient of water potential in the soil which causes a flux of water towards the roots. Owing to the poor water conductivity of the soil this gradient needs to be steep and so there must be a considerable lowering of the water potential round the roots. This is no more than saying that zones of dryness develop round the root. The results reported in this paper support the earlier conclusions discussed in the introduction, but carry us a stage further. They suggest that not only are the residual water deficits at sunrise due to soil deficits but that all diurnal water deficits (at least the substantial ones) in the plant are due to s o i l deficits which arise because of the resistance to water movement in the soil.

One secondary feature of the results is of interest. It will be noted in figure 4 that the leaf W.P.D. remained unchanged, but at the not inconsiderable value of 6 atm. It is understandable that such a value should arise due to root resistance even with the low transpiration rates during the high humidity period. But why should there be no change in W.P.D. with a change in transpiration rate ? It is suggested that the reason lies in a fall in root resistance with increased rate of transpiration. Such changes in resistance have been demonstrated by Köhnlein (1930), Brewig (1939), Brouwer (1954) and Mees & Weatherley (1957). The final tentative picture at least for Ricinus is that root resistance leads to small water deficits which remain rather constant in the face of changes in transpiration rate. The larger water deficits suffered by these plants and the well established diurnal variations in water deficit are due to W.P.D.'s which arise in the soil.

Summary

Potted or water culture grown plants *(Ricinus communis)* were subjected to predetermined patterns of change in atmospheric humidity in a climatological wind tunnel and responses in transpiration and water potential of the leaves followed. When the transpiration rate was altered by changing the humidity of the atmosphere there was an accompanying change in leaf water potential (e.g. development of a water deficit) only in the soil rooted plants. There was no change in the leaf water potential of the plants with roots surrounded with water even when the transpiration rate was increased × 3. It is concluded that the changes in leaf water potential arise from low water conductivity in the soil.

References

Barrs, H. D., Weatherley, P. E.: A reexamination of the relative turgidity technique for estimating water deficits in leaves. — Austr. J. biol. Sci. *15* : 413—428, 1962.

Brewig, A.: Die Regulationserscheinungen bei der Wasseraufnahme und die Wasserleitgeschwindigkeit in *Vicia faba* Wurzeln. — J. wiss. Bot. *82* : 803—828, 1936.

Brouwer, R.: The regulating influence of transpiration and suction tension on the water and salt uptake by the roots of intact *Vicia faba* plants. — Acta bot. neerl. *3* : 264—312, 1954.

Honert, T. H. van den: Water transport in plants as a catenary process. — Discuss. of the Faraday Soc. *3* : 146—153, 1948.

Köhnlein, E.: Untersuchungen über die Höhe des Wurzelwiderstandes und die Bedeutung aktiver Wurzeltätigkeit für die Wasserversorgung der Pflanzen. — Planta *10* : 381—423, 1930.

Macklon, A. E. S.: The soil : plant : atmosphere system; a study of plant water relations. — Ph. D. Thesis, Univ. of Aberdeen, 1962.

—, Weatherley, P. E.: In preparation.

Mees, G. C., Weatherley, P. E.: The mechanism of water absorption by roots. II. The role of hydrostatic pressure gradients across the cortex. — Proc. Roy. Soc. B, *147* : 381—391, 1957.

Weatherley, P. E.: Studies in the water relations of the cotton plant. I. The field measurement of water deficits in leaves. — New Phytol. *49* : 81—97, 1950.

—: Studies in the water relations of the cotton plant. II. Diurnal and seasonal variations in relative turgidity and environmental factors. — New Phytol. *50* : 36—51, 1951.

—, Barrs, H. D.: A climatological wind tunnel. — Nature *183* : 94—96, 1959.

—, Slatyer, R. O.: Relationship between relative turgidity and diffusion pressure deficit in leaves. — Nature *179* : 1085—1086, 1957.

Discussion

W. C. Visser: The very fine results of Prof. Weatherley confirm in a useful way our results concerning moisture flow through soil, plant and atmosphere. Have you already tried to fit your data in some formula of nonsteady capillary flow, as might be possible if the relation were entirely of a soil physical nature?

P. E. Weatherley: No. These results represent only a beginning. They certainly suggest that we must pay increasing attention to the soil : root system.

B. Slavik: Have you found any definite relation between the velocity of recovery of water potential and the difference between the two transpiration rates (after having reached an equilibrium value of water deficit during the high transpiration, of course)? This relation would give some data about the rate of water flow in unsaturated soil.

P. E. Weatherley: No, these are in a sense only qualitative data and proper quantitative treatment remains to be carried out.

J. Úlehla: Is there any explanation for the peak occurring before the end of the period of exposure in an atmosphere of 45% humidity?

P. E. Weatherley: I am not sure the rise at the end is significant. There is a general rise throughout the period presumably due to increasing W.P.D. in the soil round the roots.

J. Úlehla: The increase in water potential deficit observed by Prof. Weatherley after changing the air humidity from 90 to 45% (and the decrease in W.P.D. after changing back to 90% humidity) may be supposed to be accompanied by a corresponding elastic deformation of the root system. During the period of low air humidity a gap between the roots and the surrounding soil could, perhaps, enlarge or develop. An enlarged gap would then increase resistance to flow of water from the soil to the roots. After the switching back to high air humidity the contracted roots would gradually increase in volume and close or reduce the supposed gap, thereby increasing the transmissibility for water.

P. E. Weatherley: Well, I don't see if this is not a possibility. What I think is that the soil-root relationship must be studied in greater detail.

P. G. Jarvis: It would be expected that the development of the water potential deficit would be proportional to the intensity of potential evapo-transpiration. Because, when potential evapo-transpiration is high, a "dry-zone" of low capillary conductivity would rapidly develop around the roots and limit water uptake. But when potential evapo-transpiration is low, capillary conductivity would not fail rapidly and the roots would have access to water in the whole mass of the soil. Does Prof. Weatherley have any data on the development of WSD under different conditions of potential evapo-transpiration?

P. E. Weatherley: Not yet.

B. Slavík: The method used for water potential determination in this work was your own method originally described for osmotic potential measurement, wasn't it?

P. E. Weatherley: Yes.

WATER RELATIONS OF PLANTS IN EXPERIMENTS WITH HEAVY ISOTOPE O^{18}

B. B. VARTAPETYAN

K. A. Timiryazev Institute of Plant Physiology, Academy of Sciences of the U.S.S.R.,
Moscow, U.S.S.R.

Definite progress in the problem of the entry of water into a plant and its distribution in the organs of a plant has been achieved in a number of countries in the last five years due to the application of labelled water (Vartapetyan and Kursanov 1959, Hübner 1960 a, b, Ducet and Vandewalle 1960, Vartapetyan 1960a, Ordin and Gairon 1961, Hübner and Wetzel 1961, Biddulph, Nakayama and Cory 1961, Vartapetyan and Kursanov 1961 and others).

Our laboratory has also participated in this work. This article gives a brief account of investigations carried out in this field in the aforementioned period using H_2O^{18} and O_2^{18}.

Material and Methods

Various types of plants have been used for the main experiments, though animals (fishes and insects) as well as microorganisms (yeast) have been also used in some cases.

Heavy oxygen O^{18} in the form of H_2O^{18} and O_2^{18} was used. Its concentration in solutions and in the gaseous phase was usually not very high (1 to 4 atom per cent O^{18}); only rarely higher concentration was used (11 atom per cent O^{18}). Analysis of O^{18} was carried out by means of mass-spectrometric technique. For this purpose the samples of water obtained from the examined objects or solutions by means of lyophilisation were brought into isotopic exchange with a small amount of carbon dioxide in which the content of heavy oxygen was then determined (Cohn and Urey 1938).

More detailed information about the investigated objects and the methods can be found in the respective original papers cited below.

Results and Discussion

In the first experiments (Vartapetyan and Kursanov 1959, Vartapetyan 1960a) a study was made of the entrance of H_2O^{18} into the organs of *Phaseolus vulgaris* with the transpiration stream. The roots of the plant were placed into a nutrient solution with H_2O^{18} so that it was possible to follow the rate of substitution of the ordinary water of the plant's organs by the labelled water entering from the nutrient solution.

Fig. 1 shows that the substitution of water in the roots took place at such a great speed that in 15 to 30 minutes equilibrium was reached between the content of H_2O^{18} in roots and in the outer solution. Equilibrium was reached later in stems (in several hours) and even later in leaves.

It is interesting to note that the equilibrium in the organs occurs at different levels. The concentration of H_2O^{18} in the stem, for example, reached (or was close to) its level in the outer solution, thus indicating that all the ordinary water in the parenchymal cells of this organ (the water contained in protoplasm, vacuole, cell wall) can be substituted by the outer water.

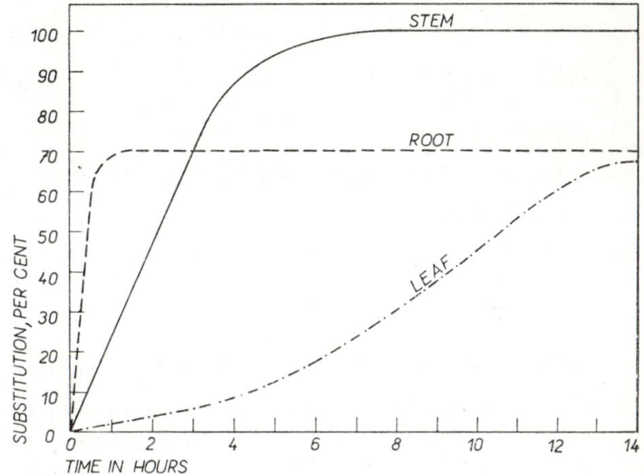

Fig. 1. Dynamics of substitution of ordinary water in *Phaseolus vulgaris* organs by H_2O^{18}.

The concentration of H_2O^{18} in the leaves and roots, on the other hand, reached only 60—70 % of that of the outer solution and showed no further increase. Similar results were obtained by Biddulph, Nakayama and Cory (1961). At that time we believed that such a distribution of H_2O^{18} in the organs of a plant is a result of the fact that a part of cell water in the roots and leaves (as distinct from the stem) is substitued by the outer water with difficulty. This opinion has gained the support of a number of other scientists (Vasileva and Burkina 1960, Lebedev 1960, Ducet and Vandewalle 1960, Biddulph, Nakayama and Cory 1961).

Hübner (1960a, b) demonstrated by using D_2O that root tissues can easily exchange the intercellular water for atmospheric moisture, and therefore a part of the labelled water may escape from the tissues after by exchange with ordinary atmospheric water. When such an exchange was excluded by

working with roots and leaves in a dehydrated atmosphere it became clear
that it was possible to obtain complete substitution of the intercellular water
by the outer water even in these organs. Thus the experiments of Hübner proved that the intercellular water can be completely substituted by the

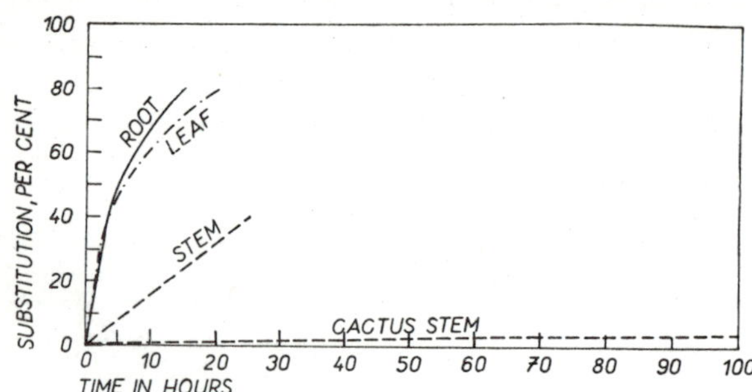

Fig. 2. Substitution of ordinary water in *Phaseolus vulgaris* organs and cactus stem by H_2O^{18} vapours
from atmosphere.

outer water not only in the
cells of the stem as had
been found in earlier ex-
periments (Vartapetyan and
Kursanov 1959, Vartapetyan
1960b) but also in the cells
of leaves and roots.

The subsequent experi-
ments (Fig. 2) revealed that
the leaves of *Phaseolus
vulgaris* also easily exchange water with atmospher-
ic vapour while the stems
are rather inactive in this respect (Vartapetyan and Kursanov 1961). Even
higher stability in exchange with atmospheric water vapour has been demonstrated in succulent organs and plants (cactuses, bulbs) due to the peculiarities of their anatomical and morphological structure which, in general,

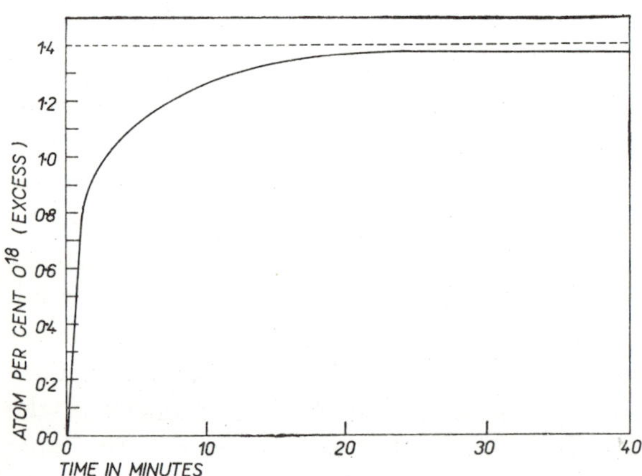

Fig. 3. Substitution of ordinary water in *Vallisneria* by
H_2O^{18}. The horizontal dotted line indicates concentration of H_2O^{18} in solution.

hamper gaseous exchange
with the outer atmosphere
(Vartapetyan and Badano-
va 1963).

As for the high rate of
exchange with the liquid
water of the nutrient me-
dium, it is necessary to point
out that this is characteristic not only for the roots
of the plants (Vartapetyan
and Kursanov 1959, Hüb-
ner 1960, Ducet and Van-
dewalle 1960, Vartapetyan
1960), but also for other
organs and organisms. For
example, the leaves of the

water plant *Vallisneria*, when placed into the solution of H_2O^{18} exchanged
more than 50 per cent of the intercellular water in one minute (Fig. 3).
The rate of water exchange of microorganisms is even greater. Yeast (*Sac-
charomyces cerevisiae*) when placed into a solution of H_2O^{18} completely ex-

changed the whole of its intercellular water in less than a minute (Vartapetyan and Kursanov 1961). The water animals, for example sea- and freshwater fishes, relatively easily and completely exchange their intercellular water (Vartapetyan 1962b).

These results are of interest as they show that it is quite easy to exchange the bound water of the cell. That is why in the following experiments we tried to find out the degree of mobility of biocolloid hydrated shells, i.e. of the most firmly bound water of the cell.

Fig. 4 illustrates schematically the structure of hydrated shells of hydrophilic colloids. According to the existing conceptions, dipoles of water are located around the molecule of the hydrophilic colloid. These dipoles of water are bound to the colloid matter, the first monomolecular layer of water, nearest to the colloid matter being bound most firmly. Making use of H_2O^{18} we tried to find whether these water shells were really not very mobile (Vartapetyan 1962a). Small starch grains of maize were chosen for the experiments.

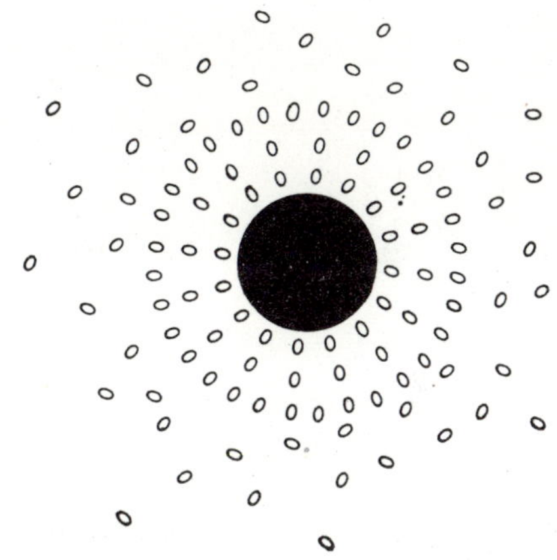

Fig. 4. Hydrate shells around particle of hydrophilic colloid.

The experimental procedure is shown schematically in Fig. 5.

Starch grains were first thoroughly dehydrated for 15—20 hours period in a lyophilisation apparatus by means of liquid nitrogen (—195·8° C) in the vacuum of $1 \times 10^{-2} — 1 \times 10^{-3}$ mm. of mercury column. Under these conditions the starch lost its hydrated layers including the first monomolecular Langmuir's layer of water (A). Then, after injection of H_2O^{18} into the flask and after suspending the starch grains in this flask (B), the hydrated layers were built anew, but now from the H_2O^{18} (C). After that the flask (a) was filled with ordinary water (D) and the starch grains were quickly precipitated by centrifugation (E). The concentration of H_2O^{18} in hydrated shells of starch (sediment) and in the supernatant was then determined by using the mass-spectrometer technique (E). The results of these experiments showed that the concentration of H_2O^{18} in the water of hydrated shells of starch and in the fluid, surrounding the starch grains, was the same (I). Consequently, the hydrated shells of biocolloid (including the first monomolecular layer of water) are not securely "bound", but are rather mobile (in respect to exchange), otherwise the content of H_2O^{18} in the starch grains should have exceeded the concentration of H_2O^{18} in the surrounding fluid (II). The mobility of hydrated shells does not decrease appreciably even when the cell is partially dehydrated due to the fact that it had been placed into an osmotically active solution.

More recently we have been occupied with determining the role of the water which appears in the plants as the result of biosynthesis from atmospheric O_2 and from the hydrogen of respiration substrate. It had already been demonstrated experimentally in our laboratory, making use of O_2^{18},

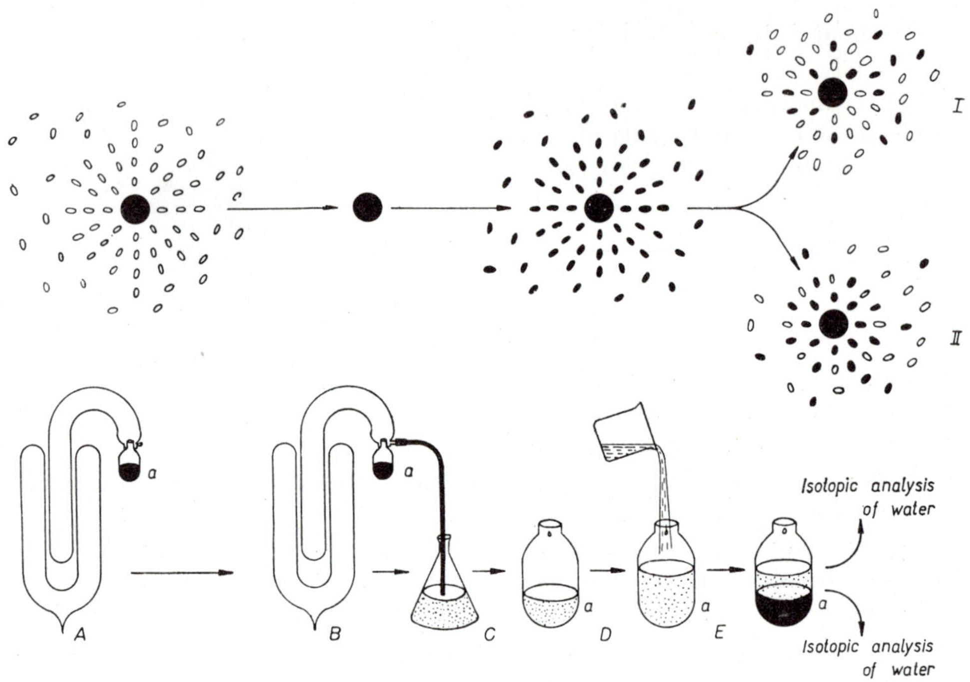

Fig. 5. Consequence of procedures during investigation of hydrate shell mobility. White points: ordinary water, full points: heavy water.

that the oxygen which is absorbed from the atmosphere during respiration is utilized for the synthesis of water in cells (Vartapetyan and Kursanov 1955). The subsequent experiments on an animal object (the silkworm *Bombyx mori*) revealed that such endogenous means of obtaining water by some organisms can play an essential role in the general water balance (Vartapetyan 1963).

Fig. 6 demonstrates the procedure of experiments with *Bombyx mori*. O_2^{18} which is obtained in electrolysis (A) from H_2O^{18} enters chamber (C) where it is used for respiration. As a result of O_2^{18} fixation in the tissues of *Bombyx mori* H_2O^{18} appears and its amount gradually increases while the initial ordinary water of the tissues is gradually lost, some of it physically (evaporation) and some of it biochemically (metabolism). The results of these experiments have shown that insects obtain a considerable amount of water endogenously and that by the end of life almost the entire amount of water in the body of *Bombyx mori* may be of endogenous origin.

Later we tried to find out the significance of endogenous water for plants, and especially for those growing in conditions of water deficit, in drought and desert areas. From this point of view the succulent plants are of interest, because they are able, similar to *Bombyx mori,* to live for a long time without obtaining water from the outside and to preserve a considerable amount of water in the tissues under these conditions.

Our experiments with O_2^{18}, however, showed that succulent organisms

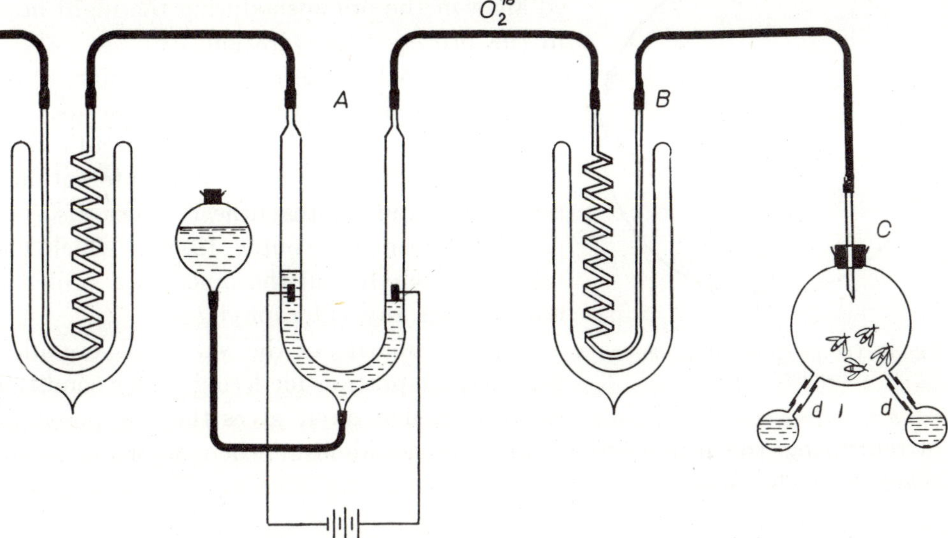

Fig. 6. Apparatus for investigation of endogenous synthesis of water.

(cactus), having been without water even for a long time obtained an insignificant amount of H_2O^{18} by means of biosynthesis from atmospheric O_2^{18}. On the basis of these findings the conclusion could be reached that the endogenous water is here of no importance. But such a conclusion might prove to be not quite correct taking into account some anatomical and physiological peculiarities of succulent plants. The thick cuticle preserving these plants from extra water expenditure must also prevent the loss of O_2 resulting from photosynthesis during day hours as well as of the respiration CO_2 accumulated at night. As to CO_2, its loss is also eliminated biochemically. It is known the succulents synthesize a considerable amount of organic acids at night as a result of carboxylation of CO_2 from respiration (Bennet-Clark 1949).

Consequently, at night there is a decomposition of water in the tissues of the succulents caused by respiration. The oxygen of this water conjugates with the carbon of the respiration substrate forming CO_2 which is later conserved in organic acids in the process of carboxylation. During the day hours the water spent by the organism in the process of respiration is again regenerated in the organism during the process of photosynthesis.

But the process of photosynthesis itself leads to the decomposition of the cell water to O_2. The O_2 of photosynthesis accumulated during the day hours in the succulent tissues is again synthesized into water during the night hours due to respiration. Thus the water spent by a cactus during the day hours as a result of photolysis is regenerated anew in the organism during the night hours in the process of respiration.

Consequently, there is a peculiar closed water cycle in the tissues of a cactus. This cycle leads to a periodic reutilization of oxygen (Fig. 7). In the light hours such interaction of respiration and photosynthesis must also take place between the outer tissues which contain chlorophyll and the inner ones, which do not contain any chlorophyll.

This peculiarity of the succulent organisms is a good adaptation for living under conditions of water deficit as it gives them a possibility of reutilizing the intercellular water by metabolism when no water is supplied from the outside.

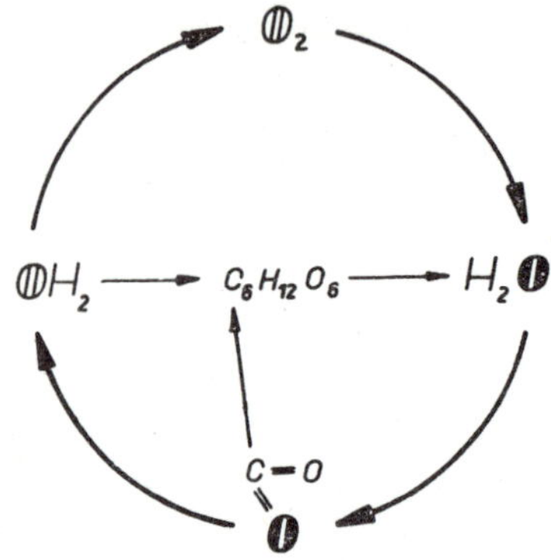

Fig. 7. Oxygen reutilization in succulent plants.

Summary

Using different kinds of labelled water it was possible to show that intercellular water, both in unicellular and multicellular organisms, undertakes rapid and complete exchange with the external water. High mobility is also peculiar to the bound structural water of the cell, including monomolecular hydrate shells of the biological colloid. The partial moving off of the water from the cell does not markedly reduce the mobility of biocolloid hydrate shells.

Plant organs' water exchanges with the vapour of the atmosphere. This exchange proceeds in nonsucculent organs with high rate while succulent organs are rather resistant in this respect. The importance of this phenomenon for the possibility of using atmospheric water vapour by plants should be specially studied.

The water stress of some organisms can be compensated in some degree by the endogenously originated water forming in the cell by the biosynthetic way from the molecular oxygen and hydrogen of the oxidation substrate.

References

Bennet-Clark, T. A.: Organic acids of plants. — Ann. Rev. Biochem. *18* : 639—654, 1949.

Biddulph, O., Nakayama, F. S., Cory, R.: Transpiration stream and ascension of calcium. — Plant Physiol. *36* : 429—436, 1961.

Cohn, M., Urey, H. C.: Oxygen exchange reactions of organic compounds and water. — J. Amer. chem. Soc. *60* : 679—687, 1938.

Ducet, G., Vandewalle, G.: [Absorption of water by excised roots.] In Russ. — Fiziol. Rast. *7* : 407—413, 1960.

Hübner, G.: Die Probenahme für die massenspektrometrische Deuteriumanalyse. — Kernenergie *3* : 888—890, 1960a.

—: Zum Wassertransport in *Vicia faba*. — Flora *148* : 549—594, 1960b.

—, Wetzel, K.: Zum Mechanismus des Wasserdurchtritts durch lebende Membranen. — Ber. dtsch. bot. Ges. *74* : 255—256, 1961.

Lcbedev, G. V.: [The state of water in the plant cell.] In Russ. — Fiziol. Rast. *7* : 398—400, 1960.

Ordin, L., Gairon, S.: Diffusion of tritiated water into roots as influenced by water status of tissue. — Plant Physiol. *36* : 331—335, 1961.

Vartapetyan, B. B.: [Participation of H_2O^{18} in the metabolism of photosynthesizing tissues.] In Russ. — Fiziol. Rast. *7* : 414—418, 1960a.

—: [Further investigation of exchange of water in plants with aid of heavy water H_2O^{18}.] In Russ. — Fiziol. Rast. *7* : 395—397, 1960b.

—: [Mobility of starch hydrate shells.] In Russ. — Dokl. Akad. Nauk SSSR *147* : 221—223, 1962a.

—: Untersuchungen des Wasseraustausches des Meerefisches *Mugil saliens* Risso unter Anwendung von H_2O^{18}. — Kernenergie *5* : 384—386, 1962b.

—: [Endogenous water of the silkworm.] In Russ. — Biokhimiya *28* : 764—768, 1963.

—, Badanova, K. A.: [Rate of water exchange in dormant plant organs.] In Russ. — Fiziol. Rast. *10* : 106—108, 1963.

—, Kursanov, A. L.: [The participation of water oxygen and atmospheric oxygen in plant respiration.] In Russ. — Dokl. Akad. Nauk SSSR *104* : 272—275, 1955.

—, —: [A study of water exchange in plants with aid of water containing heavy oxygen, H_2O^{18}.] In Russ. — Fiziol. Rast. *6* : 144—150, 1959.

—, —: [Exchange of water between plant tissues and liquid and vaporous water of the environment.] In Russ. — Fiziol. Rast. *8* : 569—575, 1961.

Vasileva, N. G., Burkina, Z. S.: [Water conditions in cell organoids.] In Russ. — Fiziol. Rast. *7* : 401—405, 1960.

Discussion

B. Slavík: As in the first experiments mentioned the shoot part of *Phaseolus* plant was in an open atmosphere, the equilibrium reached in the stem and leaves was determined by the transpiration rate and has, therefore, no absolute significance.

B. B. Vartapetyan: The equilibrium reached in the content of H_2O^{18} in the overground organs of *Phaseolus vulgaris* was due to the following factors:
1. The one-way flow of H_2O^{18} together with the transpiration stream through the root system.
2. The exchange of water in the organs of the plant with unlabelled atmospheric water vapour.
Since the rate of this exchange in the stem was small it could not basically affect the H_2O^{18} content. As against this, in the leaves the exchange greatly decreased the concentration of H_2O^{18}. Therefore in these experiments the transpiration rate actually had a relative character.

STATE OF WATER IN A PLANT CELL, MOBILITY OF WATER AND FACTORS WHICH DETERMINE IT

G. V. LEBEDEV and N. A. ASKOCHENSKAYA

K. A. Timiryazev Institute of Plant Physiology, Academy of Sciences of the U.S.S.R., Moscow, U.S.S.R.

The discovery of heavy water laid the foundation for direct investigations on the rate of its diffusion into plant and animal tissues. A whole number of studies carried out in this direction confirmed the remarkable ease with which water enters animal and plant cells (Hevesy and Hofer 1934, Wartiovaara 1944, Thimann and Samuel 1955, Kutyurin 1956, Ordin and Bonner 1956, Ordin and Kramer 1956, Lebedev 1959, 1960, Vartapetyan 1960, 1962, Ducet and Vandewalle 1960, Hübner 1960, Samuilov and Efremov 1962, Lebedev, Sabinina and Chuchkin 1963). In accordance with experimental and theoretical data it was established that when tissues are immersed in heavy water solutions or when they are transferred from heavy water to H_2O a certain water exchange occurs in them in compliance with the laws of diffusion of two-component systems.

In a work published in 1959 (Lebedev 1959) we also confirmed the above findings and simultaneously arrived at the conclusion that the "bound" water kept by cell colloids is mobile and easily exchangeable. This conclusion was based on the fact that when bean and pumpkin seeds were immersed in heavy water a complete substitution of outer medium H_2O for heavy water contained in them takes place, the rate of this process being such that after 30 minutes 50 per cent substitution of one form of water for another was observed* (Fig. 1).

Recently, a similar conclusion of fast exchangeability of "bound" water was also arrived at by Hübner (1960), Vartapetyan (1962), and Samuilov and Efremov (1962).

Simplifying as much as possible our conceptions of cell structural elements and their interactions with water let us try to envisage schematically the given process on the basis of current ideas concerning the functional dependence of the degree of colloid hydration on the value of the field of force of a hydrated particle (in this work the character of the interaction of forces is not considered).

It is well known that in a water solution around every particle (ion, mole-

---

* Determination of diffusion coefficients D_2O and H_2O^{18} showed that water penetrates a cell in the form of non-dissociated molecule (Hübner and Wetzel 1961).

cule, radical, globule) there is a zone of interaction with H_2O dipoles surrounding it. On entering this zone water molecules are subjected to a certain influence and arranged in a rather ordered state forming the polymolecular layer of "bound" water.

Forces keeping water in a "bound" state are distributed irregularly being characterized by an S-shaped curve. In this case the water layers nearest

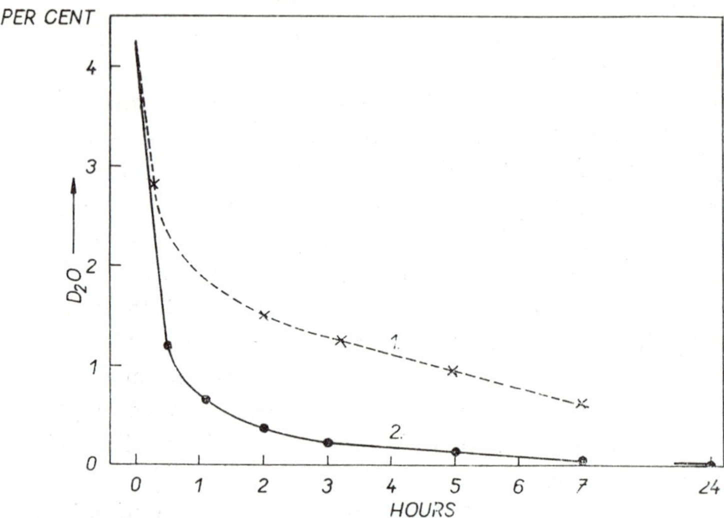

Fig. 1. Concentration of heavy water in seeds (exchange $H_2O \rightleftarrows D_2O$). 1 — Bean seeds, 2 — Pumpkin seeds.

to the particle are attracted to it more strongly than all succeeding ones the so-called Langmuir's monomolecular layer being subjected to the greatest action (Fig. 2).

In these conditions one would naturally expect a considerable decrease in the mobility of water molecules firmly bound by a colloid. A consideration of the given system in a static condition does not enable us to answer the question why all water molecules in conditions of interaction with ultimately swollen colloid possess substantial mobility and are easily exchanged with water of the surrounding solution.

This question can be answered if we consider the relation between the values of vapour elasticity and water content in the considered system. These data show that "as water content decreases, forces of water retention accounting for one molecule of water increase" (Crafts, Currier and Stocking 1949).

In all probability the characteristics of the interaction field in this case should be represented as the following isotherms (Fig. 3).

This conception about the distribution of forces in the zone of interaction is to some extent based on the consideration of vapour elasticity values and hydration degree in the zone of monomolecular layer formation and enables us to understand the changes in the character of mobility of water molecules around the hydrated particle as it becomes dehydrated.

The attached diagram shows that an increase in hydration promotes a change in the relation of forces applied to each molecule of water and, in conditions of complete hydration, Langmuir's layer in the zone of monomolecular layer formation will be held around the particle much more loosely and the entire interaction zone and, in particular, its adjacent layers will be more "friable". Therefore, the crystal lattice of water in this case should not be well ordered. Accordingly, physical and chemical parameters of Langmuir's layer as well as diffusion layer should depend on the degree of particle hydration.

In case of complete hydration of the particle there exists the highest though heterogenous mobility of all water layers around it due to the distribution of interaction forces throughout the polymolecular layer.

This high mobility of

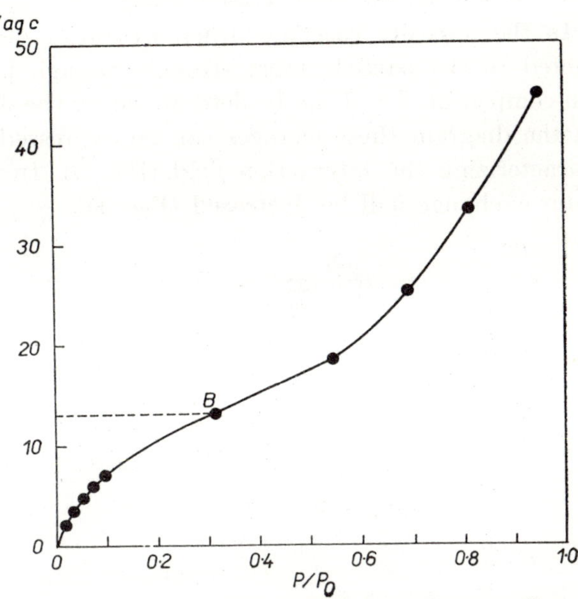

Fig. 2. Isotherm of water vapour absorption on collagen at 25° C. OB-zone of the formation of Langmuir's monomolecular layer (Pasynsky 1959).

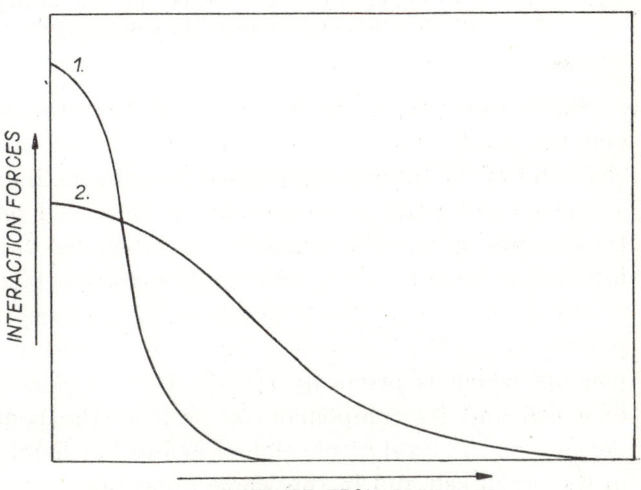

Fig. 3. Characteristics of the interaction field of a particle in conditions of variable hydration. 1 — Highly dehydrated system. 2 — Partial dehydration.

all water layers accounts for high permeability of animal and plant tissues to H_2O.

Besides this, the activity of water in this case is highest (Alekseev 1938, Alekseev and Gusev 1957, Gusev 1959).

In the opposite case, i.e. when hydration decreases, water layers are attracted to the particle more strongly though part of the system energy is non-compensated and can be determined by the difference of vapour elasticity. On the diagram these changes can be expressed in the decrease of the area characterizing the interaction field (Fig. 3). In these conditions the rate of water exchange will be decreased (Fig. 4).

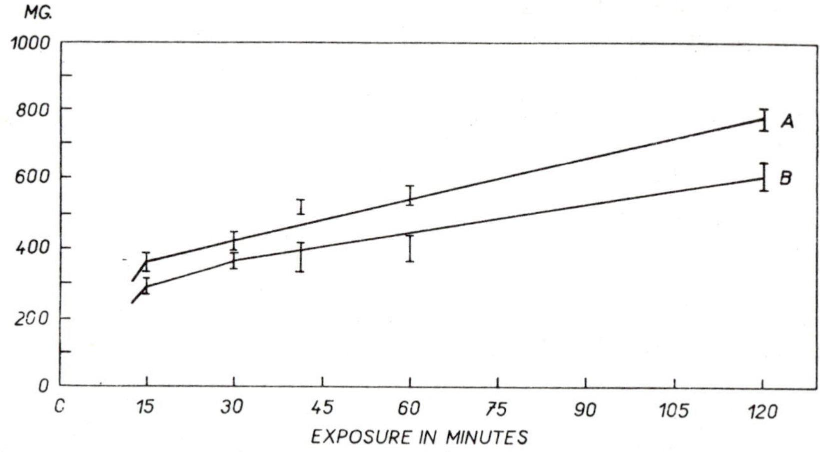

Fig. 4. Amount of exchanged water "bound" by 1 g. of dry bean seeds after their swelling. A — seeds in water, B — seeds in the mannitol solution (Lebedev, Sabinina and Chuchkin 1963).

Simultaneously, it should be noted that the size of the interaction field and the gradient of forces by which it is formed will depend both on the character of the interacting particle and the medium by which it is surrounded.

Such regularities can naturally be observed only under conditions where the interacting particle remains invariable, the value and configuration of the interaction zone changing due to dehydration factor.

When there is a system in which the particle itself can easily change, the parameters of the interaction zone will depend on the state of the both components which is primarily specific to biological systems since the structure of a cell and its components as well as the composition of a liquid phase depends on the level of physiological and biochemical processes occurring both in the given cell and in the whole organism.

A change in the energy level of the system due to its dehydration results in the development of the above-mentioned processes with the simultaneous change of interaction zones and consequently in a certain deformation of the molecular structure of interacting components. A deviation from the standard in cell structures brings about a change in the level and tendency of physiological and biochemical processes which in its turn determines a new state of functioning structures and the composition of the cell liquid phase as well as a new level of metabolism in an organism against the background of newly formed criteria of water relations. The described changes should occur both in conditions of normal and pathological physiology in processes associated with the variation of structural peculiarities of the functioning protoplasm and organoids which in the former case accounts for the rythmics of the

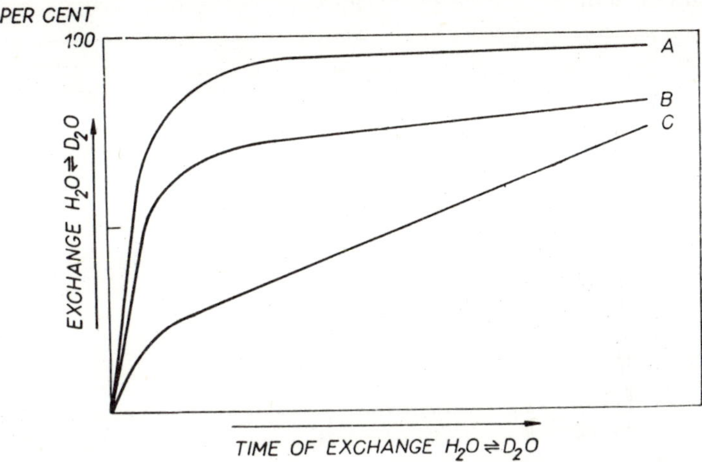

Fig. 5. A — high permeability of all cell elements to water. B — fast water exchange in cell superficial layers (cell walls, protoplasm), and slow water exchange — through biological membranes (vacuoles, nucleus, chloroplasts, mitochondria, etc.), C — slow diffusion of water through superficial cell layers.

water relations of the organism or in the latter case violates its water balance. The utilization of substances stabilizing structures should certainly affect the level of physiological and biochemical. processes and consequently influence their water relations.

From this point of view we can also consider the osmotic pressure, suction force and turgor pressure as quantities which, taken together characterize the level and tendency of the above processes.

Thus the water relations of a cell and of the whole organism in normal conditions should be considered as a s t r i c t l y   r e g u l a t e d   p h y s i o l o g i c a l   p r o c e s s   functionally dependent on the activity of the whole organism with due regard for the factors of environment.

Various cell structural elements should also differently determine the rate of water diffusion due to both their functional peculiarities and structural specificity.

In this case investigators working on problems of water diffusion into animal and plant tissues should encounter the changes shown in Fig. 5. Under conditions of adequate permeability of all cell elements to heavy water its rapid diffusion into tissues will be observed and the character of the enrichment curve will be that which is inherent in the diffusion process of the two-component system — Fig. 5A. But in the presence of retarded diffusion, for instance, because of cell membranes the process of metabolism $H_2O \rightarrow D_2O$ will be characterized by the curve with a plateau which rises comparatively slowly (Fig. 5B).

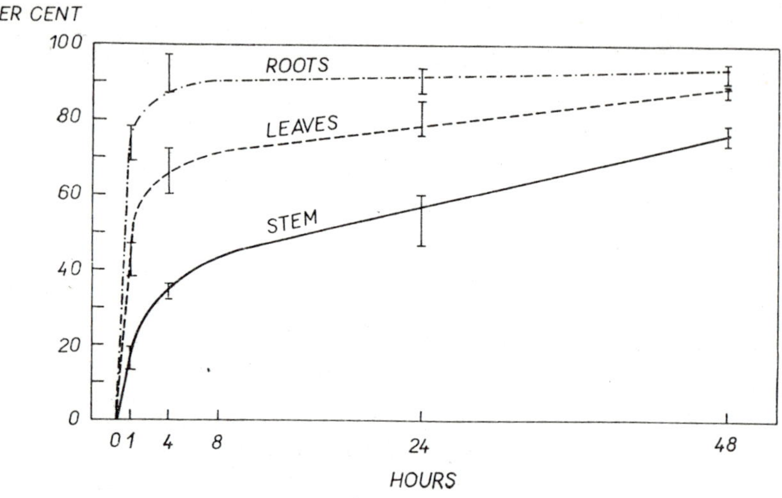

Fig. 6. Water exchange ($H_2O \rightleftarrows D_2O$) in different bean tissues (in relative percentages). $\cdots\cdots$ roots, ——— stem, $----$ leaves.

If the permeability of upper layers of a cell to water is much lower than that of the inner ones the process of diffusion will be expressed by a gradually rising curve (Fig. 5C). The other cases will be intermediate between the findings shown in this Figure.

This description of diffusion processes in the system $H_2O \rightleftarrows D_2O$ at the boundary of two media divided by biological membranes refers to all cell organoids.

Studying this problem we carried out experiments with Askochenskaya basically as follows:

Bean plants *Vicia faba* in the phase of the second formed leaf were dismembered into roots, stem, and leaves while the surface of parts was laid in paraffin. Every part of the plant was placed in a solution of heavy water (—10 atom%) for 1 minute, 24 and 48 hours. After these periods they were taken out of the solution, dried with filter paper while the paraffined ends were cut off. This operation was carried out in a dry chamber with $CaCl_2$. Plant parts and water samples in which the former were placed were subjected to analysis for determination of D_2O in them (Lebedev and Chuchkin 1962). This was repeated five times and in the case of 24 hours exposure — eight times. The data obtained were statistically treated and are given in Fig. 6. Similar experiments were carried out with maize with exposure for 1, 3, 5, and 24 hours and were repeated twice (Fig. 7).

These data enable us to draw the conclusion that the most intense metabolism $H_2O \rightleftarrows D_2O$ occurs in the first place in roots then in leaves and finally in the stem. These data also point to the varying character of the permeability of the tissues analysed. Comparing the data given in Figs. 4, 5, and 6 one can suppose that cell vacuoles perhaps also participate in the process of water passage through the root system of plants.

The process of water mobility in a cell should also be influenced by the

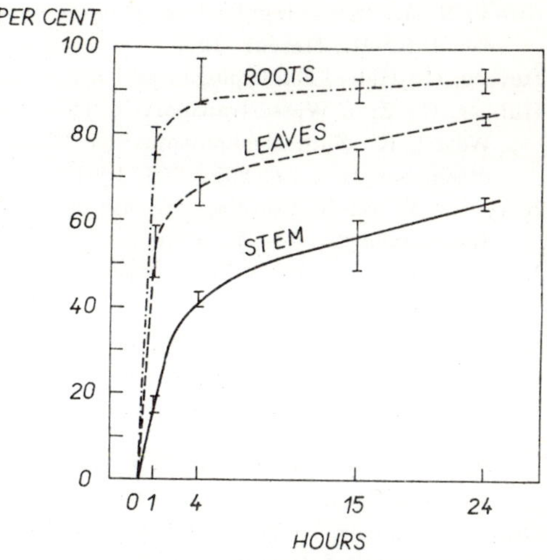

Fig. 7. Water exchange ($H_2O \rightleftarrows D_2O$) in different maize tissues (in relative percentages). · · · · · · roots, ——— stem, - - - - - leaves.

movement of protoplasm and the presence of vacuoles with contractile functions (Dangeard 1950). It is quite possible that root cells have such vacuoles and in leaves, as specialized the organs, cells do not possess these functions or possess them to a less extent. Therefore, in the latter case much more time is needed for evening up of water concentration between a cell and an outer solution.

Thus, by studying the process of water diffusion in a cell it is possible to some extent to characterize the functional state of biological membranes which is important from the standpoint of the knowledge of their role in the general functional relationship of reacting structures.

References

Alekseev, A. M.: [Water relations of a plant and the effect of drought on it.] In Russ. — Tatgosizdat, Kazan 1948.

—, Gusev, N. A.: [Influence of mineral nutrition on water relations of plants.] — In Russ. — Publ. House Acad. Sci. U.S.S.R., Moscow 1957.

Crafts, A. S., Currier, H. B., Stocking, C. R.: Water in the physiology of plants. — Chron. Bot., Waltham, Mass. 1949.

Dangeard, P.: [Cytology of plants and general cytology.] In Russ. — Foreign Lit. Publ. House, Moscow 1950.

Ducet, G., Vandewalle, G.: [Absorption of water by cut-off roots.] In Russ. — Fiziol. Rast. 7 : 407—413, 1960.

Gusev, N. A.: [Some regularities of water relations of plants.] In Russ. — Publ. House Acad. Sci. U.S.S.R., Moscow 1959.

Hevesy, G., Hofer, E.: Diplogen and fist. — Nature *133* : 495—496, 1934.

Hübner, G.: Zum Wassertransport in *Vicia faba*. — Flora *148* : 549—594, 1960.

—, Wetzel, K.: Zum Mechanismus des Wasserdurchtritts durch lebende Membranen. — Ber. dtsch. bot. Ges. *74* : 255—256, 1961.

Kutyurin, V. M.: [Concerning the rate of deuterium penetration into plant tissue.] In Russ. — Biokhimiya *21* : 50—52, 1956.

Lebedev, G. V.: [Rate of water exchange in swollen plant seeds.] In Russ. — Dokl. Akad. Nauk SSSR *128* : 632—634, 1959.

—: [State of water in a plant cell.] In Russ. — Fiziol. Rast. *7* : 398—400, 1960.

—, Chuchkin, V. G.: [Purification and isolation of isotopic water forms from biological objects by means of distillation in nitrogen flow and on inserts.] In Russ. — Fiziol. Rast. *9* : 259 — 262, 1962.

—, Sabinina, E. D., Chuchkin, V. G.: [State of water in a plant cell. Mobility of colloid water and water bound by crystals.] In Russ. — Fiziol. Rast. *10* : 108—110, 1963.

Ordin, L., Bonner, J.: Permeability of *Avena* coleoptile sections to water measured by diffusion of deuterium hydroxide. — Plant Physiol. *31* : 53—57, 1956.

—, Kramer, P. J.: Permeability of *Vicia faba* root segments to water as measured by diffusion of deuterium hydroxide. — Plant Physiol. *31* : 468—471, 1956.

Pasynsky, A. G.: [Colloid chemistry.] In Russ. — State Publ. House "Higher School", Moscow 1959.

Samuilov, F. D., Efremov, Yu. Ya.: [Study of water exchange in plants with the help of heavy water (D_2O).] In Russ. — Fiziol. Rast. *9* : 438—445, 1962.

Thimann, K. V., Samuel, E. W.: The permeability of potato tissue to water. — Proc. Natl. Acad. Sci. U.S. *41* : 1029—1033, 1955.

Vartapetyan, B. B.: [Further investigation of water exchange in plants with the help of heavy water H_2O^{18}.] In Russ. — Fiziol. Rast. *7* : 395—397, 1960.

—: [Rate of water exchange in a sea bony fish.] In Russ. — Dokl. Akad. Nauk SSSR *143* : 721—723, 1962.

Wartiovaara, V.: The permeability of *Tollypellopsis* cell for heavy water and methyl alcohol. — Acta bot. fenn. *34* : 1—22, 1944.

WATER STRESS IN CONIFERS DURING WINTER

G. HYGEN

Agricultural College of Norway, Vollebekk, Norway

I

In a northern continental climate the water supply to evergreen conifer needles is apparently cut off in the late autumn by the onset of frost. Transpiration still proceeds, however, even if the rate is reduced to a minimum. Consequently the needles must be exposed to a slowly increasing water stress. This results in a gradual reduction of the leaf water content during winter, and a corresponding increase in the osmotic potential of the cell sap. If these changes should exceed a comparatively moderate range, the needles would die. Since most of them do survive after all, we are faced with the problem of how they may avoid excessive water loss.

Let us to begin with take it for granted that freezing of soil, stems, twigs and leaves completely prevents upward water transport, and that the only opportunities for water replenishment occur during occasional thawing periods. Even in the most severe winter climate the mercury sometimes creeps above zero, although it does not stay there for long. Local thawing of frozen tissue might also occur on clear days with air temperatures a little below zero, owing to absorption of solar radiation.

Under such conditions, however, another obstacle may prevent translocation of fluid water, viz., the occurrence of air embolism. When ice melts, the dissolved gases are expelled in the form of bubbles. Scholander, Love, and Kanwisher (1955) and Scholander, Ruud, and Leivestad (1957) have shown in recent experiments with living stems that freezing and subsequent thawing cause strings of tiny air bubbles to form in the conducting xylem elements. Thereby the cohesive water columns are broken, so that the xylem elements concerned become eliminated from the conducting system, at least temporarily. Water diffusion through the heavily lignified cell walls of the tracheids could not occur to any significant degree, considering the diffusion distance and the temperatures involved.

Thus we arrive at the preliminary conclusion from current theories that after freezing has once occurred, the needles should not be able to obtain any water from the stem to balance the transpiration loss.

II

Now let us consider the magnitude of the transpiration loss. Norway spruce *(Picea abies)* may serve as an example to illustrate the relation. This is one of the several ecologically different species whose water relations have been thoroughly investigated by Pisek and his coworkers in Austria (1938, 1939, 1953). It is evident from these investigations that the spruce needles respond to water stress by closing of the stomata, so that after a while only cuticular

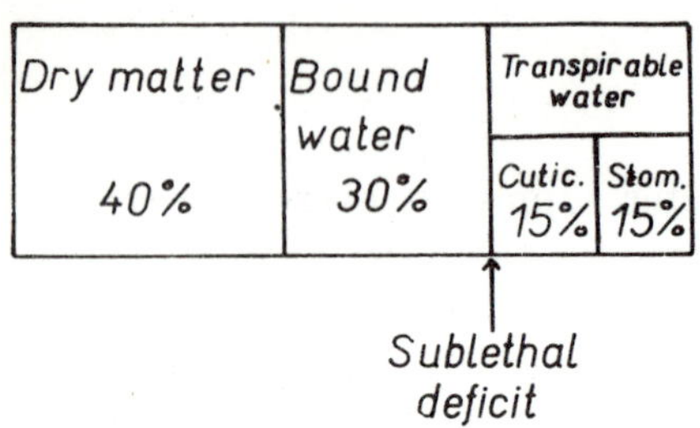

Fig. 1. Gross distribution of water in spruce needles. (From data presented by Pisek and Winkler 1953).

transpiration proceeds. The rate of water loss is hereby very markedly reduced. But if transpiration is allowed to proceed without water replenishment, the needles must sooner or later become damaged by drought.

The critical stage can be defined by the so-called sublethal water deficit. This is the average amount of water lost when 5 percent of the needles have suffered irreversible damage. The water economy of the spruce needles can be illustrated by a simple diagram which indicates the order of magnitude of the different water fractions (Fig. 1).

At saturation the total water content of the needles varies around 60 % of the fresh-weight. The sublethal level is attained when about one half of this quantity is lost. Therefore, the total amount of water available for transpiration represents about 30 % of the needle fresh-weight. Approximately one half of this amount is lost in the initial phase, before the closing movement of the stomata is completed. Thus, the water fraction which remains to be spent by cuticular transpiration, before the occurrence of irreversible damage, constitutes only 15 % of the initial needle fresh-weight, or 150 mg. per gram.

Now, if the water supply to the needles should be cut off early in winter by freezing of soil and xylem, this quantity of 150 mg. water per gram would be all the needles have to spend during the whole winter.

The endurance of this initial water capital would depend upon the rate of spending. The rate of cuticular transpiration is determined by the transpiring properties of the needles on the one hand, and by the existing meteorological conditions on the other. Pisek and Winkler (1953) estimated that spruce needles could only keep alive about 4 days without water replenishment under the conditions of their laboratory experiments. But when we try to apply these data to the radically different conditions which exist in a Norwegian spruce forest during winter, we are on uncertain grounds. To some extent we must resort to assumptions.

Unfortunately, actual measurements of spruce transpiration under natural conditions in winter are not available from Norway, but a minimum value can be estimated by taking the cuticular transpiration to be proportional to the atmospheric vapour pressure deficit at any one time. This assumption holds good only under idealized conditions, when the stomata keep closed all the time, when the leaf temperature does not rise above the air temperature, and when the transpiration rate is not influenced by wind. But any error caused by deviations from these hypothetical conditions must all go in the same direction, and lead to an underrating of the water loss, and a corresponding overrating of the life span of the needles.

The minimum water loss during a certain period can therefore be estimated from the following simple formula:

$$T_c = 20 \cdot \Sigma D \text{ mg./g.}$$

T_c is here the total amount of water lost by cuticular transpiration during the period in question, and D is the daily means of the atmospheric saturation deficit in mm. mercury. The figure 20 is a proportionality factor estimated from available transpiration data.

Bearing in mind that the maximum amount which could be spent by cuticular transpiration is 150 mg. per g., it follows that the needles would be doomed when the deficit sum attains the value of 7·5.

The atmospheric saturation deficit at any one time can be easily derived from temperature and humidity registrations. The diagram

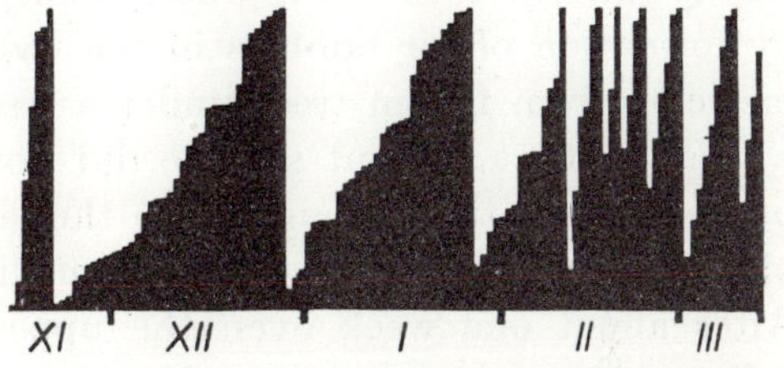

Fig. 2. Atmospheric saturation deficit in Oslo in the winter 1961/62. The following dates are marked on the abscissa: 15/XI, 1/XII, 1/I, 1/II, 1/III and 15/III. The maximum level of the ordinate corresponds to a deficit of 7·5 mm. Hg. Explanation in the text.

in Fig. 2 may serve to illustrate the relations which obtained during the winter 1961/62 in Oslo.

Here the abscissa indicates the time, and the ordinate shows the accumulated sum of the daily deficit means. Whenever this sum exceeded the critical level 7·5, the step curve of the diagram was discontinued and started anew from zero. During each of the resulting periods the needles must have spent at least 150 mg. water per gram, according to our formula. From the number of breaks in the curve it appears that the total water expenditure during winter would amount to nearly ten times the critical quantity.

III

Let us reflect a little on how the problem stands by now. On the one hand we have found that in winter the water supply to the needles should be completely blocked because of freezing and air embolism in the xylem. On the other hand it appears that the needles nevertheless spend several times their initial content of transpirable water. Indeed, the total water loss during winter would seem to exceed the initial fresh-weight of the needles by 50 %.

Obviously, these two statements cannot be reconciled. It seems that the needles must somehow have got an additional provision of water. If so, three possible channels of supply are open to suspicion.

Firstly, a fairly feasible manner of obtaining additional water would be by way of absorption of moisture from the atmosphere. Such absorption does occur during rainfall, dew and melting of snow, and probably also to a small extent from a water-saturated atmosphere.

Without accurate observations we cannot say for certain whether this source might suffice to account for the apparent water surplus, but tentative estimates do not support this explanation.

As a second possibility we must examine the theory of air embolism more closely. In order to see if water translocation becomes permanently blocked by formation of air bubbles in the xylem during thawing, we cut off spruce branches from frozen trees under a freezing mixture at various times during winter, so that the cut surface did not come into contact with free air. The branches were then brought into the laboratory and transferred to a solution of *lignin pink* in water. They immediately started to absorb dye solution. After about one week even the uppermost little twigs showed a clear red colour of the wood upon cutting. Sections at different levels showed that the major part of the xylem had been very vividly coloured, thus demonstrating an apparently unhampered conducting function of the xylem. We failed to observe significant differences between samples taken at different times during winter. All the branches behaved in a similar way (Fig. 3).

These experiments proved quite convincingly that the xylem of frozen spruce branches is able to resume water transport immediately upon thawing, in apparent contradiction to the theory of air embolism.

As a third and most unlikely channel of supply we may consider the possibility of water uptake from frozen soil and water transport through frozen stems. When wood freezes, the water content is not quantitatively transferred to ice. A certain fraction remains unfrozen. Working in Oslo under Scholander's supervision, Lybeck (1959) used a calorimetric method to determine the amount of ice formed in the wood of different species at various temperatures. Fig. 4 shows the results of her experiments with Norway spruce.

In this diagram the horizontal lines indicate the initial water content of

Fig. 3. Cross section of spruce branches cut in November, December, January, and April. Treatment explained in the text.

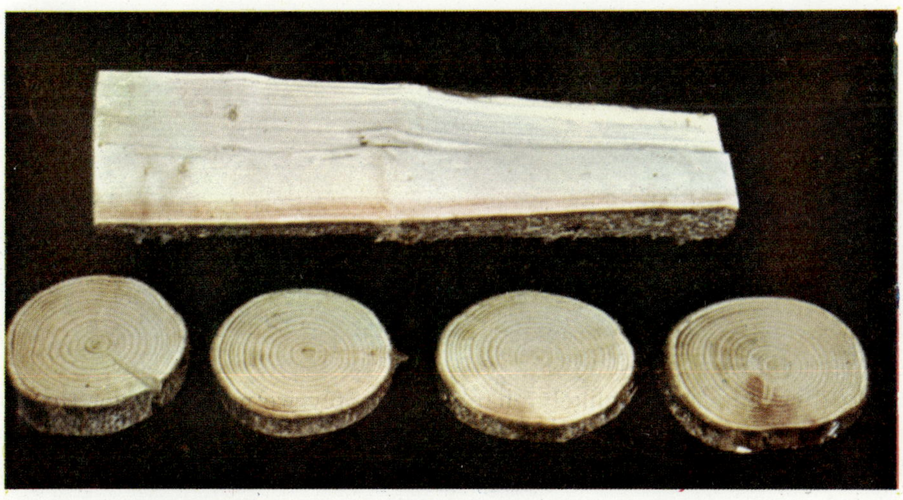

Fig. 5. Dye marks in a spruce stem, indicating translocation through frozen wood. Explanation in the text.

the wood samples, which varied between 65 % and 125 % of the dry-weight, with a mean around 87 %. The slightly curved line at the bottom indicates the amount of unfrozen water in the wood at the different temperatures.

It is evident from this curve that freezing takes place at temperatures very close to zero. When freezing has occurred the amount of water remaining unfrozen is always the same, irrespective of the initial water content. And most important of all, the amount of unfrozen water is practically independent of the temperature. The curve runs horizontally during the entire temperature range from —2° C to —20° C.

Lybeck concluded from these results that "the unfrozen water is essentially that which is bound to the cellulose walls", and that "in winter time practically all sap within vessels and tracheids freezes, forcing the dissolved gases out as bubbles".

To my mind these conclusions do not seem absolutely convincing.

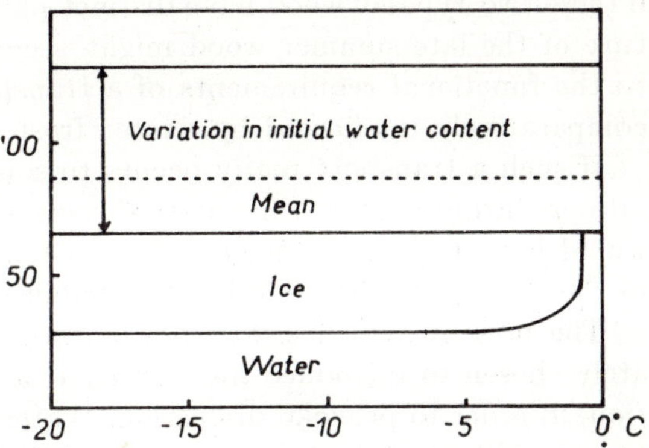

Fig. 4. Amount of water and ice in frozen spruce wood. (After Lybeck 1959.)

The presence of such a considerable amount of liquid water in the frozen wood might perhaps afford some opportunity for translocation. This idea is not original. As early as in 1914 Gates expressed the belief that even at temperatures down to —15° C a certain water transport might take place, sufficient to compensate for the transpiration loss. And Michaelis stated in 1934 that freezing of the soil would not prevent water absorption completely, even if the rate of uptake might be extremely slow.

So far we have made only a single exploratory experiment to test this possibility. A suitable vessel was fastened to the base of a healthy spruce tree last winter, immediately above the snow cover. The vessel was filled with a solution of lignin pink in a glycol/water mixture. A small hole was then bored a couple of inches into the stem under the surface of the solution. After eight days the tree was felled and cut in lengths of 50 centimeters. The stem was undoubtedly frozen through and through. Nevertheless the red colour could be traced upwards from the hole through 6 stem pieces. This means that the solution had actually passed upwards at a rate of about 1·5 cm. per hour, which corresponds to some 3 % of the ordinary flow rate in summer.

Fig. 5 shows the first stem piece above the hole and 4 cross sections taken half a meter apart. The red colour indicates where the dye solution has passed upwards. A transport rate of this order of magnitude would probably be amply sufficient to account for the slow water replenishment needed to balance the loss by cuticular transpiration in winter.

It is tempting to speculate a little further. Coniferous wood is marked by differentiation into two types of xylem elements, viz., the comparatively wide and thin-walled tracheids of the spring-wood, and the much narrower and more thick-walled tracheids of the later summer wood. We do not know if these two types of wood have distinct physiological functions. But the structure of the late summer wood might seem to be particularly well adapted to the functional requirements of a transport system which might carry on comparatively unaffected by winter frost.

If such a transport really occurs to some extent, then soil moisture conditions during winter, thickness of snow cover, and depth of ground frost would have to be reconsidered from a fresh view-point, as essential factors in the ecology and geographical distribution of coniferous forests in general.

The present evidence does not warrant such speculations. I have deliberately chosen to introduce these ideas at a much too early stage of investigation, in order to provoke discussion. Water uptake from frozen soil and water transport through frozen stems must still be considered as the most unlikely of the three alternative channels of water supply discussed above. However, to cite a proverbial statement by Michael Faraday: "In Nature nothing is too strange to be true". And we may add: The stranger things seem to be, the more urgently do they call for investigation.

Summary

The water balance of leaf parenchyma cells depends upon the relation between the rates of water loss and water uptake of the whole leaf. The leaves of evergreen conifers survive prolonged frost in a northern winter climate, when their water supply is cut off by freezing of soil and stems. Under such conditions transpiration still proceeds, and even if the rate may be reduced to a minimum, an increasing water stress must result.

According to the cohesion theory occasional brief thawing periods would not be expected to improve the situation, because of the expelled air bubbles in the tracheids blocking normal water translocation.

However, tentative calculations indicate that the total amount of water lost by transpiration from spruce needles during winter may be several times greater than their initial content of transpirable water. If this is correct, water replenishment must take place to a certain extent even during mid-winter. Such replenishment might occur either by surface absorption or by water movement in the xylem. A few preliminary experiments were made to test the latter alternative.

Frozen spruce twigs were brought into the laboratory at various times during winter, under conditions preventing air exposure of the cut end surface. Water absorption and upward conduction then started immediately upon thawing, in apparent contradiction to the hypothesis of air embolism.

Even at temperatures several degrees C. below zero a certain fraction of the water content of wood remains unfrozen. An injection experiment in the field demonstrated upward conduction of dye solution in an apparently frozen spruce stem.

Although these data are both incomplete and inconclusive one is tempted to suggest the preliminary working hypothesis that water uptake from frozen soil and water conduction through frozen stems may play an essential part in leaf water balance and survival of evergreen conifers in areas with a severe winter climate.

References

Gates, F. C.: Winter as a factor in the xerophily of certain evergreen ericads. — Bot. Gaz. *57* : 445—489, 1914.

Lybeck, B. R.: Winter freezing in relation to the rise of sap in tall trees. — Plant Physiol. *34* : 482—486, 1959.

Michaelis, P.: Ökologische Studien an der alpinen Baumgrenze. IV. Zur Kenntnis des winterlichen Wasserhaushaltes. — Jahrb. wiss. Bot. *80* : 169—247, 1934.

Pisek, A., Berger, E.: Kutikuläre Transpiration und Trockenresistenz isolierter Blätter und Sprosse. — Planta *28* : 124—155, 1938.

—, Cartellieri, E.: Zur Kenntnis des Wasserhaushaltes der Pflanzen. IV. Bäume und Sträucher. — Jahrb. wiss. Bot. *88* : 22—68, 1939.

—, Winkler, E.: Die Schliessbewegung der Stomata ökologisch verschiedenen Pflanzentypen in Abhängigkeit vom Wassersättigungszustand der Blätter. — Planta *42* : 253—278, 1953.

Scholander, P. F., Love, W. E., Kanwisher, J. K.: The rise of sap in tall grapevines. — Plant Physiol. *30* : 93—104, 1955.

—, Ruud, B. and Leivestad, H.: The rise of sap in a tropical liana. — Plant Physiol. *32* : 1—6, 1957.

Discussion

W. Larcher: In the German literature the winter-drought problem has played a certain role since the end of the last century. Although we are not sure that each effect of winter damage described as a drought effect, has been so in fact. To distinguish winter-drought damage one must control the water relations of the plant continuously. Winter-drought damage in Norway spruce in the Austrian Alps occurs, as our experience has shown, near the tree line and, in normal winters in the lowlands too, in spruces slightly damaged by air pollution or affected by parasitic fungi, so that transpiration could not be reduced efficiently enough. But also in cases of efficient reduction of transpiration we must assume a water supply to needles during the winter. As Prof. Hygen has pointed out, there is some water transport throughout the vessels. There is also some water translocation from the vessels. There is also some water translocation from stem to branches to twigs and, finally, to the transpiring needles. We have seen in experiments with intact trees at the tree line and with excised branches exposed at the ordinary place that the water saturation deficit increases first at the cost of the stem, then in the thicker branches, then in the twigs, then the older leaves and last in the youngest needles.

G. Hygen: I thank you for these interesting comments and I am glad to hear that your investigations seem to tie in quite nicely with my own in the main point, viz., in the conclusion that a certain water transport through the xylem to the needles must occur during winter.

P. E. Weatherley: Has Prof. Hygen considered the possibility of movement of water through the walls of the tracheids rather than through the lumina? Demonstration of movement by dyes might indeed be wall movement only — one recalls an experiment by Greenidge in which dye movement occurred up the stem of a detopped tree — where movement in the lumina might not be very likely. The movement into the needles was extremely slow and might by explicable in terms of wall movement through tracheids blocked by ice or air emboli.

G. Hygen: So far we have not made precise calculations on the possible rate of water diffusion through the cell walls under the temperature conditions in question. My guess is that the observed translocation rate is far beyond the order of magnitude which could be obtained by diffusion through the heavily lignified walls of the xylem. I agree, of course, that experiments with dyes such as lignin pink do not allow a distinction between translocation in the cell lumen and the cell walls. Our primary aim has been to find out to what extent translocation actually occurs during the winter. The question of the relative importance of the possible channels of such movements must be left open for future research.

P. E. Weatherley: I should have thought that cell wall movement would be mass flow rather than diffusion. Mass flow could presumably take place in the intermicellar spaces of the wall in response to differences in hydrostatic pressure or by capillarity.

G. Hygen: Certainly mass flow through the walls is a mechanism to be considered, but owing to the lignification of the walls the flow resistance is likely to be much higher in the xylem than in leaf or root parenchyma. We would have to know much more about the magnitude of this resistance and the gradient of water potential in order to say anything more definite.

G. F. Makkink: The summarized moisture deficit of the air may be too high if taken from observations made in the meteorological screen. On the surface of the trees, in the canopy, conditions might be less severe according to snow or melting water.

G. Hygen: The deficit values were obtained from our Meteorological Institute and were derived from the ordinary standard data. They represent, therefore, only a rather rough approximation to the actual conditions in a forest.

G. F. Makkink: It would be worth while to study the microclimate of the conifer canopy in winter.

G. Hygen: Certainly.

P. G. Jarvis: I would like to ask Prof. Hygen about the formula $(T_c = 20 \cdot \Sigma D)$ on the blackboard because it plays an important part in his argument. How was it derived ?, and does it take into account the fact that the loss of water through the cuticle decreases as the water content decreases ?

G. Hygen: I am glad you raised this point because the estimate of the proportionality factor 20 in my formula is the weakest point in the whole argument. As I stated, we lack actual measurements of cuticular transpiration of spruce needles under natural winter conditions in Norway. As a substitute I have used data from the investigations of Prof. Pisek and his coworkers, which admittedly were obtained under quite different conditions and had to be "translated" by comparison of temperature and humidity measurements. Thereby we had to resort to a comparison between cuticular transpiration rates in Norway spruce and Scotch pine. At moderate water deficits the cuticular transpiration rate in *Pinus* appears to be approximately proportional to the atmospheric saturation deficit under otherwise constant conditions and temperature equilibrium. A similar relationship was provisionally assumed to exist for *Picea*. We hope in this way to have arrived at a figure of the right order of magnitude. However, the effect of a more severe dehydration in causing deviations from the initial proportionality has not been taken into account and I am in perfect agreement with Dr. Jarvis when he considers this to be a serious objection. We hope, however, to get more raliable data as a basis for making the appropriate corrections in experiments planned for this winter.

P. G. Jarvis: It may be of interest that we found critical levels for cell damage in spruce needles of 40% relative turgidity or 95 atm. water potential deficit (in Prof. Weatherley's terminology). The critical level of water potential deficit is the highest we have found in our own studies or in the literature. I think it essential that, in the continuation of this work, estimates of water potential deficit through the winter should be made. In the laboratory (25° C temperature, high light) we found, with excised shoots, that relative turgidity decreased at a rate of 1·2% relative turgidity per hour when the tissue was water saturated, but that after about 6 hours this rate was appreciably less — about half as far as I remember.

G. Hygen: I agree that measurements of the water potential would be of considerable value in elucidating the mechanism of the water balance during winter, and I am thankful for your suggestion.

W. Larcher: Pisek and Winkler (1953) measured the cuticular transpiration of the conifers in summer after saturating the twigs with water. Thus their transpiration rates are rather high. As Pisek and Berger (1938) had shown, the cuticular transpiration rates decline — under controlled conditions — from day to day, especially in the first 2—3 days of drying. I have found the same effect with slowly drying twigs of *Picea abies* and *Pinus cembra* in winter. Some 3 to 4 days after the beginning of the experiment cuticular transpiration rates were about $^1/_3$ of the initial cuticular transpiration rates. Thus, the coefficient in Prof. Hygen's equation may be somewhat too high.

G. Hygen: The question of the effect of dehydration on the rate of cuticular transpiration seems to be a rather controversial one, and I think more experimental data required to clarify the situation with regard to Norway spruce. I am, however, fully aware of the preliminary and approximative character of my so-called proportionality factor, which I have been careful not to call a constant. It is perfectly possible that the value I have arrived at may prove to be too high, but I do not think the possible error would change the main features in the rather sketchy picture of the water balance I have tried to outline.

As I emphazised in my concluding remarks, my intention was to present problems for debate rather than definite results. Your remarks have strengthened my own conviction that it is worth while to follow up this line of investigation, and I am thankful for the many valuable suggestions I have got.

SERIAL STUDIES ON THE HYDRATION OF FOREST PLANTS

W. SCHEUMANN

Institute for Forest Plant Breeding, German Academy of Agricultural Sciences, Graupa/Pirna, German Democratic Republic

One of the main tasks of forest research at the present time is the improvement of the genetic basis of our woods for the purpose of increasing the economy and productiveness of stands. This requires preliminary studies on the site suitability of the breeding material, expecially of frost and drought resistance. However, site suitability in general, as well as drought and frost resistance in particular, are such complex properties that investigation of the individual factors of those properties can only be of practical use if a great many of them are studied simultaneously. Therefore, applied plant physiology strives to investigate reactions and characteristics which essentially influence the properties of interest or result from a multitude of such individual factors.

The "Wasserzustand" — i.e. hydration in the sense of Walter (1931) — is such an integral involving water regime factors. Hydration is effect and cause together: on the one hand it is a result of idiotype and environment and on the other a decisive assumption for physiological capacity, effective vitality and productiveness. Proceeding from a dynamic approach, the comparative investigation of the course of hydration under various environmental factors may result in most valuable information for plant breeders and foresters as to the different, genetically conditioned "reaction norms" of their material.

Method

Since breeding of forest plants — with the exception of certain poplars — does not deal with clones but with more or less heterogeneous populations of alien pollinators, there is a prevailing demand for mass investigation methods. Moreover, reiterated sampling of small plants requires methods necessitating only minimal quantities of samples. The cryoscopic method generally used for the determination of the osmotic pressure as an expression of hydration is quite unsuitable for this purpose. That is why we have used r e f r a c t o m e t r y instead of cryoscopy for many years. In forest objects the relations between Δ_t and n_D are very close. The correlation coefficient, determined for more than hundred value pairs of 6 tree species at different periods, was $r = + 0 \cdot 95$ (Scheumann 1960). Slavík (1959) also found similar correlations in the leaves of sugar beets. Compared with the dispersion of individual samples this correlation seems to us close enough to speak of hydration. In recent papers also Walter and his school of thought admit a certain practical significance of refractometry in this sense (Walter 1962/63; Kreeb 1957, 1961). For technical reasons we even go further, using in the refractometer the "International Sugar Scale of 1936" and thus we obtain data in per cent dry matter, though the cell sap does not represent a pure sugar solution. Refractometry is an ideal method for mass investigations. Two trained workers are able to investigate over 500 samples on one day, using sample quantities reduced to only $0 \cdot 1$ g. fresh leaf substance.

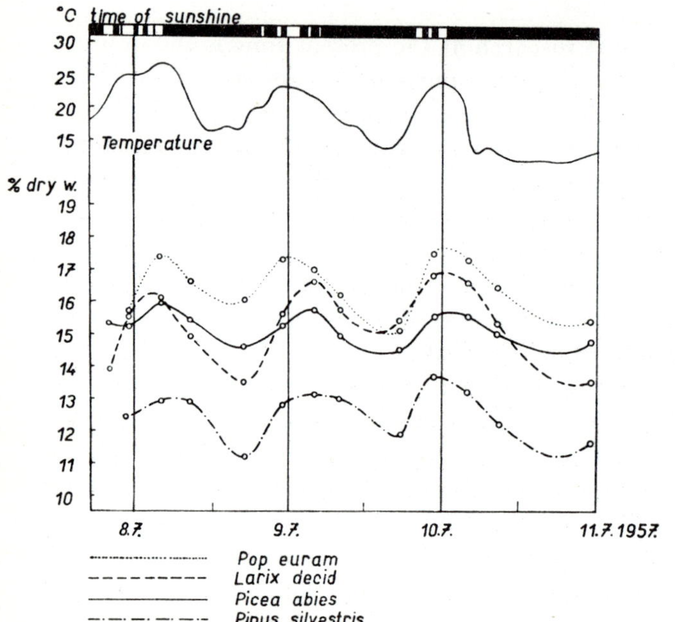

Fig. 1. Diurnal fluctuations in the cell sap concentration of various tree species (Scheumann 1960).

Results

The following example of 3 day-courses demonstrate the sensitiveness of the method: Fig. 1 shows the course of hydration in 4 tree species from July 8th to 11th, 1957. The periodic changes in press sap concentrations of leaves and needles, respectively, parallel well the course of temperature as an expression of meteorogenic factors affecting the water regime of the plants. Thus, for instance, the temperature maximum on July 8, is reached in the afternoon and the culmination point of cell sap concentration lies in the afternoon as well. On the contrary, the temperature maximum on July 10, occurs just at midday after a rather late but all the more sharp rise, and also the course of cell sap concentration reveals a rather late and steep increase with the culmination point at noon. Furthermore, it is interesting to note that the absolute values in Scotch pines are relatively low while they are the highest

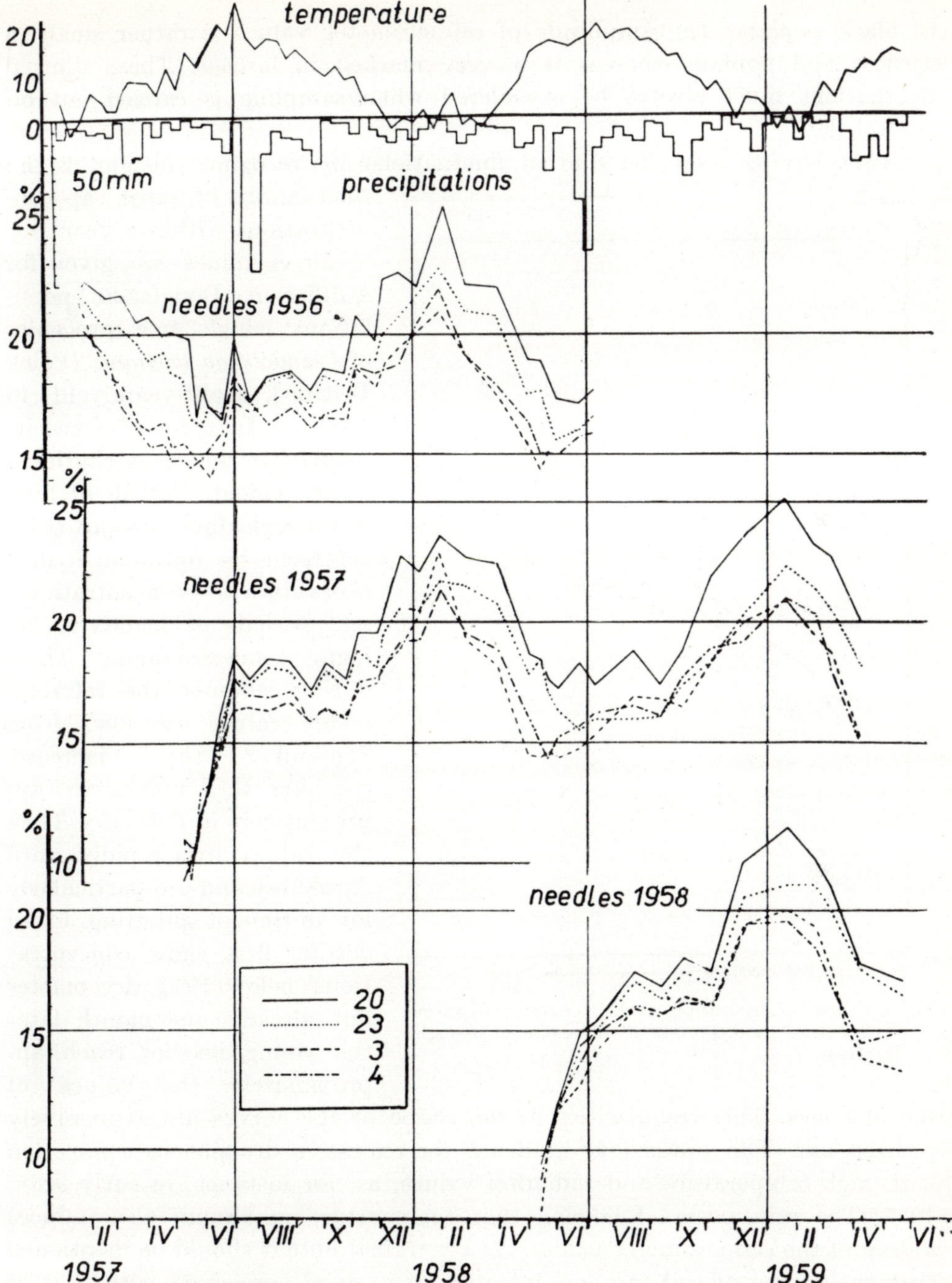

Fig. 2. Temperature course (mean of decades) and precipitation (total of decades) as well as course of cell sap concentration (mean values) of needles from 4 different single tree progenies of the green Douglas-fir [*Pseudotsuga taxifolia* (Poir.) Britton var. *viridis* Aschers. et Gröbner] (Scheumann 1960).

in black poplars. The amplitude of refractometer values is rather small in spruces and poplars whereas it is very marked in larches. These diurnal fluctuations must always be considered when sampling is carried out on several days.

After having seen the diurnal fluctuations we recognize in Fig. 2 the fluctuations of press sap concentrations within a year.

Mean values are given for 4 different Douglas-fir populations [single tree progenies of *Pseudotsuga taxifolia* (Poir). Britton], eight-years old in 1957. — In spite of some irregularities the annual rhythm is quite evident. Periods of high physiological activities are characterized by optimum hydration values (low concentrations) and periods of inactivity by high concentrations. Thus, during summer the refractometer values are low, from September they increase, reaching the maximum approximately in February. Then the values drop rapidly until April/May and are particularly low in time of sprouting. Fresh needles first show concentrations below $10^0/_0$ dry matter but already one month later the young needles reach approximately the values of

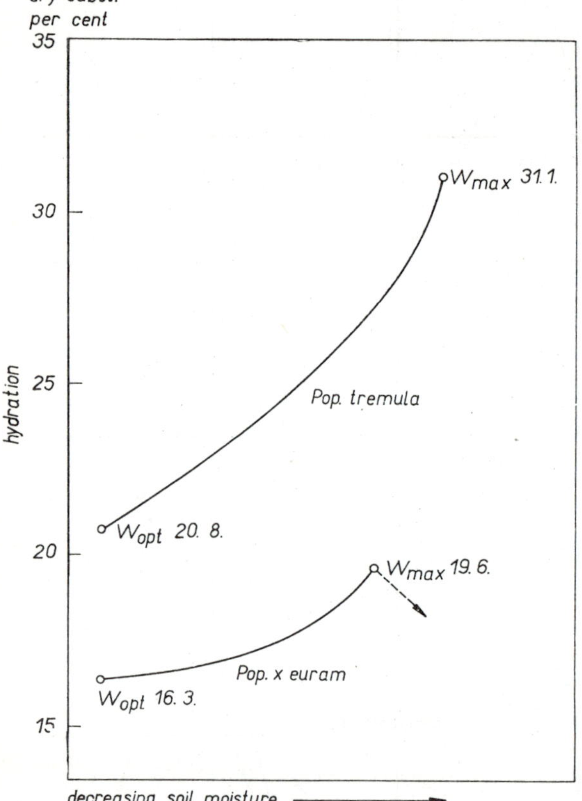

Fig. 3. Course of hydration of *P.* × *euramericana* cv. *"robusta"* and *P. tremula* in extreme drought stress.

the old ones. The irregularities in the shape of the curves are expressively pointed out. The peaks give evidence for excessive drought in connection with high temperature and radiation values, as, for instance, in early July, 1957. The rain period following that temperature maximum also induces a drop of the refractometer values. As a marginal note it should be mentioned that the course of cell sap concentration is in good agreement with that of frost hardiness in Douglas-firs. During the winter half-year the mean correlation of the four populations under study was r = + 0·74 (Scheumann 1960).

Fig. 3. shows that *Populus* × *euramericana* cv. "*robusta*" and *Populus tremula* reveal quite different hydration reactions to stressed water conditions. Similar differences, though not so marked, are also found between individuals of the same tree species. Here the cell sap concentrations of both poplar species are determined between the cardinal points W_{opt} and W_{max}, representing the limiting values of hydration between (approx.) optimum water conditions of the plant and drought stress provoking the first irreversible injuries. In our experiment increasing soil suction forces impaired and finally stopped the water uptake. Refractometer values of *P. tremula* leaves rose rapidly during intensive radiation and finally the leaves desiccated. However, there was only a slight increase of cell sap concentration in *P.* × *euramericana* leaves. Soon after, proteolytic processes set in; the leaves yellowed and fell off. Occasionally, a drop in refractometer values could be observed during that decomposition.

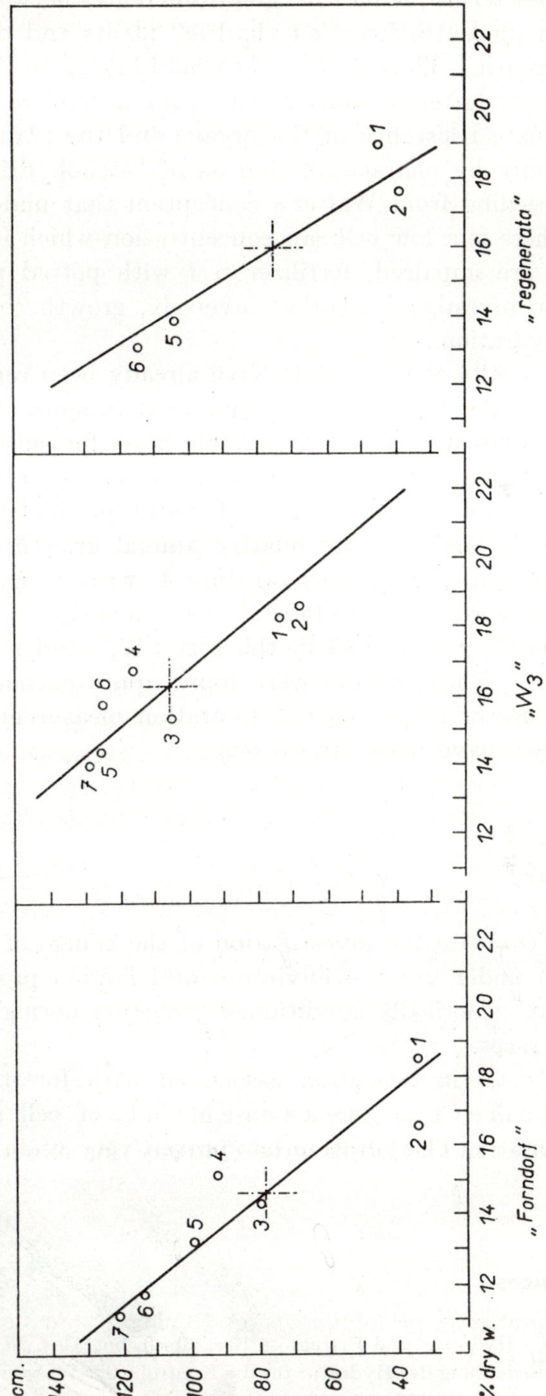

Fig. 4. Relations between hydrature (measured on July 4) and final heights of one-year old poplar shoots (mean values). Fertilization variants: 1 = O; 2 = -PKMg; 3 = N-KMg; 4 = NP-Mg; 5 = NPKMg (standard complete fertilization); 6 = N_3PKMg. (Fritzsche and Scheumann 1962.)

Walter terms plants with great differences between optimum and maximum cell sap concentrations "euryhydric" plants and those with small differences "stenohydric". Thus, *P. tremula* would belong to the former and *P. euramericana* to the latter. It must be emphasized, however, that it is not yet possible to evaluate resistance of the organs and the plant, respectively, to drought stress only by characterization as of "stenohydric" or "euryhydric" types.

Proceeding from Walter's conception that under optimum growth conditions there is a low cell sap concentration which increases when growth conditions are impaired, fertilizer test with potted plants were carried out in order to investigate whether inversely, growth conditions could be deduced from hydration.

The results of those tests have already been reported in part (Scheumann and Fritzsche 1962). It was found that in spite of some deviations refractometer values may provide valuable hints for suitability with regard to site. Fig. 4 shows correlations between refractometer values and growth heights of three poplar sorts in seven fertilization variants. Our example made it possible to conclude the relative annual growth capacity expected already from the values measured on July 4, with a significance of 73—93%. The following correlation coefficients were found: $r = -0.96$ in the poplar sort "Forndorf", $r = -0.86$ in the sort "W_3" and $r = -0.97$ in the sort "regenerata". These results were found quite encouraging and further studies on the use of refractometric hydration measurements for evaluation of site suitability have been carried out.

Summary

The comparative investigation of the course of hydration (in the sense of Walter) under varying environmental factors provides new aspects for the different, genetically conditioned "reaction norms" of species and their ecological races.

For mass investigation associated with forest plant breeding the r e f r a c t o m e t r i c   m e a s u r e m e n t   of cell sap concentration for the determination of hydration has proved very useful.

References

Kreeb, K.: Hydratur und Ertrag. — Ber. dtsch. bot. Ges. *70* : 121—136, 1957.

—: Die Bedeutung der Hydratur für die Kontrolle der Wasserversorgung bei Kulturpflanzen. — Beitr. Biol. Pfl. *36* : 57—89, 1961.

Scheumann, W.: Untersuchungen zur Entwicklung rascharbeitender Selektionsmethoden für die Frostresistenzzüchtung bei Waldbaumarten. — Inaugural-Dissert., Math.-Nat. Fak. d. Univ., Rostock, 1960.

—, Fritzsche, K.: Hydratur und Wachstum von Pappeln in Abhängigkeit von der Nährstoffversorgung des Bodens. — Züchter 32 : 179—184, 1962.

Slavík, B.: Gradients of osmotic pressure of cell sap in the area of one leaf blade. — Biol. Plant. 1 : 39—47, 1959.

Walter, H.: Die Hydratur der Pflanze und ihre physiologisch-ökologische Bedeutung. — Jena, 1931.

—: Zur Klärung des spezifischen Wasserzustandes im Plasma und in der Zellwand bei der höheren Pflanze und seine Bestimmung. Part II. — Ber. dtsch. bot. Ges. 76 : 54—71, 1963.

Discussion

W. Larcher: Irrespective of the great practical significance, refractometry can also be a useful auxiliary method for the simultaneous determination of hydrature in investigating the relationship between water deficit and photosynthesis, since it requires only a small amount of material. For this at least a rough assessment would be necessary of whether the refractometric value rose because the cell sap was concentrated by water loss, or whether it occurred because starch was hydrolysed to sugars. The product of "osmotic value × water content" gave the possibility of making such an assessment. Have you tried using the product "refractometric value × water content"?

W. Scheumann: If we observe the water relations in plants, i.e. their degree of hydration, then the question of how the changes are produced has a secondary significance. In so far as the product "water content × osmotic value" is important for the special problem, the product "water content × refractometric value" could allow similar conclusions since the relationship between cryoscopy and refractometry is, in our experience, unusually close. We ourselves have not worked with this product.

W. R. Müller-Stoll: Simple refractometric determinations can be used to establish hydration relations with sufficient reliability only when we are working with the same species of plant and with samples as far as possible in the same physiological condition (e.g. with leaves of the same age). If we compare cryoscopic and refractometric determination in a greater number of species we get big differences in the relationship of the two values, which are due to the composition of cell sap (own unpublished results). Since refractometry gives priority recording to the high molecular components of cell sap (e.g. sugars) it is physiologically inaccurate as compared with the cryoscopic method. The replacement of this method by refractometry is possible only in special circumstances and after careful investigation of the significance of the values obtained with each material investigated and for a certain relationship.

W. Scheumann: Cryoscopy is not suitable for widescale investigations within the framework of forest-tree selection, since it requires a great deal of material and time. The results of cryoscopy are only an expression of hydrature (water vapour tension). Cryoscopic results are not equal to refractometric results and it is correct when determining hydrature by refractometry to ascertain the extent to which these values are in agreement with cryoscopic values. In our case this agreement, as stated in the report, was very good (r = 0·95). The scatter of values of hydrature within one population, one plant and even one leaf, can be very great. It would be impossible to use cryoscopy for the purpose given above thereby limiting ourselves to a very small number of samples. With the aid of refractometry we can determine hydrature better under these conditions.

B. Slavík: The reaction between the refractive index and osmotic pressure of the cell sap also changes in individual plants due to ontogenetic development connected with changes in the composition of cell sap. This has been shown and discussed already (B. Slavík, Biol. Plant. *1* : 48—53, 1959).

M. Rychnovská: Why was the concentration of cell sap (besides other ecological factors) only compared with the temperature curve ? One should have used, for example, evaporating power or the hydrothermic quotient.

W. Scheumann: Further meteorological factors were measured. In the paper the temperature curve was used to express factors affecting the water relations of plants for simplification.

H. Polster: As Dr. Larcher has just stressed, increased refractrometric values can be caused both by decreased water content and by increased content of osmotically active substances. It will be necessary to determine quantitatively the ratio of the two factors using biochemical analysis of concentrated osmotically active substances, including the paper chromatography of sugars. This would at the same time make it possible to investigate more deeply into the cause of drought and frost resistance. We intend to continue along these lines of work at Graupa (G.D.R.).

WATER RELATIONS OF SOME STEPPE PLANTS INVESTIGATED BY MEANS OF THE REVERSIBILITY OF THE WATER SATURATION DEFICIT

MILENA RYCHNOVSKÁ

Botanical Institute, Czechoslovak Academy of Sciences, Brno, Czechoslovakia

In July 1962, during our studies of the water relations of some steppe plants in Hungary, I and my colleague were also interested in the sublethal value of the water saturation deficit. In our investigations we applied a method based on water inhibition measurements in plants dried up so that they reached various degrees of water deficit. The measurements were made on segments of grass blades. We were able to establish two types of saturation curves, with a characteristic difference between them:

1. Up to a certain limit the plants fully recovered from their water deficit by restoring their original water content.

2. An increase in the water deficit brought about a certain degree of increased water retention capacity of the leaf tissue. The water content in leaves saturated after the previous partial desiccation was found to be higher than that in the initial fully turgescent conditions, so that we may be justified in speaking about "oversaturation" with water.

Measurements collected from various sites revealed that the first reaction was common among all tested individuals of the species *Stipa pulcherrima, S. joannis, S. stenophylla, S. capillata* and also of *Melica ciliata* and *Andropogon ischaemum*. The second type of the saturation curve was found among *Chrysopogon gryllus, Agropyrum intermedium, Koeleria gracilis* and also among some mesophytic grasses, such as *Dactylis glomerata* and *Brachypodium silvaticum* (Rychnovská and Květ 1963).

In the present work I have tried to ascertain whether this latter property is constant and characteristic for certain species, or whether it is the result of physiological adaptation of the plants; such adaptation would be due to ecological factors, and would be found among various species under specific conditions.

Materials and Methods

During the entire vegetative periods of 1962 and 1963 I studied the variations in the sublethal water deficit of some xerophilous grasses found in a steppe of the Křížová mountain (Křížová hora) near Moravský Krumlov (southwestern Moravia). The study included the following plants: *Stipa pulcherrima, S. joannis, S. stenophylla, S. capillata* and *Bromus erectus*. As *Bromus erectus* was stationed on a shady wood edge, as early as 1960 several tufts were transplanted on the steppe among the other plants of the species *Stipa*, employed in the experiment. The necessary background was supplied by my studies of the saturation curves of psammophilous vegetation near Bzenec, of xerothermic grasses from the Mohelno steppe, of mesophytic grasses from a Brno park, and of hygrophytic vegetation from southern Bohemia. The same methods were used as described in my preceding papers (Rychnovská 1963): The collected plants (without roots) were left in an atmosphere saturated with water vapour till they reached full turgescence, their bases being submerged in water for 12 hours. Subsequently their leaves were kept in a thermostat at 28—30° C for one to eight hours so that they reached various degrees of water deficit. Then they were cut into segments and water-saturated polyurethane was applied to the cut areas to enable them to absorb water. The water content before and after absorption was determinated gravimetrically (see also Čatský 1960, Rychnovská and Bartoš 1962).

Results

The results of the experiments are expressed in graphs (Figs. 1 to 7). The water content is expressed as a percentage of the water content of full turgescence at the beginning of the experiment. The abscissa indicates the values of water deficit at which the plants were saturated with water.

The graphs show that under certain conditions all species employed give saturation curves of both the first and the second type, although some of the species tend to the first and others to the second group. From the steppe grasses the most reactive species seems to be *Stipa stenophylla* (Fig. 1), while *Stipa capillata* and *S. dasyphylla* are the least reactive of all (Fig. 5). In *Stipa joannis* the second type of curve was noted only exceptionally (in 1963) (Fig. 2). *Bromus erectus* from the steppe locality is reactive to a very small degree, while that from the station on the wood edge tends to "oversaturation" after a period of water deficit (Fig. 3, 4). From the psammophilous grasses *Corynephorus canescens* and *Festuca dominii* were analysed and their reactions found to be different (Fig. 7). The last graph (Fig. 6) illustrates the saturation curves of some mesophytes and hygrophytes.

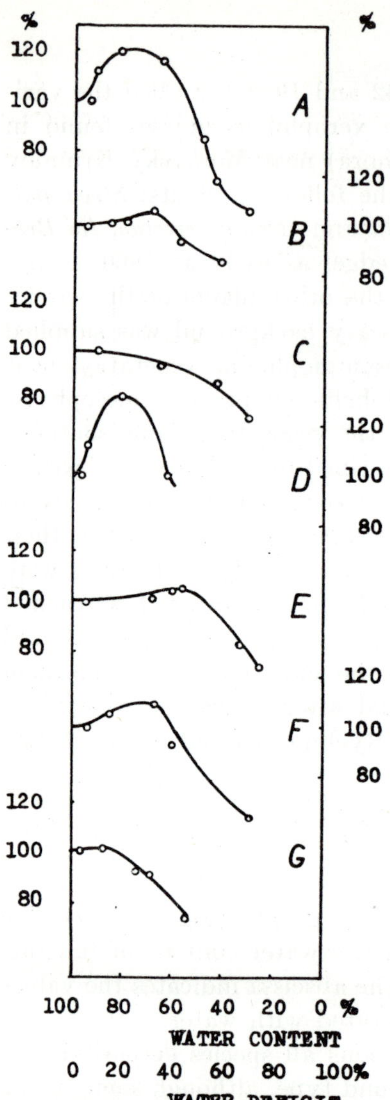

Fig. 1. Water saturation curves obtained with leaf segments of *Stipa stenophylla*. — A (25. 7. 61), B (27. 4. 62), C (6. 5. 62), D (4. 6. 62), E (4. 5. 63), F (17. 5. 63), G (16. 8. 63) The initial water content of the leaf segments has been taken as 100 %. Abscissa: water content of the leaf segments expressed as percentage of their dry weight (upper scale) and the corresponding water saturation deficit (lower scale). Ordinate: water content of the leaf segments after saturation of 2 hours expressed as percentage of the initial water content.

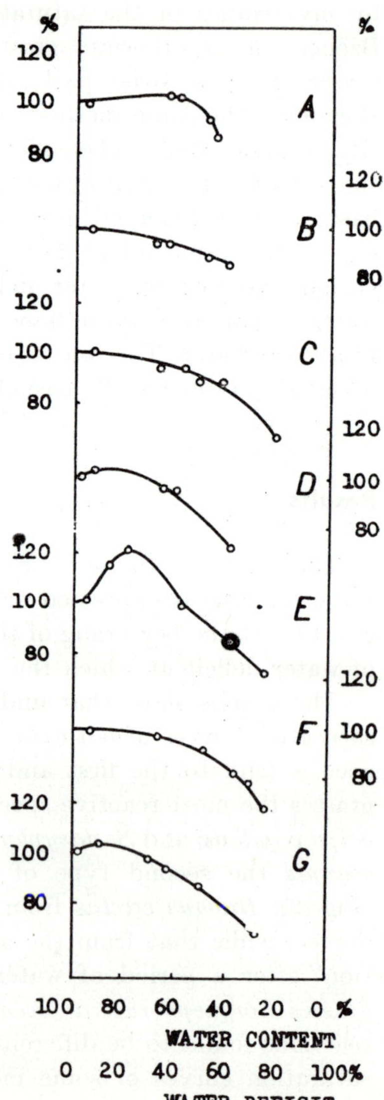

Fig. 2. Water saturation curves obtained with leaf segments of *Stipa joannis*.
A (25. 7. 61), B (27. 4. 62), C (6. 5. 62), D (4. 6. 62), E (4. 5. 63), F (17. 5. 63), G (16. 8. 63) Explanation as for Fig. 1.

Discussion

The above results point clearly to answers to the two questions put forward in the introduction to this paper; the existence of a constant property characteristic for certain plant species is entirely out of question. The materials

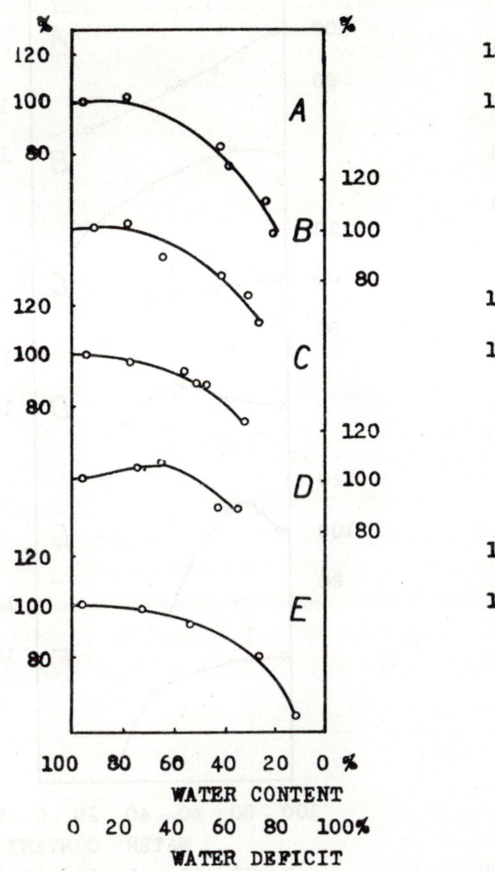

Fig. 3. Water saturation curves obtained with leaf segments of *Bromus erectus* — steppe.
A (25. 7. 61), B (27. 4. 62), C (6. 5. 62), D (4. 5. 63), E (17. 5. 63)
Explanation as for Fig. 1.

Fig. 4. Water saturation curves obtained with leaf segments of *Bromus erectus* — wood border.
A (25. 7. 61), B (27. 4. 62), C (4. 5. 63), D (17. 5. 63), E (16. 8. 63)
Explanation as for Fig. 1.

used in my studies and numerous experiments with mesophytic plants indicate that the "oversaturation" with water following the water deficit is a property — under certain conditions — of both typically xerophilous plants (species of genus *Stipa*) and mesophytic plants *(Bromus erectus, Dactylis glomerata)*, and even typical hygrophytes *(Alopecurus aequalis, Eleocharis palustris* etc.). The results also lead to the conclusion that most probably the oversaturation does not depend on the anatomical adaptation of the mesophyll, such as the size of the intercellular spaces. If the oversaturation

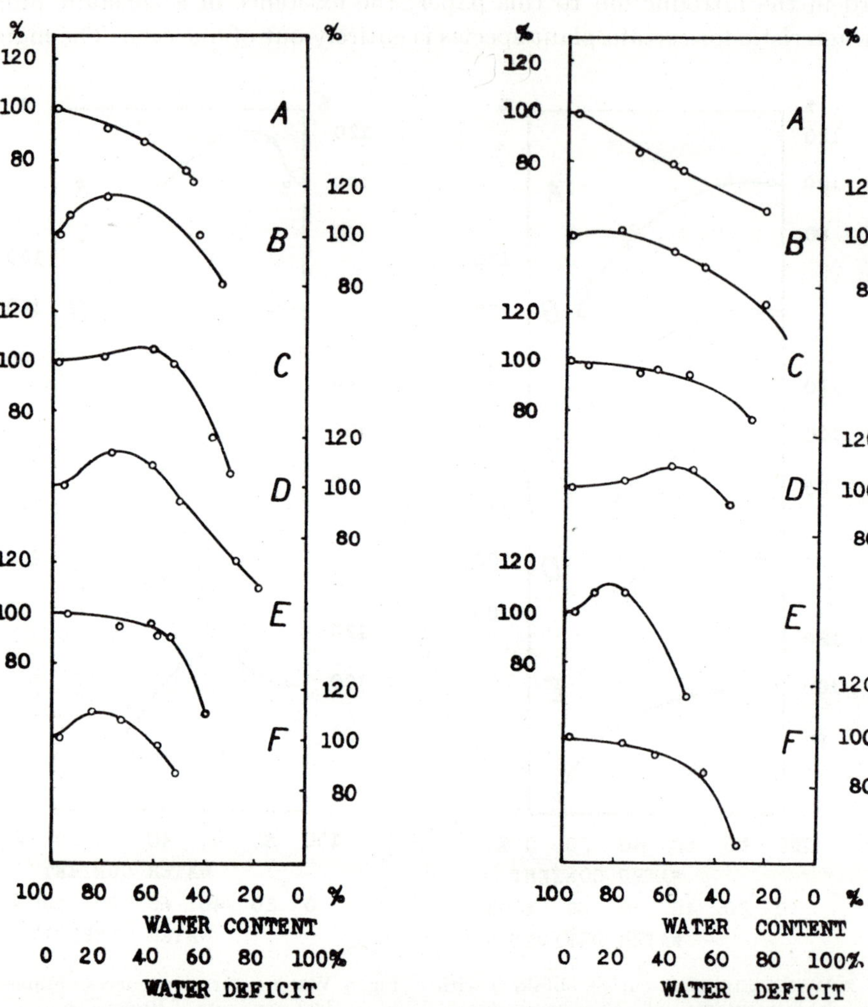

Fig. 5. Water saturation curves obtained with leaf segments of *Stipa pulcherrima* (A, B), *Stipa dasyphylla* (C, D) and *Stipa capillata* (E, F)

A, C, E on 25. 7. 61; B, D, F on 17. 5. 63
Explanation as for Fig. 1.

Fig. 6. Water saturation curves obtained with leaf segments of *Dactylis glomerata* (A), *Festuca arundinacea* (B), *Trisetum flavescens* (C), *Eleocharis palustris* (D), *Alopecurus aequalis* (E), *Carex vesicaria* (F)

A, B, C on 6. 5. 63; D on 10. 7. 63; E, F on 28. 5. 63
Explanation as for Fig. 1.

were due to a certain type of anatomical structure no changes in the type of the saturation curve during one vegetative period would be possible (cf.

Stipa, Bromus etc.). As all experiments were concluded in exactly the same manner, as concerns the collection of samples, the laboratory work and the time of the day, the possibility of a chance phenomenon due to experimental treatment is excluded.

The possibility that the results are due to the endogenic rhythm of the plant, in connection with the developmental phases, is highly improbable in the light of my results. Thus only one explanation can be offered: this property depends on the ecological conditions of the habitat and on the reaction of the species concerned to these conditions. The above results suggest that the water relations of the habitat may be the most important factor. Experiments with mesophytic grasses, conducted in spring (Fig. 6), and most of the results obtained from the steppe vegetation in the first half of 1962, with much precipitation both in this country and in Hungary, reveal the predominance of the first type of the curve. On the other hand, results from the year 1963, which was characterised by arid conditions in the steppe localities, show the predominance of the second type.

If the plants under study are mutually compared, *Stipa stenophylla* appears to be the most reactive species. This is in accord with its ecological amplitude, although there is no substantial phytogeographical difference between this plant and *Stipa joannis* and other continental grasses. As far as I could trace the occurrence of *Stipa stenophylla* in Czechoslovakia, the plant always appeared in habitats of most pronounced mesophytic character among the other steppe grasses. In these areas it also formed vegetational cover of a very progressive type. Martinovský in his monograph (1963) considers *Stipa stenophylla* to be the least xerophytic species of all our feather grasses. From a similar point of view we can also compare psammophytic grasses: *Corynephorus canescens* presents the oversaturation type of curve, while the curve of *Festuca dominii*, belonging to the same plant community, is of the saturation type only. A comparison of ecological requirements and phytogeographical characteristics also points to the more pronounced moisture demands of *Corynephorus* and thus to the ability of this plant to compensate

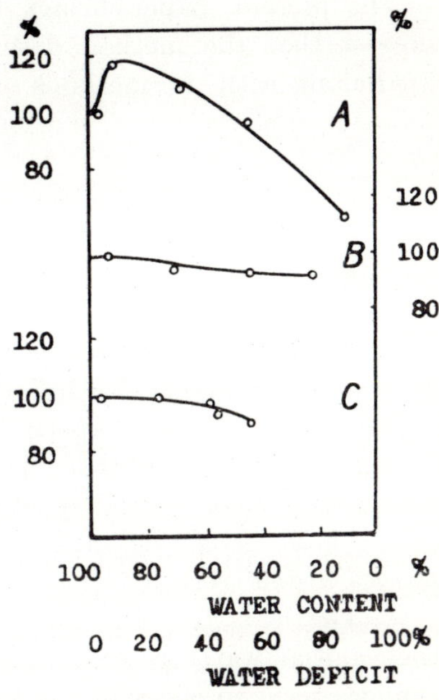

Fig. 7. Water saturation curves obtained with leaf segments of *Corynephorus canescens* (A), *Festuca dominii* (B) and *Carex hirta* (C)
A, B, C on 21. 8. 63
Explanation as for Fig. 1.

the physiologically unfavourable moisture conditions of the habitat more rapidly.

There are still some aspects which remain to be solved. The mechanism of this phenomenon, most likely due to biochemical changes, must be ascertained and its connection with the ecological conditions of the habitat verified in more cases. Both aspects are the subject of further research.

The present paper brings, however, a sufficient amount of material to suggest that the method described here constitutes a new criterion to be used along with the methods of physiology and ecology in the effort to offer a causal explanation of the natural distribution of individual plant species.

Summary

Reversibility of the water deficit in leaves of some characteristic steppe grasses was studied. The intact leaves of species tested were left to dry up under exactly defined experimental conditions so that they reached various degrees of water saturation deficit. Segments cut from these water-deficient blades were then saturated with water. The initial water content, the water deficit and its compensation were determined gravimetrically. The steppe species tested were found to fall into two groups:

1. The leaf segments fully recover from their water deficit by restoring their initial water content, but beyond a certain water deficit irreversible changes seem to occur in the leaf tissue and prevent the segments from fully compensating their water deficit.

2. An increase in the water deficit brings about an increase in the water retention capacity of the leaf tissues. In general, the water content of the leaf segments after saturation is higher than it was initially. As with the previous type, irreversible changes seem to take place, with the same consequences, if the water deficit reached a certain value.

It was shown that the different types of the saturation curve do not constitute a characteristic feature of certain plant species, as under certain ecological conditions they can both be found among all groups of plants, whether the xerophytes, or the mesophytes, or even the hygrophytes. Among the steppe plants under study the second type of the curve is most frequent in *Stipa stenophylla*, which stands — relative to the other feather grasses — closest to the mesophytes. Moreover, in the case of the psammophytes the data obtained were compared with their phytogeographical distribution and their ecological amplitude.

References

Čatský, J.: Determination of water deficit in disks cut out from leaf blades. — Biol. Plant. 2 : 76—78, 1960.

Martinovský, J. O.: Naše kavyly. [Our *Stipa*-species.] In Czech — Ochrana přírody 18 : 45—48, 1963.

Rychnovská, M.: Study of the water saturation deficit as one of the methods of causal phytogeography. — Biol. Plant. 5 : 175—180, 1963.

—, Bartoš, J.: Measurement of photosynthesis by the dry weight increment of samples composed of leaf segments. — Biol. Plant. 4 : 91—97, 1962.

—, Květ, J.: Contribution to the ecology of the steppe vegetation of the Tihany Peninsula. III. Estimation of drought resistance based on the saturation of water deficit. — Manuscript for Acta Botanica Hungarica, 1963.

Discussion

V. M. Sveshnikova: Why was water deficit chosen as one of the main elements of the water relations in evaluating the botanical and geographical extension of steppe grasses?

M. Rychnovská: The natural vegetations of Central Europe lie at the borderline of a continental and maritime type of climate. These vegetations are affected by alternating good and bad ecological conditions according to the type of climate of the given year. Therefore the water regime has a decisive role, particularly in our xerothermic communities, and the plants are forced to adapt themselves to climatic variations, i.e. mainly to variations in the water regime. This method can serve for assessing the reaction of plants to changes in the water relations.

P. E. Weatherley: Have you any physiological explanation of the oversaturation phenomenon? Is it due in some way to cell wall relaxation or something like that?

M. Rychnovská: I stated in my paper that this question will be the subject of my future work. I assume that oversaturation is not due to any physical changes since it is most marked at a water deficit of 10% and I would not expect any physical changes under these conditions. It will probably be due to biochemical changes, e.g. to the way of increasing the osmotic pressure, etc. Only future experimental work can bring a clear answer to this question.

J. Úlehla: What was the time of resaturation?

M. Rychnovská: Always two hours.

W. Larcher: With reference to Dr. Rychnovská's studies on resaturation it may be interesting that in experiments I carried out with Prof. Oppenheimer in Innsbruck, when the resaturation point (i.e. the highest saturation deficit which just permits complete resaturation) was exceeded, the photosynthesis activity of the leaf was no longer able to recover. One must, however, be very cautious in applying the results of this useful physiological test method in interpreting ecological problems.

M. Rychnovská: I look upon all these tests in my autecological studies as one stone in a whole mosaic. The autecological evaluation of the species under investigation is based on a large number of field measurements, e.g. transpiration curves *in situ*, water saturation deficit *in situ* and detailed comparative microclimatic studies, etc. There is thus no danger of interpreting ecological problems on the basis of several isolated laboratory tests.

J. Úlehla: It was observed that in segments of barley leaves in bad health condition, saturation with water proceeded at a higher rate than in segments from normal leaves. Where the saturation in the experiments of Dr. Rychnovská did not proceed to equilibrium in two hours, the oversaturation could perhaps be simulated by differences in rates of water absorption.

CHANGES IN THE WATER STATUS IN PLANTS UNDER DIFFERENT EXTERNAL AND INTERNAL CONDITIONS

N. A. GUSEV

Institute of Biology, Kazan Filial of the Academy of Sciences of the U.S.S.R.,
Kazan, U.S.S.R.

The division of water in plants into "free" and "bound" is now well recognised in plant physiology. This concept of the state of water played an important role in the development of the study of plant water regime. Nevertheless, this concept has some weak points, the main being the influences affecting the division of water into these two fractions due to the absence of sharp boundary between them. The ratio of free and bound water depends upon the amount of the water-removing force applied (Tumanov 1940, Alekseev 1948) and is thus an arbitrary value.

Moreover, for removing free water strong forces are generally required (from 34 to 135 atm. and more). That is why not only free water but also a part of bound water are considered to enter the removed water, the amount of this part of bound water increasing with the strength of the removing force.

It should also be borne in-mind that many authors do not accept the presence of free water in plants (Crafts, Currier and Stocking 1949, Sabinin 1955, Okuntsov and Tarasova 1952).

Therefore, we consider it more reasonable to characterize different water fractions in plants by their retaining forces registered by checking the amount of water removed by a number of gradually increasing water removing forces. Such forces may be various temperatures of freezing (Tyurina 1957) and sucking forces (S_x) of solutions of different concentration (Gusev 1962). This makes it possible to exclude the so-called conditional nature of dividing water into fractions, to observe the changing state of water balance and give a sharp differentiation of capacity (the amount of bound water) and strain (water-retaining forces) and to define their relation.

Methods and Results

For your years we have been using the technique of continuous characterization of the state of water in plants for studying the changes in the state of water in spring wheat leaves (Hordeiforme 189, Staratovskaya 29) caused

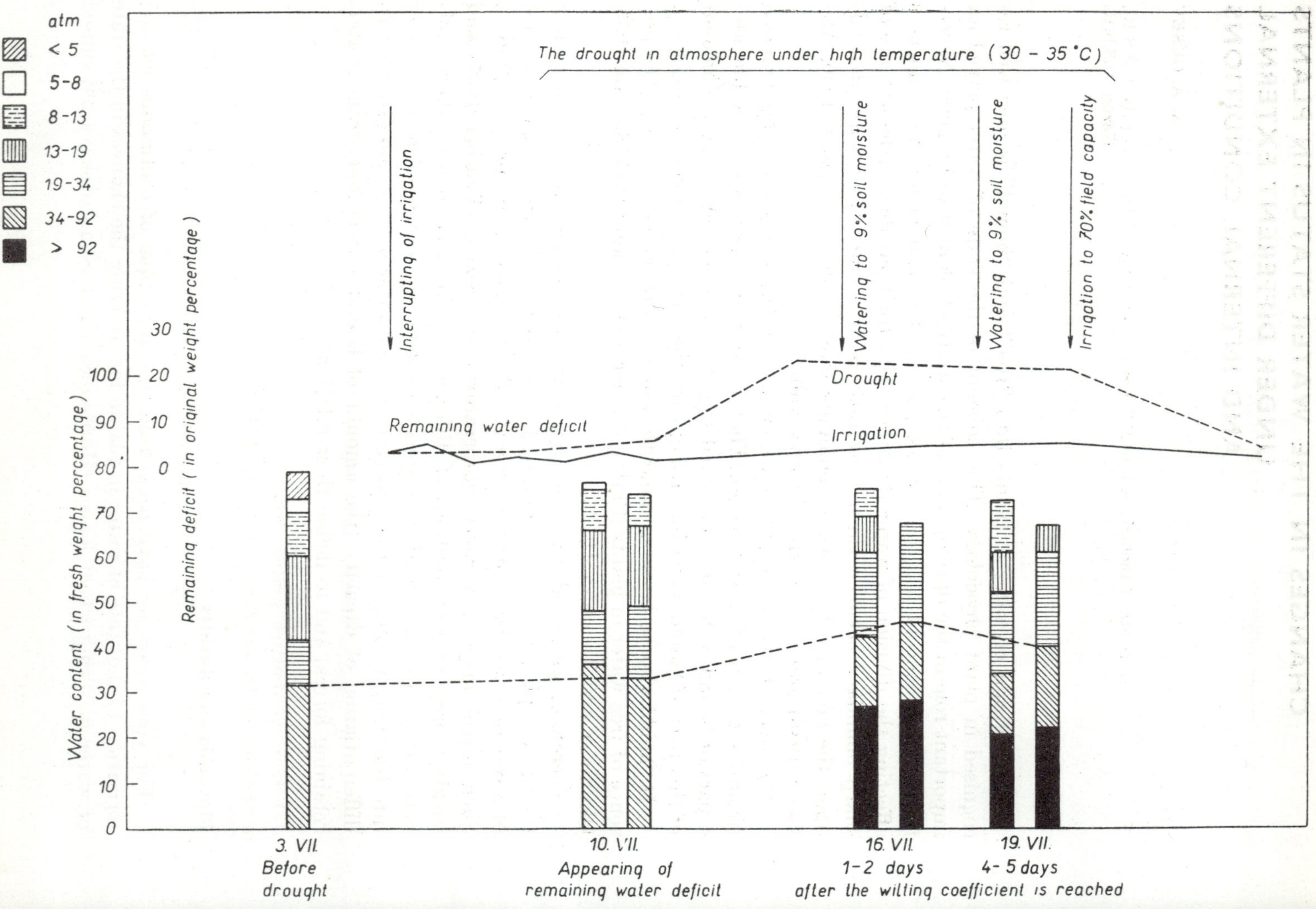
atm
< 5
5-8
8-13
13-19
19-34
34-92
> 92
The drought in atmosphere under high temperature (30 – 35°C)
Interrupting of irrigation
Watering to 9% soil moisture
Watering to 9% soil moisture
Irrigation to 70% field capacity
Drought
Remaining water deficit
Irrigation
Remaining deficit (in original weight percentage)
30
20
10
0
Water content (in fresh weight percentage)
100
90
80
70
60
50
40
30
20
10
0
3. VII.
Before drought
10. VII.
Appearing of remaining water deficit
16. VII.
1-2 days
19. VII.
4-5 days
after the wilting coefficient is reached

by environmental conditions. The wheat was grown in a vegetative vessel on grey forest slightly eluated soil widely represented in the Middle Volga Region.

Fertilizers were added in the following amount: 1·50 g. P_2O_5; 0·37 g. N; 0·75 g. K_2O for every 6 kg. of soil. Nitrogen nutrition was added to half the vessels during the period of shooting. Watering was done up to the 70% of total moisture capacity. Drought was caused by not watering. The plants were subjected to high temperatures in a glass thermochamber.

The experimental results made it possible to establish the correlation of the state of water and the water supply, nutrition and temperature. In the present paper attention is focused on the changes induced by the insufficient water supply.

Dehydration of plants during drought (Fig. 1) was first due to the loss of the most weakly bound water (retained by forces up to 5—8 atm.). As the period and the degree of the drought increased the more strongly bound water fractions began to be wasted (those retained by forces up to 13—19 atm.). Hence the dehydration of plants during drought will be less if they contain relatively less weakly bound water and much more strongly bound water. But the drought affected the amount of strongly bound water as well. At the beginning of drought (Fig. 1) the amount of water retained by the forces more than 34 atm. can somewhat drop and, on the contary, it increased during drought (from the moment the wilting coefficient was reached). In this case the increasing of the amount of water retained by maximal forces (more than 92 atm.) were most marked. With further progress of the drought the amount of more strongly bound water again decreased (as a result of intensifying hydrolytic processes under the influence of waste of water and reheating of plants).

The change in fractional composition of water is due to the change in water-retaining forces. This concept, however, should be made more exact before we start analysing the change in water-retaining forces. When comparing the plants of several series of experiments by the amount of some water fractions (i.e. by the capacity factor) some definite water-removing force (i.e. the strain factor) is taken as a constant. The plants are compared by the amount of water removed by this force (say a force of 34 atm. corresponding to a 30% saccharose solution).

When comparing plants by their water-retaining forces (i.e. by strain factor) it is necessary, on the contary, that some definite amount of removed water (capacity factor) be taken as a constant. The plants in this case should be compared by the forces to be applied to them for removing the same

Fig. 1. Changes in the water content in wheat leaves during drought (1963, earing — flowering).

amount of water. As the definition is made at the moment of setting a dynamic equilibrium between the plant water and the solution the plant is submerged into, the water-removing force of the solution may be considered to be equal to the water-retaining force of the plants under these conditions.

The increase in the amount of strongly bound water observed under drought was always connected with an increase in the water-retaining forces of the plants.

Thus, in one of our experiments in 1961 (Fig. 2) a water retaining force at 10% water remove (S_{10}), increased under the influence of drought by 42% (17 atm. compared to 12 atm. in control plants) and at 20% water remove (S_{20}) by 45% (29 atm. as compared with 20 atm. in the control).

However, a drought is often attended by a high temperature that has quite a different effect on the water-retaining forces and the state of water in plants. In the earlier experiments we observed that a high temperatures (35°—38° C) causes a decrease in the amount of strongly bound water and, on the contrary, an increase in the amount of weakly bound water (Gusev 1957, 1959, 1959a, Alekseev, Gusev and Belkovich 1953). These changes can be explained by a negative temperature coefficient of chemohydration and by a positive coefficient of structural binding and osmotic uptake of water (Lipatov 1943, Dumansky 1948, Alekseev 1950).

The method of continuous characterization of the state of water made it possible to give a thorough specification of this fact. Thus, the experiments of 1962 showed that only the amount of water retained by the forces up to 12·4 atm.

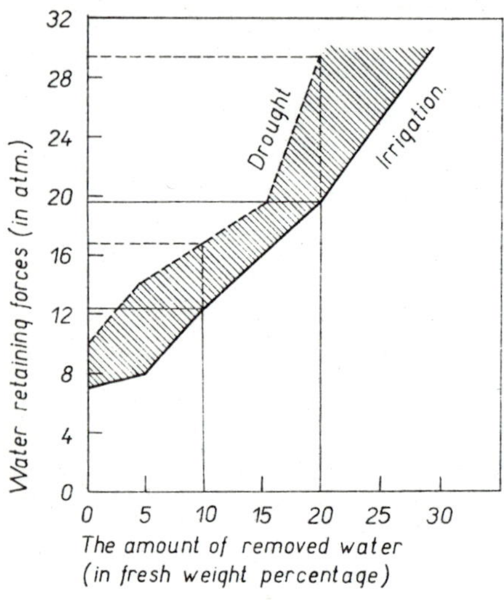

Fig. 2. Water-retaining forces in wheat leaves at different water supply (1962, flowering)

in atm.	irrigation	drought
S_0	7·1	9·0
S_5	8·2	14·7
S_{10}	12·3	16·9
S_{15}	16·1	19·3
S_{20}	19·8	29·1
S_{25}	25·1	—
S_{30}	30·5	—

(Fig. 3) can be constantly increased under the influence of high temperatures but the amount of water more strongly bound decreases, the amount of most strongly bound water (retaining by forces more than 34 atm.) decreasing.

The decrease in water-retaining forces was the reason for the decrease in the amount of strongly bound water. Thus, S_{15} decreased by 5·9 atm. (15·2 atm. compared to 21·1 atm. in the control) S_{30} — by 9·3 atm. (24·9 atm. compared to 34·2 atm. in the control). Similar results were obtained in other experiments.

A high temperature thus stimulates a greater waste of water in two ways: firstly, it increases the difference in water activity in plants and in the atmosphere (Alekseev, Gusev, and Belkovich 1963) and secondly, it decrease water-retaining forces in plants, both leading to increase in transpiration, i.e. increase in the loss of water.

The above means that the increase in water-retaining forces is one of the main factors increasing drought resistance in plants particularly at a high air temperature.

This can be achieved by special treatment of the seeds before sowing (hardening by Genkel's method, treating seeds with microelements) and by organising a special nutrition regime.

Our investigations showed that the increase in water-retaining forces of summer wheat leaves was due to rich phosphorous nutrition in the first part

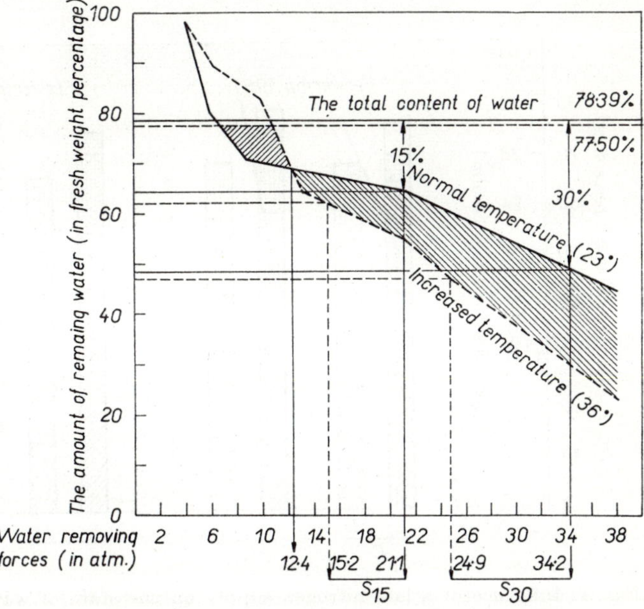

Fig. 3. The state of water and the water-retaining forces in wheat leaves at different temperatures (1962, shooting). Normal temperature (23° C) ...... $S_0 = 6\cdot3$ atm. Increased temperature (36° C) .... $S_0 = 9\cdot6$ atm.

of vegetative period and to rich nitrogen nutrition in the second part (beginning with earing). The results of the experiments made in 1961 and 1963 (Fig. 4) may serve as a good example. The addition of extra nitrogen in the period of shooting increased the amount of strongly bound water (retained by forces more than 34 atm.). This increase was higher under drought. It depended upon the increase in water-retaining forces. Thus, S_{20} increased by an average of 22% (31—35 atm. compared with 25—29 atm.) in plants without extra nitrogen nutrition. This resulted in retaining a still higher total amount of water in plants that underwent extra nutrition under drought (by about 2% compared with the plants in the control group i.e. plants without extra nitrogen nutrition).

The effect of extra nitrogen nutrition on the water-retaining forces and the state of water in wheat leaves was retained even at a high temperature (Fig. 5). Thus the amount of strongly bound water (retained by the forces more than 34 atm.) in the plants that had extra nutrition decreased by 17 to 20% only, while in control plants it decreased by 40—55% of the original value.

The effect of nutrition conditions (as well as other environmental conditions) on the state of water in plants is effected by means of changing metabolism. The dependence of the degree of hydration of highly polymeric proto-

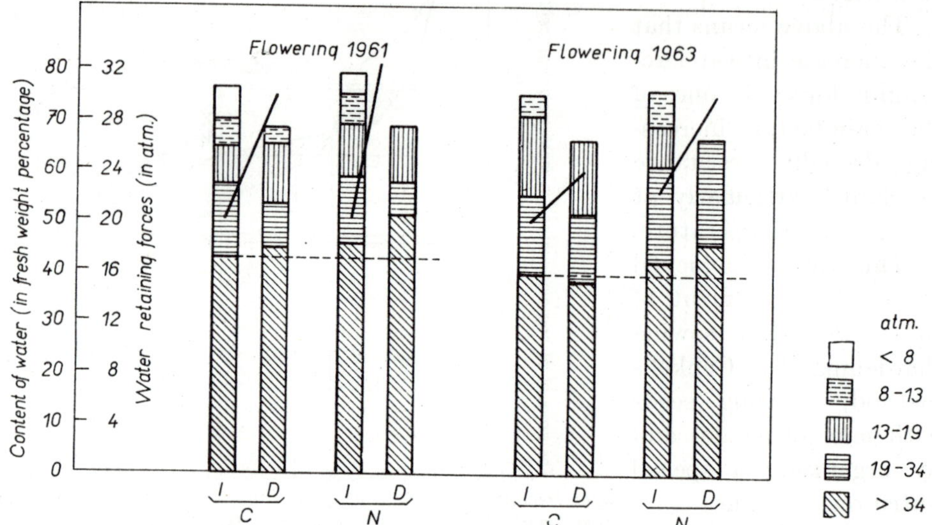

Fig. 4. Influence of a late nitrogen supply on the state of water in wheat leaves at different water supply.
I — irrigation, D — drought, N — nitrogen supplied, C — control, ——————— S_{20}

plasm compounds upon their composition and state was theoretically based and experimentally proved by the laboratory of plant physiology of the Kazan Branch of the Academy of Sciences (Alekseev 1948, 1950, Alekseev and Gusev 1957, Alekseev, Vasileva and Startseva 1959, Gusev 1959 and others).

The total degree of hydration of these compounds proved to be positively connected with the amount of proteins (watersoluble and nonextractable in particular) and of organic phosphorous compounds (especially nucleoproteids).

Using the method of continuous characterization of the state of water showed (Fig. 6) that the amount of watersoluble and non-extractable proteins is in positive correlation only with the amount of strongly bound water (retained by the forces more than 34 atm.), correlation coefficients between these two values being as much as +0·72 to +0·99.

With the weakening of the forces typical for retaining water, a positive correlation with the watersoluble and non-extractable proteins weakens and disappears (Fig. 6). A negative correlation of these values (from —0·54 to —0·99) appears and increases instead. Some rearrangement of water may seem to occur with the change in the amount of watersoluble and nonextractable proteins: the amount of strongly bound water increases at the expense of a decrease in the weakly bound water with increase in the amount of proteins.

The changes take the opposite direction when the amount of proteins decreases. Thus, an indirect negative correlation of the amount of weakly bound water with the amount of proteins arises. The increase in this cor-

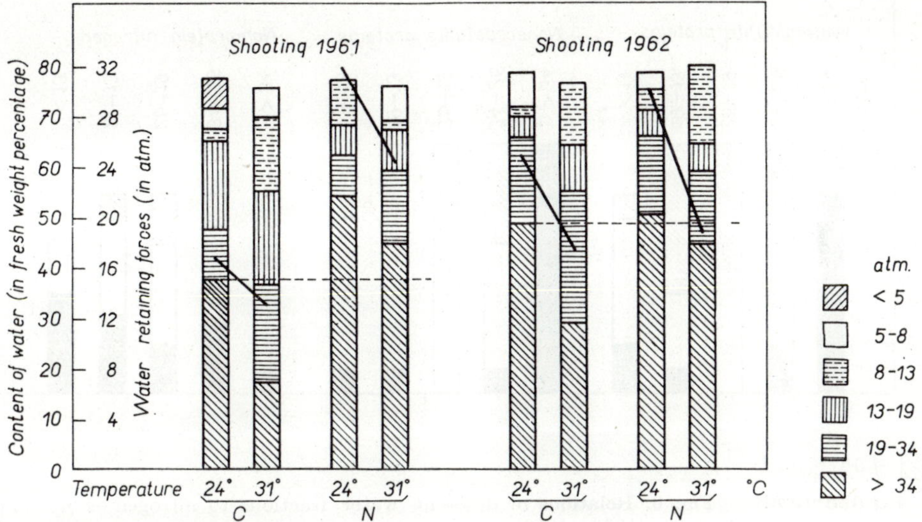

Fig. 5. Influence of a late nitrogen supply on the state of water in wheat leaves at different temperatures.
N — nitrogen supplied, C — control, ——————— S_{20}

relation with the decrease of water-retaining forces (Fig. 6) shows that the weaker the water is bound the easier it may be used in such a rearrangement.

A more weakly bound water (retained by forces less than 13 atm.) does not, however, usually exhibit a correlation with the amount of proteins (Fig. 6). It may be so mobile that its amount depends upon the immediate effect of environmental factors (temperature, air humidity, wind, light), the indirect dependence on the amount of proteins playing a second role. In addition, a positive correlation of the amount of this water with the amount of non-protein nitrogen (r from +0·66 to +0·97) is usually observed.

From the above we may draw the conclusion that the more strongly bound water is bound by the highly polymeric compounds (proteins) the more weakly bound water is bound by the low-polymeric compounds (by the non-protein nitrogen compounds) and the latter probably belong to the osmotic uptaken water (Alekseev 1950, 1960).

From this point of view the above changes of the state of water in plants,

depending upon high temperatures, drought and nutrition conditions can be
clearly understood.

It was repeatedly stressed in the literature that high temperatures cause
the decomposition of proteins in plants. (Khlebnikova 1934, 1937, Altergot
1936, 1937, Zauralov and Kruzhylin 1951, Gusev 1957, 1959a, 1959b and
others.) The thorough investigations of Altergot (1963) showed that under

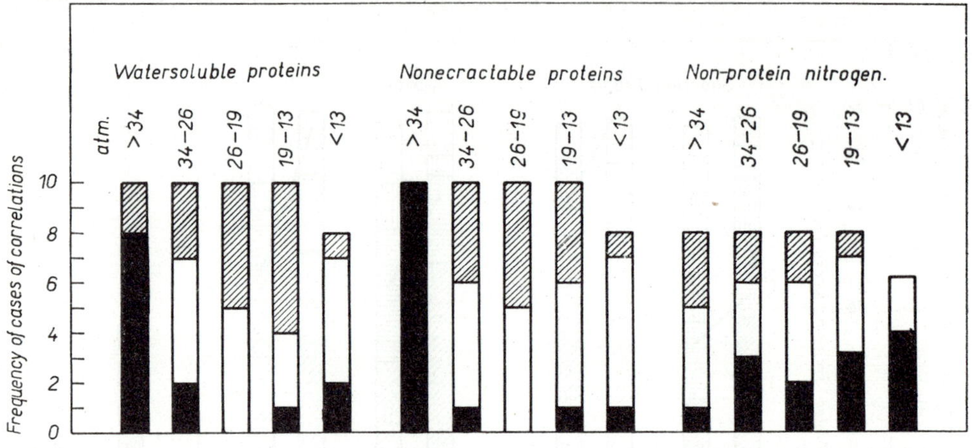

Fig. 6. Relations of different water fractions to nitrogen compounds
(1960—1962).

high temperatures, synthetic processes in metabolism stop and depolymerisa-
tion of complex organic compounds and plasmatic structures occur. These
facts were confirmed by the experiments considered.

Thus, the experiments of 1962 (Fig. 7) showed that the ratio of protein
nitrogen to non-protein nitrogen $\left(\dfrac{p}{n}\right)$ in wheat leaves decreases with high
temperature (36° C) two-fold (2·0—2·5 compared to 3·7—4·9 in the control).

This decrease had a stronger effect upon the state of water than the increase (though much weaker) in the ratio of more hydrophilic proteins
$\left(\dfrac{m.h.p.}{l.h.p.}\right)$ that is the sum of watersoluble and non-extractable proteins to the
sum of saltsoluble and alkalisoluble proteins that took place simultaneously.
A decrease in water-retaining forces and in the amount of strongly bound
water in leaves resulted. In addition a high temperature could effect the
state of water directly, causing partial dehydration of highly polymeric compounds of protoplasm and thus decreasing the amount of strongly bound
water (Dumansky 1948).

The influence of drought upon the nitrogen metabolism was more complex. A long but not very severe drought (Fig. 7, 1962) nearly doubled $\frac{p}{n}$ (5·0—5·9 compared to 3·0—3·3 in the control). The change in $\frac{\text{m.h.p.}}{\text{l.h.p.}}$ in this case was

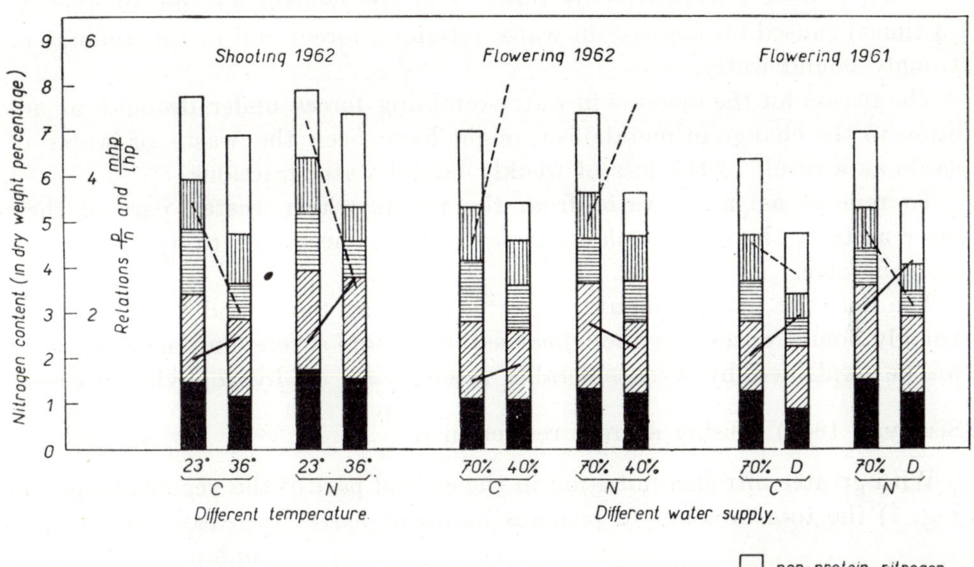

Fig. 7. Influence of a late nitrogen supply on nitrogen metabolism in wheat.
N — nitrogen supplied, C — control, D — drought, W — normal water supply.
Shooting 1962:

$$- - - - - \left(\frac{p}{n}\right)_{36} = 53\% \left(\frac{p}{n}\right)_{23} \qquad \text{————} \left(\frac{\text{m.h.p.}}{\text{l.h.p.}}\right)_{36} = 139\% \left(\frac{\text{m.h.p.}}{\text{l.h.p.}}\right)_{23}$$

Flowering 1962:

$$- - - - - \left(\frac{p}{n}\right)_{D} = 174\% \left(\frac{p}{n}\right)_{W} \qquad \text{————} \left(\frac{\text{m.h.p.}}{\text{l.h.p.}}\right)_{D} = 101\% \left(\frac{\text{m.h.p.}}{\text{l.h.p.}}\right)_{W}$$

Flowering 1961:

$$- - - - - \left(\frac{p}{n}\right)_{D} = 77\% \left(\frac{p}{n}\right)_{W} \qquad \text{————} \left(\frac{\text{m.h.p.}}{\text{l.h.p.}}\right)_{D} = 130\% \left(\frac{\text{m.h.p.}}{\text{l.h.p.}}\right)_{W}$$

slight and had no definite direction. Here again the change in $\frac{p}{n}$ played a decisive role and led in this case to increase in the water-retaining forces and in the amount of strongly bound water.

On the contrary, under a severe but short period of drought (Fig. 7, 1961) $\frac{p}{n}$ dropped, but the change was not as great as in the preceding cases and, therefore, could not exceed the influence of changes in $\frac{\text{m.h.p.}}{\text{l.h.p.}}$ the increase of which (2·0—2·7 compared to 1·5—2·3 in the control, i.e. on an average 1·3 times) caused an increase in water-retaining forces and in the amount of strongly bound water.

The reason for the increase in water-retaining forces under drought in addition to the change in metabolism might have been the waste of water of plants as a result of the loss of weakly bound water fractions.

To remove as much water from the plants which wasted part of their water as from the control plants it seems to be necessary to apply a much stronger force.

The decrease in the amount of water-retaining forces and the amount of strongly bound water (see above) caused by a more severe and longer drought can be explained by a considerable intensifying of hydrophylic processes (Sisakyan 1940) causing a great reduction in $\frac{p}{n}$.

With greater nitrogen nutrition in the second part of the vegetative period (Fig. 7) the total amount of proteins in wheat leaves increases (2—15 % as against the corresponding amount in control plants). $\frac{\text{m.h.p.}}{\text{l.h.p.}}$ increased as well (under optimal conditions of water supply by 23—59 %, under drought by 13—34 %, at a high temperature by 72 % of the corresponding amount the control plants). In the plants with nitrogen nutrition, the increase in hydrolytic processes at high temperatures was less. The increase in the absolute amount of proteins (by 15%) as well as the increase in $\frac{p}{n}$ (23 % as against the corresponding amount of the control plants) prove this.

A higher yield of corn of the plants with nitrogen nutrition under drought (2·0 g. compared with 1·3 g. per one vessel in the experiment of 1962) is a final result that proves the improvement of drought resistance in these plants.

Thus, directed changes can be induced in the state of water in plants that favour better survival in drought by affecting the course of metabolism using special nutrition conditions.

References

Alekseev, A. M.: [Water relations in plants and drought affecting it.] In Russ. — Tatar Publ. House, Kazan, 1948.

—: [On the water relations in plants.] In Russ. — Problemy Botaniki *1* : 298—320, 1951.

—: [Some results of the study of the water relations and problems to be studied.] In Russ. — Theses, Guest Session, Biol. Sect., Kazan, Publ. House Univ., Kazan 1960.

—, Vasileva, I. M., Startseva, A. V.: [Physiology of metabolism of the red clover.] In Russ. —. Publ. House Acad. Sci. U.S.S.R., 1959.

—, Gusev, N. A.: [Mineral nutrition affecting the plant water relations.] In Russ. — Publ. House Acad. Sci. U.S.S.R., 1957.

—, —, Belkovich, T. M.: [The diurnal dynamic of water relations in plants.] In Russ. — *In*: Plant Hardeness, State Publ. House of Light Industry, Moscow 1963.

Altergot, V. F.: [On the causes of the decay of plants at high temperatures.] In Russ. — Izv. Akad. Nauk SSSR, Ser. biol. *1* : 79—88, 1936.

—: [Selfpoisoning of plant cell at high temperatures.] In Russ. — Trudy In-ta Fiziol. Rast. im. K. A. Timiryazeva *2*, 1937.

—: [The effect of high temperatures on plants.] In Russ. — Izv. Akad. Nauk SSSR, Ser. biol. *1* : 57—73, 1963.

Crafts, A. C., Currier, H. B., Stocking, R.: Water in the physiology of plants. — Chronica Botanica Co., Waltham, Mass., 1949.

Dumansky, A. B.: [Colloidal study.] In Russ. — The State Chem. Publ. House, Moscow 1948.

Gusev, N. A.: [The effect of dry wind on water relations in spring wheat.] In Russ. — Fiziol. Rast. *4* : 305—311, 1959.

—: [Several laws of plant water relations.] In Russ. — Publ. House Acad. Sci. U.S.S.R., Moscow 1959a.

—: [High temperatures affecting plant water relations.] In Russ. — Izv. Akad. Nauk SSSR, Ser. biol. *1* : 79—86, 1959b.

—: [The characteristic of the state of water in plants.] In Russ. — Fiziol. Rast. *9* : 432—437, 1962.

Khlebnikova, N. A.: [Physiology of fruit trees and vegetables. 1. Hardiness of fruit trees and vegetables to high temperatures under conditions of Astrakhan zonal station.] In Russ. — Trudy Komisii po Irrigacii Akad. Nauk SSSR *3*, 1934.

—: [Chemical nature of plant organism hardiness to the influence of temperature factor.] In Russ. — Trudy In-ta Fiziol. Rast. im. K. A. Timiryazeva Akad. Nauk SSSR *1* (2), 1937.

Lipatov, S. M.: [Highpolymeric compounds: lyophilic colloids.] In Russ. — Tashkent, 1943.

Okuntsov, M. M., Tarasova, E. N.: [Water status in plants.] In Russ. — Dokl. Akad. Nauk SSSR *83* : 315—317, 1952.

Sabinin, D. A.: [Physiological grounds of plant nutrition.] In Russ. — Publ. House Acad. Sci. U.S.S.R., 1955.

Sisakyan, N. M.: [Biochemical characteristics of the plant drought hardiness.] In Russ. — Publ. House Acad. Sci. U.S.S.R., 1940.

Tumanov, I. I.: [Physiological grounds of plant frost-hardiness.] In Russ. — Agric. Publ. House, Leningrad, 1940.

Tyurina, M. M.: [Studies of frost-hardiness in plants in mountain conditions of Pamir.] In Russ. — Trudy In-ta Bot. Akad. Nauk Tadzh. SSR *57*, 1957.

Zauralov, C. A., Kruzhilin, A. S.: [The change of nitrogen metabolism in cabbage leaves at high temperatures.] In Russ. — Dokl. Akad. Nauk SSSR *77* : 733—736, 1951.

THE EFFECT OF RELATIVE AIR HUMIDITY ON THE DEVELOPMENT OF WATER SATURATION DEFICIT IN THE CUT LEAVES OF TWO VARIETIES OF BARLEY

L. NÁTR and IVANA KOUSALOVÁ

Cereal Research Institute, Kroměříž, Czechoslovakia

The study of the water relations is carried out at our Institute with a view to its application in plant production management. From the point of view of plant physiology, the whole of plant production is conceived as essentially an accumulation of the products of photosynthesis by the plant. It is known that among numerous factors influencing photosynthesis, the water relations have a particularly important significance. The indirect effect of water deficit on the photosynthetic rate is especially striking. This indirect effect acts by the hydroactive closing of the stomata with increasing water deficit, causing restricted up to complete halting of CO_2 uptake (Pisek and Winkler 1956). The fact is even more important since the highest water deficit occurs most often on warm sunny days when all other factors, such as light, temperature and CO_2, are very favourable for high net photosynthesis. Under these conditions — relatively small change in water deficit can lead to relatively big changes in photosynthetic rate.

The unfavourable balance of the water relations of plants on these warm sunny days arises as a result of excessive water loss, since even with optimal conditions of soil moisture, plants cannot replace loss of water arising from rapid transpiration. In our region a period of drought before the end of the vegetative period is the most frequent cause of a big decrease in grain yield. This is certainly not only a question of the effect of reduced soil moisture but obviously also of decreased air humidity. We, therefore, wished to obtain quantitative measurements of the effect of air humidity on changes in tissue saturation and thereby on the photosynthetic rate. In this paper we report the effect of different relative air humidity on the development of water saturation deficit in the leaves of spring barley after stopping water uptake.

Material and Methods

The experiments were made on the leaves of two varieties of spring barley, Valtický and Selekční hanácký. The plants were cultivated under artificial lighting (40 W neon lamps) at a temperature of 25° C in the day and 15° C at night. The measurements were made on cut leaves aged 8—12 days. Immediately they had been cut the leaves were placed in a simple plexiglass cuvette in such a way that the lower part of the leaves was left sticking out of the cuvette and the cut surfaces immersed in water (Fig. 1). After 30 min. the leaves were removed from the cuvette.

The parts which had been placed in the cuvette were cut off and immediately weighed. After weighing these parts (marked A on Fig. 1) they were replaced in the cuvette and weighed at 10-minute intervals for one hour. For determining the amount of water required for the complete saturation of the tissue, the leaves were then placed in vessels in the dark and weighed after 20 hours. Dry weight was then determined and water deficit of the leaves calculated in the usual way in percentages. The intensity of irradiation was 6.49×10^4 erg.cm.$^{-2}$sec.$^{-1}$ in wave lengths 380 up to 780 mμ, air temperature 25° C, air speed 24 l.hr.$^{-1}$. The CO_2 content was kept constant at 0·035 vol. %. Constant air humidity was maintained by the known method of saturating the air with water vapour at normal or raised temperature and then cooling to a given temperature to precipitate the excess of water vapour. The water saturation deficit was determined at relative air humidities of 35%, 60% and 85%.

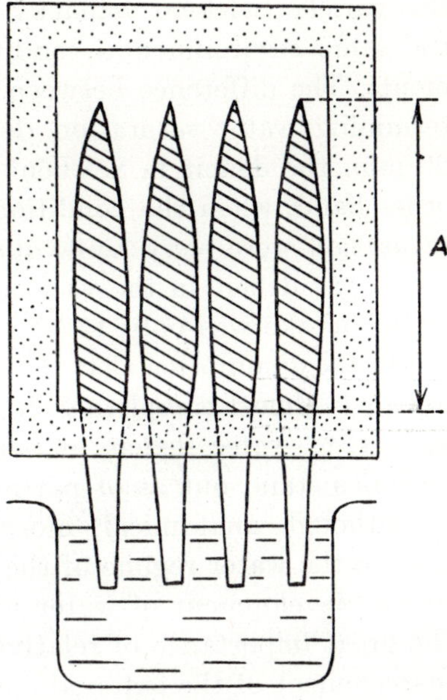

Fig. 1. Diagram of leaves placed in simple assimilation chamber on saturating with water. Sector A denotes part of leaves in which the development of water saturation deficit was determined.

Results and Discussion

The curve of water saturation deficit in the variety Valtický is given in Fig. 2. Each point is the mean of 10—15 measurements. At relatively high air humidity (85%) water saturation deficit amounted to about 12·5% one hour after stopping water uptake, whereas in dry air the deficit amounted to 19%. According to preliminary results the CO_2 uptake of the leaves of the two varieties under experiment was almost completely halted at a water deficit of about 12%. From this point of view the differences in the developing of water deficit in the leaves at different air humidity are particularly important: at 85% air humidity net photosynthesis falls to zero one hour after the complete stopping of water uptake, at 60% air humidity CO_2 uptake stops after 22 min. and at 35% air humidity already after 13 min.

The values of the water deficit curves of the other variety, Selekční hanácký, are similar but the absolute values differ (Fig. 3). The results of experiments by Zemánek at the above institute showed that water loss in the variety Selekční hanácký is less than in the variety Valtický under the same environmental conditions. This was also observed in our experiments. The difference between the two varieties was relatively small on determining water saturation deficit at an air humidity of 85%, when after 60 min. the deficit in Selekční reached 11·3% and in Valtický 12·5%. On the other hand, when the conditions for maintaining a good level of water tissue saturation were very unfavourable, i.e. at a relative air humidity of 35%, the water deficit in the variety Selekční hanácký 60 minutes after stopping water uptake was only 15·5%, while in Valtický it was 19%.

On the assumption that the uptake of CO_2 by the leaves from the surrounding air is almost completely stopped by the hydroactive closing of the stomata at a 12% water deficit in both varieties, the differences between the two varieties stand out as of particular importance.

Although anatomical, morphological, biochemical and other factors influence the water regime of the given plant, the quantitative results obtained in the development of water deficit on stopping water uptake demonstrated the great importance of relative air humidity. The use of these results in the management of the external environment of crops to obtain raised and stable yields would mean that the photosynthesis rate could be greatly influenced through the water regime of the plant by the regulation of air humidity. Since the study of non-root nutrients of plants and increasing the concentration of CO_2 in the air has also shown a favourable effect by increasing net photosynthesis (Mortimer 1959, Bezděk 1962 etc.), the simultaneous application of all three factors would be of great physiological advantage. It is probable that the use of very dilute solutions of mineral nutrients with possible CO_2 enrichment with increased air humidity would be physiologically more effective and more economical than the existing methods of irrigation.

Summary

Determinations were made of the effect of differing air humidity on the creation of water saturation deficit after stopping water uptake in the cut leaves of two varieties of spring barley grown on sand culture with artificial irradiation.

The results, shown in the figures, confirm the great influence of the water

9*

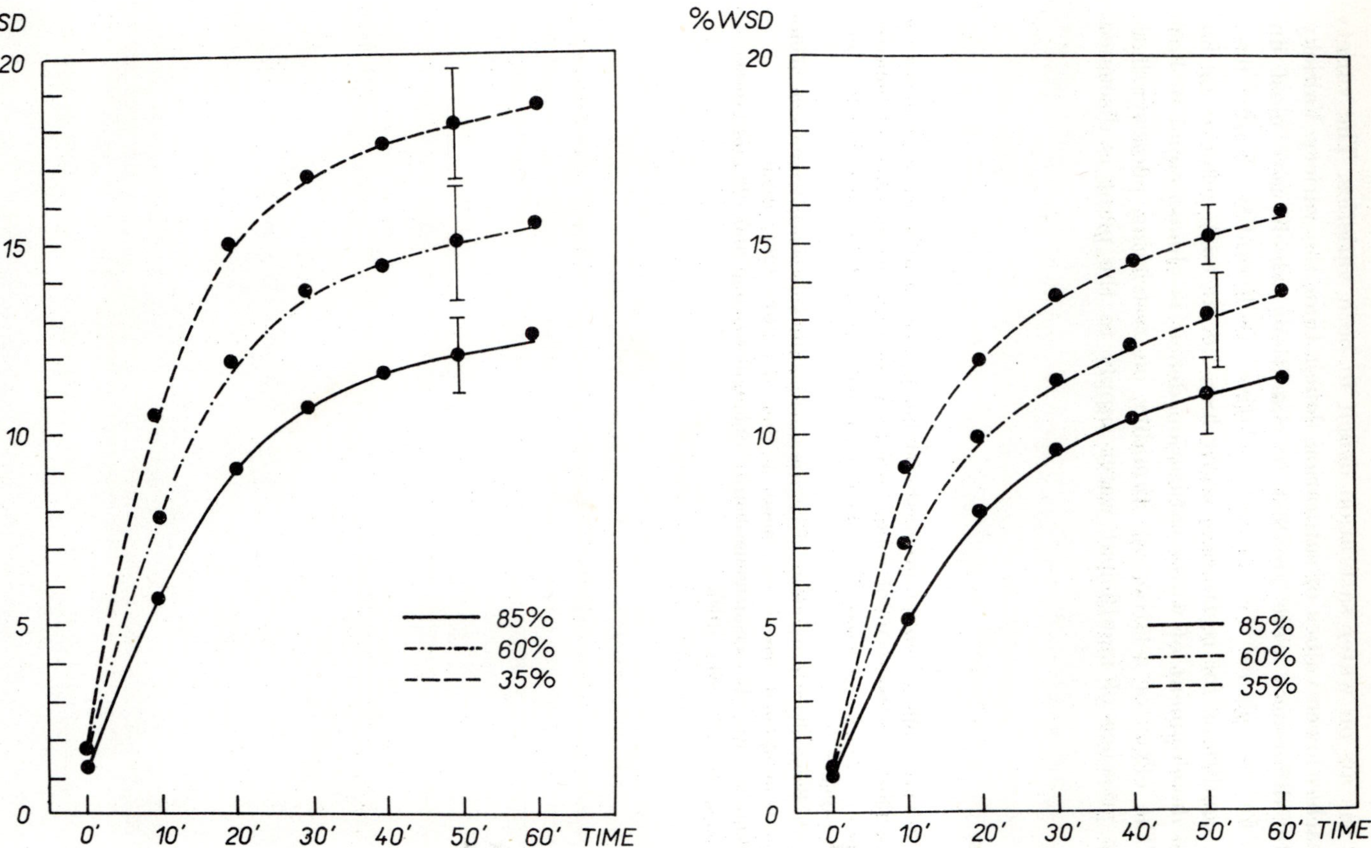

Fig. 2. Curve of development of water saturation deficit (W.S.D.) in cut leaves of spring barley, variety Valtický, at different relative air humidities after stopping water uptake. Perpendicular lines denote reliability intervals of $\pm\, t_{0,025} \cdot s\overline{x}$.

Fig. 3. Curve of development of water saturation deficit (W.S.D.) in cut leaves of spring barley, variety Selekční hanácký, at different relative air humidities after stopping water uptake. Perpendicular lines denote reliability intervals of $\pm\, t_{0,025} \cdot s\overline{x}$.

content of the air on water saturation deficit in both varieties. The variety Selekční reaches lower values of saturation deficit than the variety Valtický. The varietal differences were greatest in conditions which were most unfavourable for the plants, i.e. at 35% relative air humidity. Under more favourable conditions for restricting water loss, i.e. at 85% relative air humidity, the development of water saturation deficit is almost equal in both varieties. The effect of relative air humidity on restricting photosynthetic rate in the presence of insufficient water supply to the plant is discussed.

References

Bezděk, V.: Možnosti aplikace roztoků průmyslových hnojiv postřikem na list u ozimé pšenice. [The possibilities of treatment of winter wheat leaves with the solutions of industrial fertilizers.] — Věd. Práce Výzk. Úst. obiln. v Kroměříži [Sci. Papers of Cereal Res. Inst. Kroměříž] 1962.

Mortimer, D. C.: Some short-term effects of increased carbon dioxide concentrations on photosynthetic assimilation in leaves. — Canad. J. Bot. *37* : 1191—1201, 1959.

Pisek, A., Winkler, E.: Wassersättigungsdefizit, Spaltenbewegung und Photosynthese. — Protoplasma *46* : 597—611, 1956.

Discussion

H. Polster: The determination of photosynthetic rate in the stand of poplars showed that high air humidity can almost counterbalance great dryness of the soil. Despite considerable dryness of the soil (14-day period of drought) a high photosynthetic rate was determined, but in the presence of low air humidity almost no assimilation was found. This findings confirm the results described.

L. Nátr: The communication of Dr. Polster significantly confirms our results that relative air humidity is an important factor in the regulation, not only of water relations, but also indirectly of the photosynthetic rate of plants.

G. Meinl: We confirmed the results of Dr. Nátr in potatoes when investigating photosynthesis. We have not found, however, significant species differences. Question: Were the plants investigated from species with the same time of ripening, that is, were the organs investigated of the same physiological age? In order to obtain comparable results we make relative comparisons of the same vegetation periods in potatoes and thus of the individual phases of development (maximum total mass = 100, this time point also = 100). Only thus, in our opinion, is it possible to make a direct comparison of the values measured.

L. Nátr: Unfortunately, I cannot reply because I do not know. From experiments carried out in our institute by Eng. Zemánek it was shown that out of the large number of varieties investigated, the greatest differences were found between Valtický and Selekční. We therefore worked with these varieties. The experiments were made on the first leaves of about 12-day plants, which had been cultivated under constant conditions. We are, therefore, of the opinion that in this case possible differences in the duration of growth and a different course of ontogenesis could not cause interference. We see the solution to the problem of the impossibility of comparing values obtained when measuring physiological qualities of varieties with different length of vegetative period at the same time intervals, only in the possibility of completely regulating external conditions throughout the entire vegetation period and in the repeated measurement of the corresponding physiological processes. This, of course, requires good technical equipment which at the present time is accessible only to some laboratories. The calculation of relative times which is used by Dr. Meinl, is in my opinion a good idea. However, even in this case we cannot exclude changes in the external environment since measurements are made which are comparable from the viewpoint of the ontogenesis of the corresponding variety but under quite different climatic conditions.

W. R. Müller-Stoll: When reviewing results of physiological experiments with regard to their importance and applicability for agricultural practice, the same caution is needed as whenever experience of laboratory work is to be generalized in field conditions. With reference to the influence of the water situation in plants on their photosynthesis, the reaction of the test objects will be more or less different under field conditions and under controlled conditions in the laboratory. For the productivity of crop plants in the field, their specific ontogenesis, for example, is also very important. Agricultural experience has proved that after premature development under relatively dry conditions, the productive power in the mature stage may be quite different (higher) from that of premature development under relatively wet conditions. Many signs of adaptation, e.g. different formation of subterranean and overground organs, will cause modified reactions in plants and may yield different pictures in subsequent years, according to the weather conditions. In view of these facts, research on the productive power of crop plants by methods of plant physiology is a difficult, yet very interesting work, whose great importance is still frequently disregarded in agriculture.

L. Nátr: Air humidity was chosen intentionally. For high grain yield it is very important that conditions are favourable for photosynthesis at the end of swelling of the grain. Under our conditions there is often excess soil and air dryness in June and July. Our aim was therefore to carry out a basic experiment on the degree to which different relative air humidity can affect photosynthetic rate through water deficit in plants.

W. Larcher: The methods of maintaining constant external factors in physiological investigations and the working technique itself has advanced in recent years so that the testing of species, varieties etc., under standard conditions has become relatively easy. We can determine, for example, that different varieties, under optimal conditions, do not differ at all in their photosynthetic rate or we can find varietal differences. We should not forget, however, that they are all only tests and that they are influenced by the united effect of very varied situations, e.g. the past history of the plant, the degree of development, etc. Varietal differences may in fact have their base here. Despite this we should not abandon such tests because we must start from somewhere.

L. Nátr: I agree with the opinion of Dr. Larcher that it is necessary to carry out basic studies of important physiological processes and how they are affected by factors of the external environment in plant cultures and in testing varietal differences. At the present time testing appears to be most useful for practical purposes because the more important results afford reproducible observations obtained in regulated conditions and analysing the individual factors of the external environment.

B. Slavík: I rather think that varietal differences (generally) would show up clearly in many cases under unfavourable rather than almost optimal conditions.

L. Nátr: It is certainly necessary to agree strongly with Dr. Slavík's remarks. The results reported here also confirm this. Varietal differences in water deficit were greater under unfavourable conditions, i.e. at 35% air humidity, than in relatively better conditions — at higher air humidity. Of course, this need not necessarily always be the case. In our investigations of varietal differences in photosynthetic rate we made measurements under conditions closely approaching optimal, since we wished to differentiate varieties according to their highest activity. Here also we found varietal differences which bear a definite relation to activity.

In general, however, we consider, in agreement with Dr. Slavík, that the best results on the effect of the corresponding factors can be obtained if measurements are made under a wide range of conditions, i.e. from very unfavourable to the best conditions. Unfortunately, this is not always possible for technical-reasons.

H. Polster: When making physiological measurements under standardized conditions, relatively low intensity of illumination is often used. I doubt whether measurements at 7,000 lux can be compared with ecological conditions where the intensity of illumination usually reaches 20—70,000 lux. This is valid mainly if we wish to assess yield on the basis of comparison of determined photosynthetic rates.

L. Nátr: In studying photosynthesis under laboratory conditions the effect of intensity of illumination should without doubt be given primary attention. However, when investigating other factors, e.g. water deficiency, it is necessary to work with a given intensity. And it is always possible to find objections, whether we use high or low intensity. It would certainly give the best results if measuring were made under different light intensities, which however, at the present time come up against technical problems and shortage of time.

R. Zwicker: In investigating the effect of light on physiological processes, it is necessary to consider not only the intensity, but also the quality of light, in the same way as in the course of natural ecological conditions from May to September, red and blue affect the transpiration of plants with different nutritional requirements (oats) differently. It can be assumed that the photosynthesis of different varieties will be differently influenced by the quality of the light.

L. Nátr: Unfortunately, we have not investigated problems of the quality of light in connection with the study of the effect of different air humidities. If, however, the effect of different qualities of light showed itself in transpiration in the experiments of Dr. Zwicker, it is without doubt — and there are data in the world literature — it will be even more apparent in the photosynthetic rate and possibly in the quality of assimilates formed.

W. Larcher: Such investigations take up much more time than is often assumed. To obtain a statistically significant series of light and temperature curves and curves dependent on deficit in one species or variety of plant about three weeks are required, including trial runs. I am unable to work out the time that would be required to investigate these three dependencies in combination. It is not only a methodological problem but a question of the programming of the experiments. This might be solved by carrying out as far as possible, exact investigations on a small number of plants which would show us where it is necessary to carry out serial experiments.

L. Nátr: I completely agree with the opinions of Dr. Larcher. Since we had no conception of the effect of relative external humidity on the development of water deficit under conditions of inadequate water supply and on the photosynthetic rate, we had first to deal with the basic study of the relationship of relative air humidity to water deficit. The results showed us the extent to which this factor must be taken into consideration and if it is necessary to consider this factor at all in studying further factors.

It is, of course, true, that this work makes greater demands on time. If, however, we want to understand the whole complex of physiological questions connected with the production of crop plants we must necessarily gradually acquire such basic knowledge of the interaction, first of one and then of the combined effects of a large number of factors.

THE REFRACTOMETRIC METHOD OF DIAGNOSING WATER DEFICIT IN PLANTS*

L. A. FILIPPOV

Moldavia Horticulture, Viticulture and Wine-Making Research
Institute, Kishinev, U.S.S.R.**

The refractometric method of diagnosing water stress in plants was proposed by Lobov (1949) with a view to determining the watering-dates for the tomato plant as well as for other vegetables and potatoes (Lobov 1957). Later on this method was highly appreciated for diagnosing water stress in wheat (Galinskaya and Kalinin 1953, Petinov 1959), in maize (Tityov 1958), tomato (Babushkin 1959, Belik 1960), and sugar-beet (Chunosova 1963).

For several years we have carried out research to exactify and further improve the refractometric method of diagnosing the water stress first in the cotton plant and then, in the apple tree. The principal purpose of our investigation was to establish the interdependence between the cell sap concentration (CSC) in leaves as determined by refractometer and other indicators, characterizing the state of water relations in plants (the osmotic indicators of the cell sap pressure in leaves, their water content, soil moisture etc.). At the same time the changes of CSC in leaves, depending on the air temperature and humidity, as well as on the soil hydrological conditions of the place where the plants were grown and their watering were studied. Simultaneously·the principles of selecting the leaves for evaluation of the condition of water relations in the above plant species were worked out.

Material and Methods

The refractive index, or the cell sap concentration in leaves, was determined by refractometer, and expressed as a percentage of the equivalent concentration of sucrose solutions at the temperature of 20° C. The term "cell sap concentration" was suggested by Lobov (1949), and it is, as will be seen later on, very appropriate for use in practice. The determinations of CSC in leaves were carried out at different times by two types of field refractometers (RP, Kiev, and RR1, Poland) directly under orchard conditions. In laboratory

---

* This contribution was not read at the Symposium meeting.
** Present address: Sochi Experimental Station, Sochi 2, Bzugu, U.S.S.R.

investigations the determinations were made by the refractometer of the system Stoks and comparative determinations have shown similar results. The suction pressure of the leaf, or the diffusion pressure deficit (DPD) were determined by the jet-method of Shardakov (1956), the osmotic pressure of the cell sap, by the method of collodium follicles (Boyko and Boyko 1959), specific gravity of the leaf cell sap by the micromethod described in one of our works (Filippov 1958), and the dry matter content of the sap by drying 1 ml. of sap at the temperature of 100 to 105° C to constant weight. The water content in leaves and in the soil was determined by drying the samples to constant weight at the temperature of 100 to 105° C, the air temperature and humidity determined by the Assman's psychrometer.

In research under field conditions, the sap was expressed from newly plucked fresh leaves, from which the sap concentration and the criteria characteristic of the beginning of water deficit were determined. About 6,000 determinations were made with the cotton plant, and 20,000 with the apple tree.

Results and Discussion

Cotton plant *(Gossypium hirsutum)*. Table 1 gives the correlation coefficient between the CSC of leaves, on the one hand, and the soil moisture, leaf suction force and air temperature and humidity, on the other. The correlation coefficients were calculated according to the data of simultaneous

T a b l e 1.

Correlation coefficients between CSC in leaves of cotton plant, soil moisture (W), the diffusion pressure deficit (suction force) of leaves (DPD), temperature (t) and available air humidity (A).

Factors compared	Correlation coefficients	
	1955	1956
CSC × W	—0·85 ± 0·058	—0·83 ± 0·060
CSC × DPD	+0·85 ± 0·058	+0·71 ± 0·096
CSC × t	+0·15 ± 0·122	+0·16 ± 0·163
CSC × A	—0·31 ± 0·113	—0·39 ± 0·152

determinations carried out on the irrigated and non-irrigated experiment variant plots in the period between the bud-setting and ripening. The determinations were carried out at 1 to 2 p.m. once every 10 to 15 days. The CSC and the suction force of leaf were determined in well lighted leaves, the fourth from the growing point, the reduplication of determinations being sixfold. The soil moisture was determined in the soil layer of 0 to 100 cm. in 3 to 4 reduplications, the air temperature and humidity, at the level of leaves taken for analysis. There is a high and stable negative correlation between the CSC and the soil moisture. Decrease of soil moisture in the root area was accompanied by an increase in the CSC which resulted in lower values of the CSC of leaves on the irrigated variant plots, as compared to those on the non-irrigated plots (Filippov 1959). The changes in the CSC in

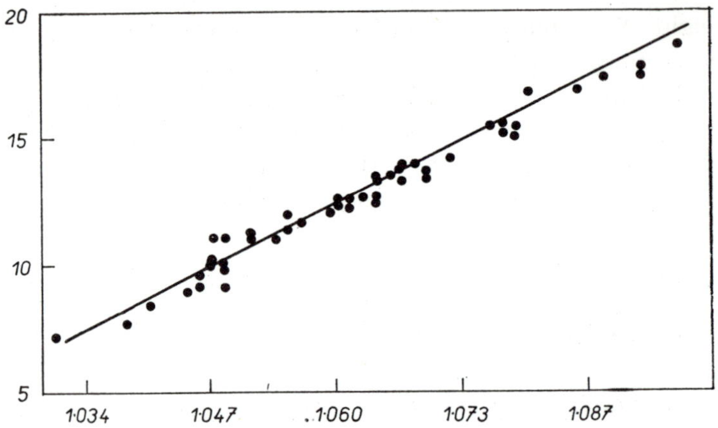

Fig. 1. Correlation between concentration and specific gravity of cell sap in leaves of cotton plant. Abscissa: specific gravity in g.ml.$^{-1}$, ordinate: cell sap concentration in per cent.

the leaf on the whole corresponded to those in the leaf DPD and show a positive correlation. The influence of the air humidity on the CSC of leaves has also been proved to be reliable.

Fig. 1 shows the interrelation between the CSC of leaves and its specific gravity. Correlation between these indicators in the wide range of their changes proved to be very high. Equally high positive correlation was found between CSC determined by refractometer and the concentration of the same sap as determined by drying to constant weight. Correlation between these values proved to be equal to 0·83 $\pm$ 0·006, the refractometric method giving higher values than the method of drying. In compliance with these data we may expect no less high correlation between CSC of leaves and their osmotic pressure which later on was demonstrated by us for the apple tree.

The range of changes in the CSC of leaves of the cotton plant, depending on the state of water relations, was within 9—10 to 24—25%. At the beginning of water deficit the CSC of leaves increased along all the tiers of the crown branches. In the lower leaves, however, the CSC was always higher than in the upper ones (Filippov 1959). For diagnosing the state of water relations in the cotton plant, as well as in other crops, it is necessary to analyse the leaves of a definite tier.

According to our data, a favourable water relations in the cotton plant is secured, if the CSC of the fourth from the growing point leaves is not over 12 to 13 %. The increase in CSC at 1 to 2 p.m. up to 14 to 15% is followed by loss of turgor in the middle of the day, and by the inhibition of the stem growth which indicates the onset of water deficit in plants and points to the necessity of recurrent watering. A permanent water deficit followed by the deep leaf wilting and the inhibition of the stem growth takes place at increase of the CSC of leaves up to 17—18 % and over. Depending on the air temperature and humidity during determinations, these values increase or decrease or 1 to 1·5 %, which must be considered when fixing the dates of cotton plant waterings.

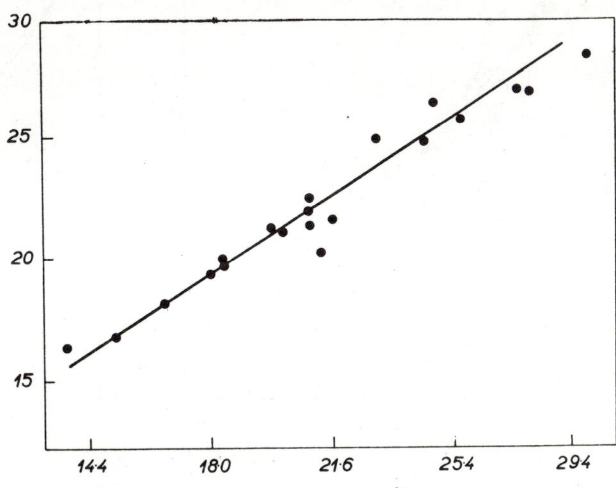

Fig. 2. Correlation between concentration and osmotic pressure of cell sap in leaves of apple tree. Abscissa: osmotic pressure in atmospheres, ordinate: cell sap concentration in per cent.

The apple tree *(Malus)*. When working out the refractometric method for diagnosing a water deficit in the apple tree we found the same high correlation between the CSC of leaves and other indicators, characterizing the state of the water relations in the leaf.

Fig. 2 shows the correlation between the CSC of leaves on the fruit-bearing twigs and the osmotic pressure of their cell sap. The correlation between CSC, its specific gravity and the content of dry matter was determined by correlation coefficient as approaching +1, notwithstanding the great number varieties, the differences in the age of trees, the position of leaves in the crown, the state of their water relations and the time of selecting the leaf samples for analysis. The content of dry matter in the cell sap of the apple tree leaves, as determined by drying (real concentration), coincided quantitatively with the value of CSC, as determined refractometrically.

Fig. 3 shows the correlation between the CSC of leaves and the whole content of water in them for a group of apple tree varieties. Comparative determinations were carried out during July—September with the leaves

completely developed from the morphological point of view. Samples were
taken from the different parts of crown in the trees of different age and grown

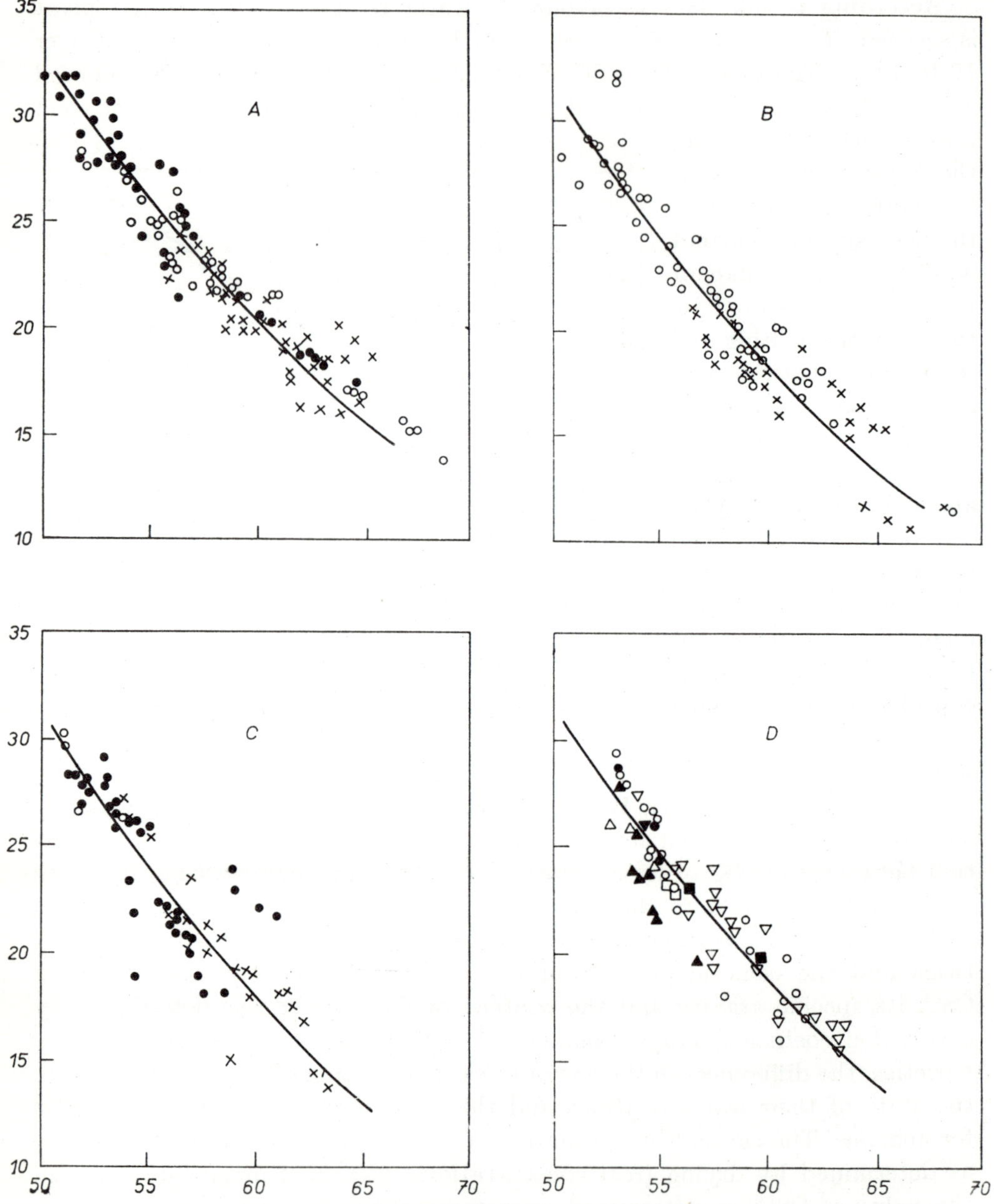

Fig. 3. Correlation between cell sap concentration and water content in apple tree leaves.
A: Wagner's prize, B: Calvil snowy, C: Renet champansky, D: a group of six varieties.
○: yielding trees, ●: not yielding trees, x: young trees and budlings in nursery. The
other markings in Fig. D signify the different varieties of apple tree. Abscissa: water
content in per cent, ordinate: cell sap concentration in per cent.

on various rootstocks, fruit-bearing and not fruit-bearing. The quantitative correlation between these indicators proved to be high and nearly equal, in spite of great variability of the objects analysed. The negative correlation coefficients between them range within 0·89—0·94.

Fig. 4 shows the changes in CSC of the leaves of the apple tree variety "Renet Champansky", depending on the changes in the store of available moisture in the layer of soil to the depth of 150 cm. during the vegetative season. The lowering of the moisture store in the soil was followed by regular increase in the CSC of leaves. In a variant experiment plot, where·the inter-rows were planted to perennial grasses, thoroughly drying the soil, the CSC of leaves was higher than in a variant with the inter-rows kept fallow. These data show that the refractometric method is sufficiently reliable for diagnosing the water deficit in the apple tree.

The range of changes in CSC of the apple tree leaves, depending on the state of their water relations, was within from 13—14 to 34—35 %. When comparing the results of the numerous determinations of CSC in leaves in its relation to the soil hydrological conditions of growth, in the great number of varieties, we fixed the criteria for the state of their water relations. Under conditions of favourable water relations in the apple tree, the CSC of well lighted leaves of the fruit-bearing twigs at the crown periphery does not exceed 21 to 22 % in the middle of day.

The favourable water relations

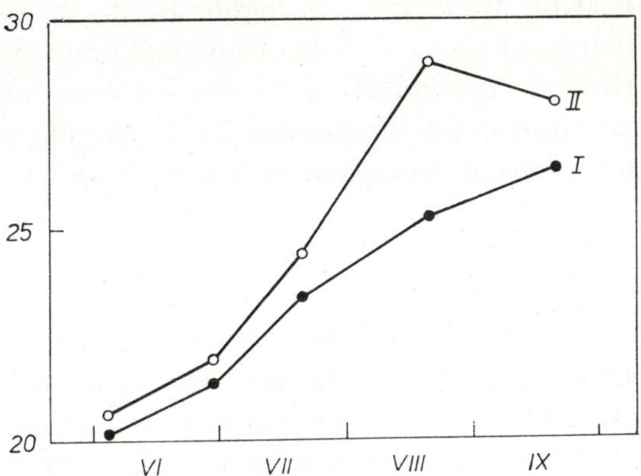

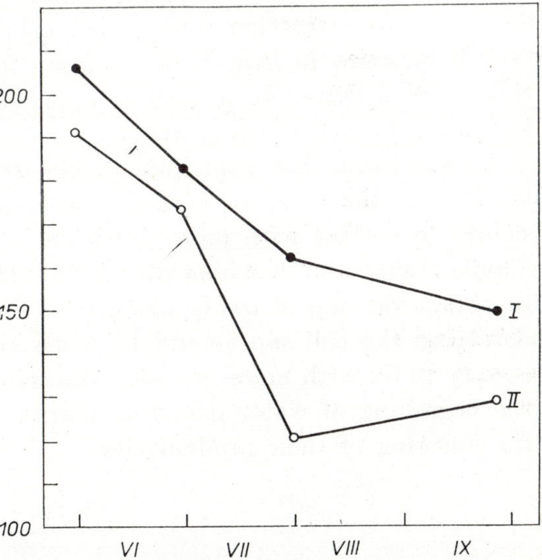

Fig. 4. Correlation between changes in cell sap concentration in leaves of apple tree variety "Renet champansky", and store of available moisture in soil with different keeping of interrows. I: fallow, II: perennial grasses. Abscissa: months, ordinate: cell sap concentration in per cent, and store of available soil moisture in mm., resp.

of fruit-bearing apple trees in the irrigated orchards of Moldavia is secured by 3 to 4 waterings.

A permanent water deficit is characterized by increase in CSC in leaves at the crown periphery up to 25—26 % and more. Every year we observed such conditions of water relations in non-irrigated orchards of Moldavia during the arid periods of the second half of summer. Besides, in the irrigated, as well as in non-irrigated orchards, the CSC of leaves in the periodically fruit-bearing apple trees is higher in the fruitless years than in good years. The increased water deficit of not yielding trees results from great moisture losses through transpiration at the expense of more vigorous growth of foliage. A criterion for diagnosing the beginning of water deficit, and of the necessity for orchard irrigation is an increase of CSC of leaves in the fruit-bearing trees up to 23—24 % in the middle of day.

By using the refractometric method in our research we have been enabled to bring out the high correlation between CSC of leaves and other indicators, characterizing the state of water relations in the cotton plant and apple tree. Positive results were also obtained by other authors when applying this method in their investigations on a series of other cultures. These facts prove that the refractometric method is quite reliable for diagnosing the water deficit in a great number of plant species. This method is very simple and efficient for investigation under field conditions, and it may be successfully used in practice in irrigational farming for fixing the dates of watering. But this method demands a further improvement in compliance with the biological properties of every crop.

In particular the principles of the selection of leaves, considering their location on the stem in annual plants, and in the crown in perennial ones, should be settled with more precision. This problem has also pointed to by Slavík (1958) and Kozinka and Nižnánsky (1963). Where it is not possible to obtain the sap of newly plucked leaves from some plants, the methods of obtaining the cell sap should be made more accurate. And finally, it is necessary to fix with more precision the critical values of CSC in leaves, showing the beginning of water deficit in plants, the inhibition of their growth, and the lowering of their productivity.

Summary

The relation between the refractive index of cell sap in the leaves of the cotton plant *(Gossypium hirsutum)* and apple tree *(Malus)* and other indicators determining the state of the water relations of the above plants, was studied. The refractive index was expressed as a percentage of the equivalent concentrations of solutions of sucrose, and is termed "cell sap concentration" (CSC) of leaves. A high negative correlation between the CSC of leaves and the soil moisture, and a positive relation between CSC and its osmotic indicators, were established. The changes in CSC in the leaves of the cotton plant and apple tree corresponded to those in the state of their water relations, and were, on the whole, determined by the soil hydrological conditions of the locality where the plant was growing. The critical values of CSC in leaves, characterizing the beginning of the water deficit were established. For the cotton plant the critical CSC values amount to 14—15% and for the apple tree to 23—24 %. These indicators are recommended for diagnosing water deficit in the above mentioned plants and for determining their watering dates. The refractometric method is considered as quite reliable, being simple and effective when used in diagnosing the water deficit of many species of crop plants and in determining their watering values.

References

Babushkin, L. N.: [On diagnosing the watering requirements in vegetables by means of determining their cell sap concentration.] In Russ. — Fiziol. Rast. *6* : 480—483, 1959.

Belik, V. F.: [Development of tomatoes and cell sap concentration in their leaves as related to soil moisture.] In Russ. — Bot. Zhurnal Akad. Nauk SSSR *45* : 1063—1066, 1960.

Boyko, L. A., Boyko, L. A.: [The method of determining the osmotic pressure of the cell sap by means of collodium follicules.] In Russ. — Fiziol. Rast. *6* : 630—631, 1959.

Chunosova, V. N.: [Diagnosis of the watering dates in sugar beet by means of determining cell sap concentration.] In Russ. — Fiziol. Rast. *10* : 234—237, 1963.

Filippov, L. A.: [Cell sap concentration in cotton leaves as related to leaf age and hydration.] In Russ. — Fiziol. Rast. *3* : 393—398, 1956.

—: [On problem of cell sap concentration in leaves as physiological indicator of hydration.] In Russ. — Byul. Fiziol. Rast. Akad. Nauk USSR [Bull. Plant Physiol. Acad. Sci. Ukr. S.S.R.] No. *3* : 60—66, 1958.

—: [Cell sap concentration of leaves as physiological indicator of the state of water relations in cotton.] In Russ. — Fiziol. Rast. *6* : 85—88, 1959.

—: [The refractometric method for determining hydration in apple tree leaves.] In Russ. — Fiziol. Rast. *8* : 138—140, 1961.

Galinskaya, M. S., Kalinin, F. L.: [Determining watering dates in spring and winter wheat by means of physiological indicators.] In Russ. — In: Vopr. obmena veshchestv s.-kh. rast. [Problems of metabolism in crop plants.] — Publ. House Ukr. S.S.R., Kiev 1953.

Kozinka, V., Nižnánsky, A.: Biometric analysis of the relationship between the osmotic pressure of the cell sap and its refractive index. — Biol. Plant. *5* : 77—84, 1963.

Lobov, M. F.: [On the determination of water consumption by plants during watering.] In Russ. — Dokl. Akad. Nauk SSSR *66* : 277—280, 1949.

—: [Diagnosing watering dates in vegetables by cell sap concentration.] In Russ. — In: Biol. osnovy orosh. zemledel. [Biological Principles of Irrigation in Agriculture]: 147—156, Publ. House Acad. Sci. U.S.S.R., Moscow 1957.

Petinov, N. S.: Fiziologiya oroshaemoy pshenitsi. [Physiology of the irrigated wheat.] In Russ. —
Publ. House Acad. Sci. U.S.S.R., Moscow 1959.

Shardakov, V. S.: [Determination of watering dates in cotton by means of measuring suction
force in leaves.] In Russ. — Publ. House Acad. Sci. Uzb. S.S.R., Tashkent 1956.

Slavík, B.: The relation of the refractive index of plant cell sap to its osmotic pressure. — Biol.
Plant. *1* : 48—53, 1959.

Tityov, G. M.: [Diagnosing watering dates in maize by cell sap concentration in leaves.] In
Russ. — Vestnik Akad. Nauk Kaz. SSR, No. 12, 1958.

WATER STRESS AND PHYSIOLOGICAL ACTIVITY

Wednesday, October 2, and Thursday, October 3

Chairmen: *P. E. Weatherley, G. Hygen* and
W. R. Müller-Stoll
Secretary: *Z. Šesták*

STOMATA REACTIVITY IN LEAVES AT DIFFERENT INSERTION LEVEL DURING WILTING

JARMILA SOLAROVÁ

Institute of Experimental Botany, Czechoslovak Academy of Sciences,
Prague, Czechoslovakia

The influence of the external environment as a primary factor on the movement of stomata has been the subject of many papers by different authors. Much less attention has been paid to the effect of internal factors, i.e. the genotypes of plants, their development and growth. The conditions of the environment have an indirect effect in this case through the plant organism, certain qualities of which are determined by them.

Of the variable internal factors, perhaps the most important role is played by water saturation deficit, whose absolute value and development depend both on the plant itself and the immediate effect of the environment. The quantitative relationships between stomata opening and water saturation deficit have received much less attention than qualitative relationships. From the publications it is seen that the reactivity of stomata is dependent on the age of the leaf and that the opening of stomata is not only dependent on the numerical value of water saturation deficit, but also in its direction, rate of development and the duration of its action (Stålfelt 1929, 1932, 1955, 1961, Milthorpe and Spencer 1957, Pisek and Winkler 1953 and Yemm and Willis 1954).

Stålfelt (1955), for instance, differentiates water saturation deficit (W.S.D.) according to its effect on stomata and states that maximal opening occurs at optimal W.S.D., when the other conditions are favourable, while at very low suboptimal deficit there is a passive compression of the pores by epidermal cells. On the other hand, with a high supraoptimal deficit, the epidermal cells distend the pores of stomata leading to their passive opening.

Pisek and Winkler (1953) point out the differences in the hydroreactivity of stomata in leaves of different age and also the different degrees of hydroreactivity of stomata caused by interaction with photoreactivity, i.e. the dependence of hydroreactivity on light intensity. Milthorpe and Spencer (1957) consider the rate of change in deficit as much more important than its absolute value and further point to the diminished sensitivity of the stomata to other factors under the influence of W.S.D., both when it is acting and also at the time when it is decreasing.

The aim of the present work was to determine the different hydroreactivity of stomata to the development of permanent W.S.D. caused by soil moisture stress.

Material and Methods

The experimental material was potted plants of fodder cabbage (*Brassica oleracea* L., convar. *acephala* (DC.) Alek., var. *medulosa* Thell., variety Markstammkohl), 90—100 days old. The number of leaves per plant varied between 6 and 8 when leaves longer than 6 cm. were included in the experiment as the youngest leaves. Soil moisture fell only due to water loss by transpiration; evaporation from the soil was prevented by sheets of polyethylene. The actual experiments were carried out in a greenhouse at a temperature of $25 \pm 2°$ C with artificial illumination 1×10^5 erg.cm^{-2} sec.$^{-1}$ which corresponds to about 20,000 lux. Determinations were made after one hour light exposure after 12 hours in the dark. Since in preliminary experiments stomata determined by the reprint method appeared as open even with high deficits, two methods, the reprint and the porometric were used simultaneously in this experiment. A simple type of porometer, similar to that of Leick (1928) as modified by Gloser (unpublished), was used with brief exposure of the leaves in the porometric chamber. The one-side forceps chamber which was sealed to the leaf with the Leick seal (Leick 1928) had an active area of 0·5 cm.². Low underpressure was produced in the chamber by a falling mercury column weighing 1·35 g. in a connected capillary and which produced a suction of 0·5 cm.³ of the air on the leaf.

Immediately after porometric measuring, reprints of the leaf surface were made with chloroform solution of methylmethacrylate, for comparison. Reprints were taken from the leaves with the aid of transparent adhesive bands and later measured microscopically.

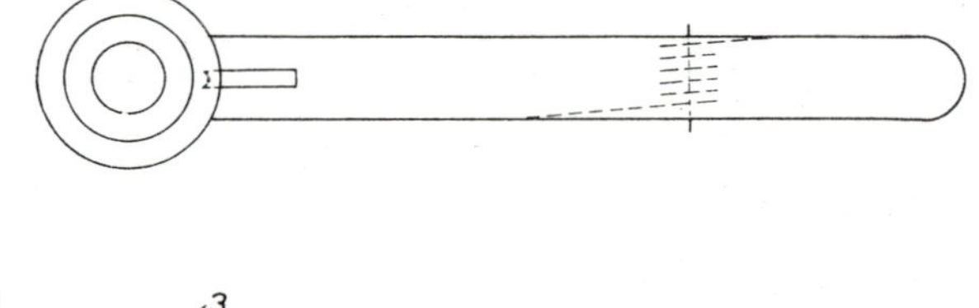
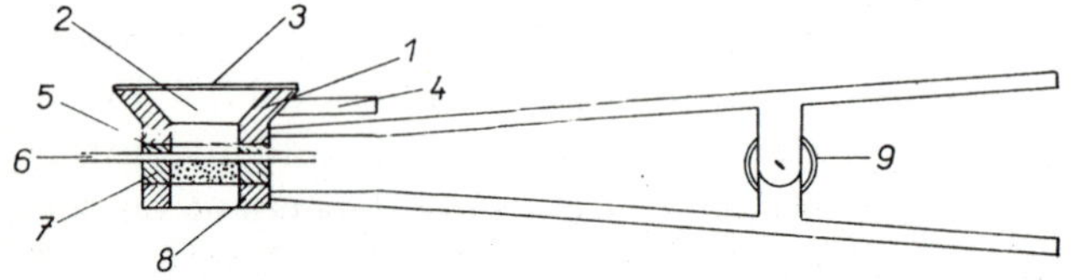

Fig. 1. Porometric chamber

1 — upper part, 2 — conical chamber, 3 — glass, 4 — air out-put, 5 — rubber seal, 6 — leaf, 7 — ring of polyurethane foam, 8 — lower part, 9 — tap with compressed (reverse) spring.

W.S.D. of the individual leaves was determined according to Čatský (1960). Disks of leaf tissue 8 mm. in diameter were punched out of the experimental leaves immediately after porometric determination beside the place where the porometric chamber was fixed. The disks were saturated in openings in

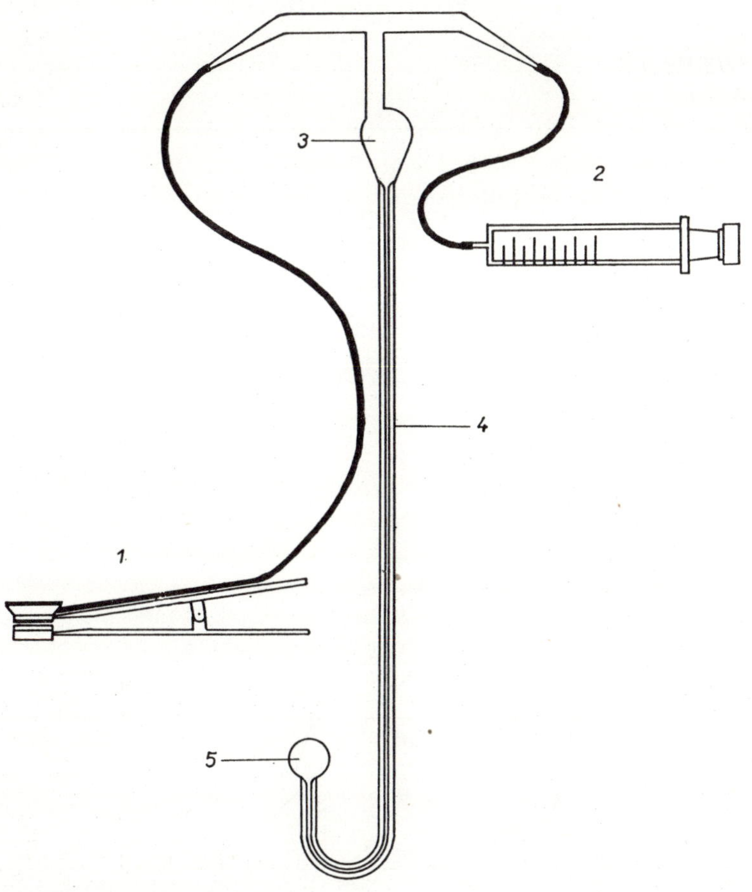

Fig. 2. Diagram of porometer

1 — porometric chamber, 2 — injection syringe for raising mercury column, 3 — safety tank, 4 — capillary with mercury column, 5 — lower safety tank.

a moist plate of polyurethane foam and weighed at the beginning of determination, after 3 hours and after 6 hours of saturation. W.S.D. was calculated according to the formula:

$$\% \,\text{W.S.D.} = \frac{2 \times \text{weight after 3 hr.} - \text{weight after 6 hr.} - \text{initial weight}}{2 \times \text{weight after 3 hr.} - \text{weight after 6 hr.} - \text{dry weight}} \cdot 100$$

Results and Discussion

Porometric determinations showed that the state of the stomata after photoactive opening was dependent on W.S.D. in a different way in leaves of different age. Maximal diffusibility was not reached in any leaves at full turgidity. In young leaves maximal diffusibility was reached at lower W.S.D.,

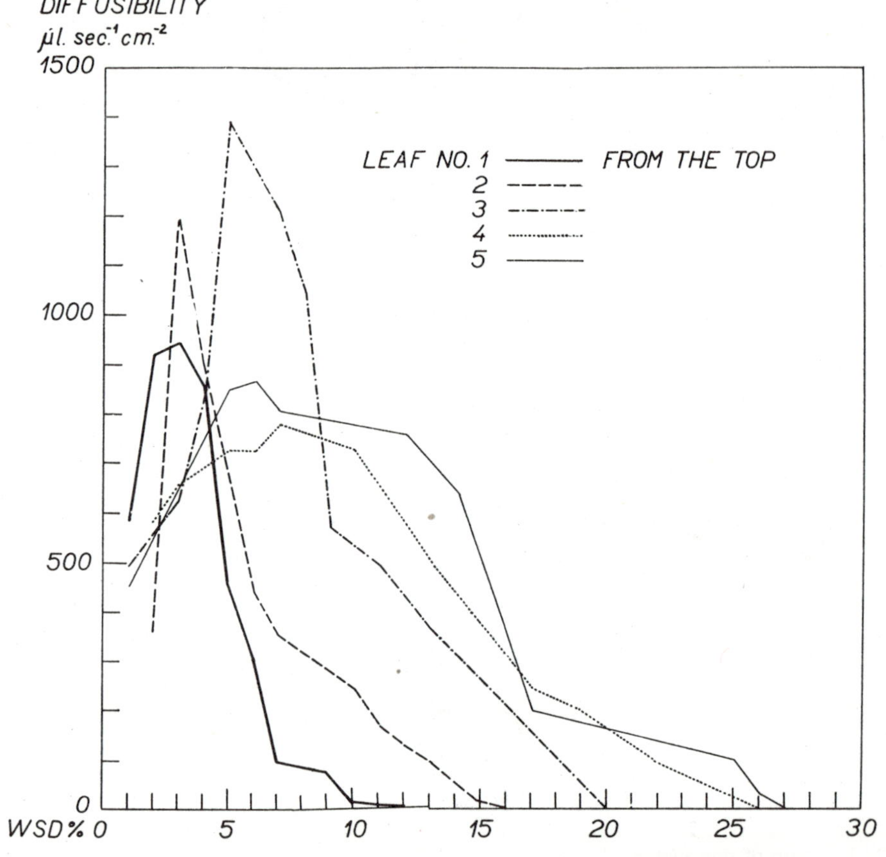

Fig. 3. Relationship between diffusibility of leaves of different age and W.S.D.

in our case 2—3%. It was reached in mature and old leaves at 5—6%. At the same time, absolute maximal diffusibility was reached by the tissue of middle leaves, whereas in both directions, towards young and ageing leaves, maximal diffusibility falls. Practically complete closing of the stomata occurred in young leaves at 12% W.S.D. At this deficit the stomata of young leaves cannot be opened by photoactivation. In old leaves there is an increase in the deficit value at which the stomata are capable of a photoactive reaction.

The microscopic measurement of reprints of stomata showed that, in agreement with porometric measurements, the pores close at low deficits. But complete closure occurs only very rarely in the reprints. At higher deficits, the

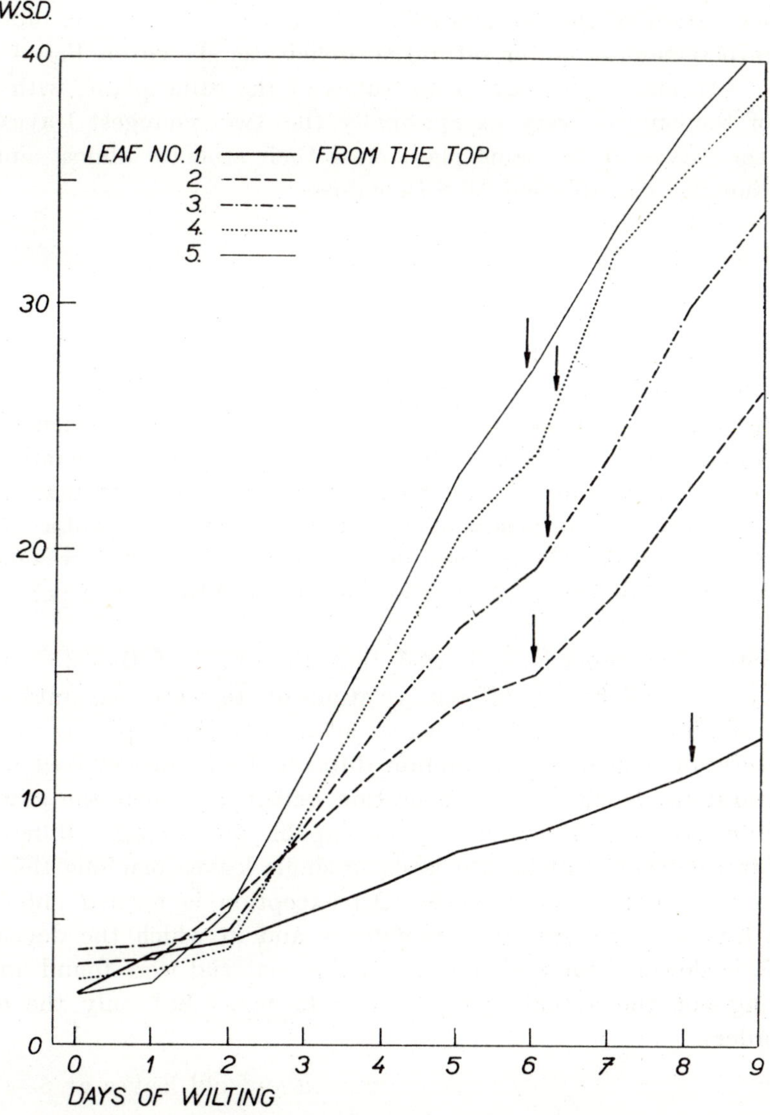

Fig. 4. Development of W.S.D. in leaves of fodder cabbage at different insertion levels.

pores, as reproduced by the reprints, do not close further but, on the contrary, they open to one-fifth up to one-half of their maximum size, while diffusibility measured porometrically remains at zero.

From the above it follows that in the material used, reprints do not reproduce the actual state of the stomata pores but only their superficial part, i.e. the upper borders. It is thus probable that at higher deficits movement does not occur of the stomata pore itself i.e. of its middle part, but only hydropassive separation of the borders.

The development of water saturation deficit, as shown on Fig. 4 reaches critical W.S.D. simultaneously in all leaves of the same plant, with the exception of the one, or very exceptionally the two youngest leaves. Thus, most of the leaves of the same plant close their stomata almost simultaneously, although at a different W.S.D. values.

Summary

Investigations were made on the hydroreactivity of stomata in leaves at different insertion levels of cabbage plants aged 90—100 days, during continuous wilting by decrease in soil moisture due to transpiration. Stomata opening was determined by direct porometry using a one-side chamber sealed every time to leaves only during measurement (one minute maximum) as well as microscopically by means of reprints made with a chloroform solution of methylmethacrylate. The increase in water saturation deficit was measured by the saturation of leaf disks cut out from leaf blades in a polyurethane foam plate.

Porometric determinations showed that the state of the stomata after photoactive opening had a different relationship to water saturation deficit in leaves of different age. Young leaves attain maximal opening at a lower water saturation deficit and minimum diffusibility is also evoked by lower water saturation deficit. Water saturation deficit at which the stomata of leaves of different insertion level are not capable of opening is different, but their natural development in situ leads to single leaves reaching this critical deficit on the whole simultaneously. An exception is formed only by the youngest leaves which are given preference and in which the development of W.S.D. is slower. Stomatal reprints in this material were found unreliable reproducing not the actual state of stomata pores but only the external upper borders.

References

Čatský, J.: Determination of water deficit in disks cut out from leaf blades. — Biol. Plant. *2* : 76—78, 1960.
—: Water saturation deficit in the wilting plant. The preference of young leaves and the translocation of water from old into young leaves. — Biol. Plant. *4* : 306—313, 1962.

Leick, E.: Ein neues Universal-Doppel-Porometer. — Ber. dtsch. bot. Ges. *45* : 43—59, 1928.

Milthorpe, F. L., Spencer, E. J.: Experimental studies of the factors controlling transpiration. III. The interrelations between transpiration rate, stomatal movement and leaf water content. — J. exptl. Bot. *8* : 413—437, 1957.

Pisek, A., Winkler, E.: Die Schliessbewegung der Stomata bei ökologisch verschiedenen Pflanzentypen in Abhängigkeit vom Wassersättigungszustand der Blätter und vom Licht. — Planta *42* : 253—278, 1953.

Stålfelt, M. G.: Die Abhängigkeit der Spaltöffnungsreaktionen von der Wasserbilanz. — Planta *8* : 287—340, 1929.

—: Der stomatäre Regulator in der pflanzlichen Transpiration. — Planta *17* : 22—85, 1932.

—: The stomata as a hydrophotic regulator of the water deficit of plant. — Physiol. Plant. *8* : 572—593, 1955.

—: The effect of water deficit on the stomatal movements in a carbon dioxide-free atmosphere. — Physiol. Plant. *14* : 826—843, 1961.

Yemm, E. W., Willis, A. J.: Stomatal movements and changes of carbohydrates in leaves of *Chrysanthemum maximum*. — New Phytol. *53* : 373—396, 1954.

Discussion

P. E. Weatherley: The material you used for taking the stomatal prints was made up in chloroform. Did you find that this in itself had any effect on the stomata — chloroform being a toxic substance?

J. Solarová: The solution dries in 20—30 sec. To prevent the influence of chloroform on further measurements I used each leaf for only one measurement.

G. Hygen: I would like to ask a question about the course of the curves in your last figure, where the arrows indicate the moment stomatal closure is supposed to be completed. After stomatal closing one would have expected a significant decrease in the transpiration rate and a corresponding slowing-down of the rise in the water saturation deficit. In contrast to this expectation your curves show a more or less markedly inclining slope after the time indicated by the arrow. I wonder how this apparent contradiction could be explained.

J. Solarová: The increase in water saturation deficit of individual leaves is not dependent only on the transpiration rate, but also on the supply of water to each of them. In adult and older leaves the increase in W.S.D. continues with declining soil moisture even when the transpiration rate is greatly decreased, due to water transport in favour of younger leaves. Therefore the slope of the curve of time dependence of W.S.D. is not a direct expression of transpiration rate in these leaves.

S. J. P. K. Bezuidenhout: In these types of experiments on the influence of water deficit one should consider the CO_2 concentration of the air too, because already near wilting the stomata become sensitive to an increase in CO_2 concentration which causes them to close.

J. Solarová: In our experiment the CO_2 concentration was not controlled and it varied within the range of 300—400 p.p.m.

WATER DEFICIENCY IN THE PROCESS OF PATHOLOGICAL WILTING

O. MAJERNÍK and C. PAULECH

Botanical Institute of the Slovak Academy of Science, Department of Pathophysiology,
Bratislava, Czechoslovakia

In diseases of the wilting type, there is a primary disturbance in water movement. Dimond (1955) characterized the initial course of this type of disease, among other things, by increased transpiration.

It is known (Majerník 1958) and it has again been proved (Majerník 1961) that apoplectic wilting of apricot trees under natural conditions is accompanied by correlated disturbances, represented mainly by destruction of tissues in some parts of the roots, stem or branches. If there is a question of a partial pathological wilting, the correlated disturbance is in the part under and above the centre of destruction. If pathological wilting of the whole tree takes place (total apoplexy), the relation between the roots and the crown of the tree has been disturbed.

In one of their previous studies (Majerník at al. 1962) the authors drew attention to the fact that in the case of pathological wilting of apricot trees, the increased water loss is due to increased opening of stomata. This was demonstrated experimentally, by artificial wounding and by means of artificial infection with the mycelia of phytopathogenic fungi; at the same time the intensity of stomatal transpiration of leaves was observed by the method of weighing.

The opening of stomata was followed simultaneously by the adhesive method. A number of results were subjected to statistical evaluation.

It was found that, compared with the control plants, 24 hours after wounding and after infection, under the same conditions, the wounded plants lost more water and the infected plants transpired the largest quantity of water during the same period of time. The enormous water loss was related to the state of the stomata. In the cases where the greatest loss of water was observed, the stomata were most markedly open.

In further experiments more information was obtained in connection with the dysfunction of stomata. In presence of fusarine acid, and in other variants in the presence of ammonia respectively, the uptake of ^{32}P was significantly increased as compared with the control groups where no toxins were present (Majerník, Janitor 1963). The increased uptake of radioactive phosphorus may be ascribed to the influence of toxic substances which stimulate increased transpiration. With decrease in water uptake and water evaporation (as a result of intoxication), there was also a decrease in ^{32}P uptake.

We concluded from the above that wounding and toxic agents may provoke dysfunction of stomata and thus stimulate transpiration.

The problems of water uptake and transpiration, observed so far only generally, were submitted to closer observation. The aim of these studies was to investigate the enormous movement of water after wounding and after infection at shorter intervals paying special attention to changes in water deficiency.

Materials and Methods

Water movement was studied on small branches of apricot trees (*Armenica vulgaris* L.) and leaves of tomato plants (*Solanum lycopersicum* L.). Branches of two-years-old samplings with 10 leaves were wounded and infected. The experiment was divided into four variants.

In the first group the wound was performed as a cut. In the second group the cut was infected with *Monilia laxa*. The third group was infected with the fungus *Fusarium solani*. The fourth group of untouched experimental branches served as a control. From each variant six selected branches were placed 10, 15 and 20 hours after infection, into a watery solution of active ^{32}P with H_3PO_4 carrier of 57 mg. (Ml); 130 μc (l). Measurements were made after six hours.

For the measurement of radioactivity disks 1 cm. in diameter were taken from the basal, middle and extreme top leaves. The measurements were made by means of a GM-tube connected to a counter of Czechoslovak make, type Optima; voltage 1,500 V. The uptake of ^{32}P was calculated from the impulses measured on the surface of the disks and from this the rate of water movement could be determined.

Repeating the test without radioactive phosphorus water deficiency was observed. The disk method of Čatský (1959) was used to determine water saturation deficit.

Results

The experiment in the way it was performed, represent certain extreme conditions for the plants, because the branches were cut off from their root system and immersed in a solution. All the variants were tested under such conditions.

The next step, i.e. a deep cut into one third of the perimeter (in the case of wounded and infected branches) may be considered a further extreme condition. This appeared, as was later found, in the different intensity of radioactive phosphorus as indicator for observing water movement. In these experiments the cut was made (Fig. 1) automatically and put about one third of the most important water conducting elements out of action. It was but logical that in these experiments a decrease in water intake would be expected.

However, as soon as the first results from the branches immersed in solution were known, it was necessary to revise this opinion.

Ten hours after wounding, the intensity of water uptake in the apricot branches was seven times higher than in the control branches, the data of which were taken as 100% (Fig. 2). It was not known whether this intensity was optimal or maximal. For this purpose it was necessary to investigate water movement at shorter intervals. It remains, however, beyond doubt that there is proof here of a difference in the intensity of water movement in comparison with the control groups.

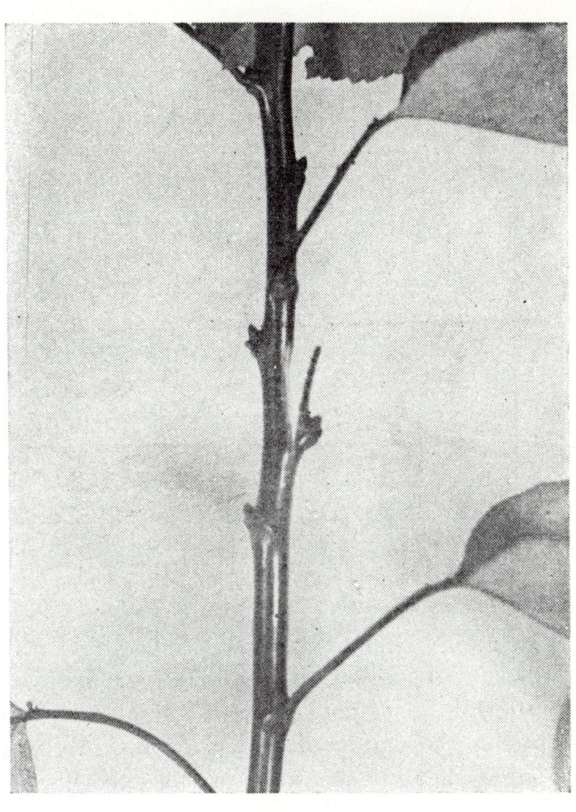

Fig. 1. Performance of the cut on the branches.

Fifteen hours after wounding, the intensity of water uptake, according to ^{32}P, dropped considerably remaining, however, higher than the values registered in the control branches. The intensity of water movement 20 hours after wounding remained at approximately the same level.

Variants infected with the fungus of *Monilia laxa*, had an even greater water uptake than the wounded variant. Ten hours after inoculation water uptake reached as much as ten times the value of the control group. In the case of the variant infected with *Fusarium solani* absorption of ^{32}P reached five times that of the control.

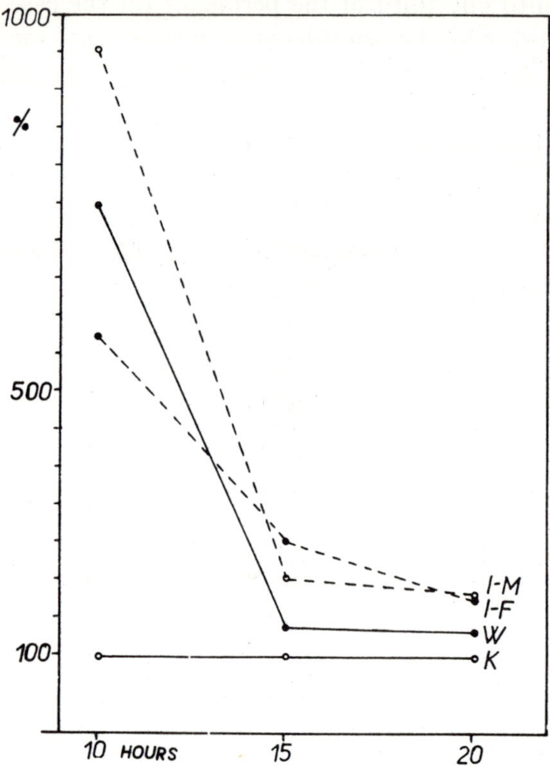

Fig. 2. The intensity of ^{32}P uptake in apricots. The data in control branches were taken as 100%.

After 15 hours even in the infected plants there was a decrease in the intensity of water movement.

The actual values of counts per minute were of the same order as represented graphically on Fig. 4. Table 1, representing the total of the average values of impulses measured (from 10 to 20 hours after infection and after wounding) shows the decreasing tendency of the movement of radioactive phosphorus.

In these experiments tomatoes showed a somewhat different intensity of water movement, as compared with apricots. The ratios of values of the individual variants on Fig. 3 indicate that 10 hours after wounding and infection none of them reached so high a figure as in the case of the apricot plant. Similarly, the values of wounded and infected groups were not so marked in tomatoes, in relation to the control, as in the case of apricot plants. The line in Fig. 3 represents the course of water uptake in a wounded tomato plant. It may be thought that in the course of the experiment, water uptake reached a certain

T a b l e 1.

Counts per minute on 1 cm. disk.

Variants	a) Apricot			b) Tomato		
	after 10 hr.	after 15 hr.	after 20 hr.	after 10 hr.	after 15 hr.	after 20 hr.
Infected-M	229	51	45	419	287	252
Infected-F	138	61	44	497	302	289
Wounded	180	138	35	406	327	215
Control	24	20	19	375	231	232

maximum. As a matter of fact, the actual data given in Fig. 4 explain this graphic representation when they show a decreasing tendency without any maximum in all variants.

We may point to another difference between the tomato and the apricot which follows from the data given in Table 1.

The counts per minute measured in apricot variants (Table 1) appear at first sight to show that the values are on the whole lower in apricots than in tomatoes. At the same time the extremes of radioactive phosphorus uptake in individual variants are much more marked in the case of apricots, being in places as much as tenfold.

From the above it may be judged — in so far as the speed of water movement may be ascribed to the rate of transpiration — that the transpiration rate is much higher in tomatoes than in apricot plants. According to this, the apricot is more sensitive to disturbance such as wounds and infections. From the methodic point of view the apricot is more suitable for certain experimental purposes as it reacts sensitively to interference and its changes in metabolism are more marked.

Water deficiency in apricot plants appeared in a simmilar way (thought not so expressively) as water movement (Table 2).

24 hours after infection the values of water saturation deficit were lower than in control vari-

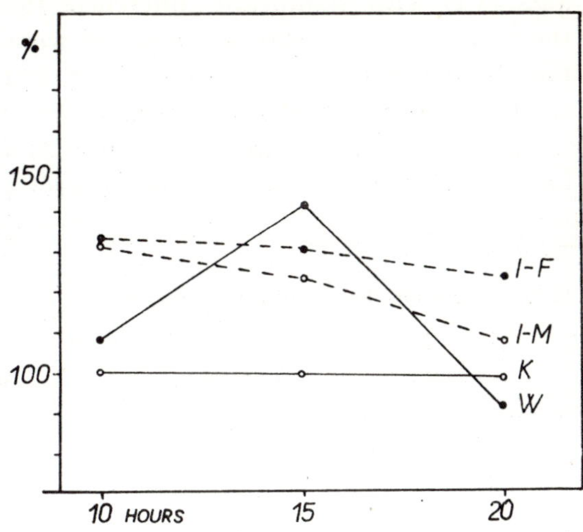

Fig. 3. The relative values of the intensity of ^{32}P uptake in tomatoes

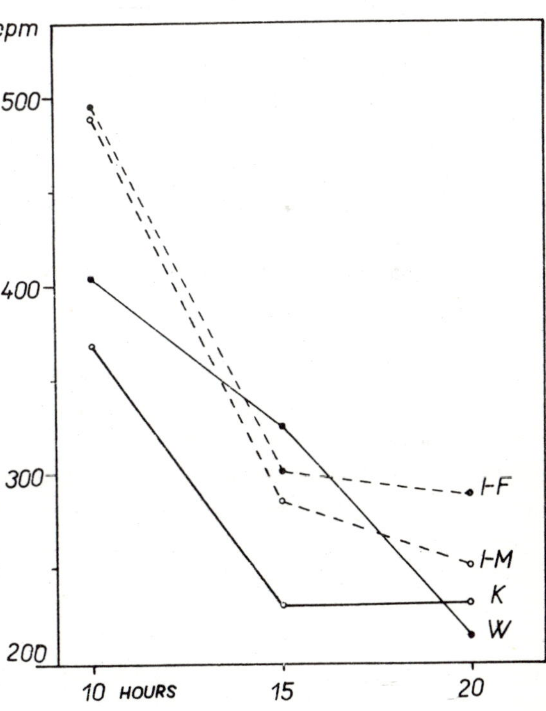

Fig. 4. The actual data of the intensity of ^{32}P intake in the tomatoes in counts per minute per disk

ants (V1) with significant differences (t > t 0·001). 48 hours after infection the values of water saturation deficit were higher than in control plants (V 2) but the difference is not so marked (t < t 0·05).

In the course of repeated tests it was found that water saturation deficit in the case of active infection (after 24 hours) was sometimes significantly higher than in the case of the control plants. (V 3, t 0·001 < t < t 0·01.)

T a b l e 2.

Water saturation deficit in apricot plants

Variants	Control x ± σ	Infected x ± σ	t-test value
1	8·863 ± 0·407	7·160 ± 0·569	7·70415
2	8·552 ± 0·423	8·795 ± 0·509	0·63586
3	6·780 ± 0·732	7·600 ± 0·407	3·63033

Variants: 1   24 hours after infection
         2   48 hours after infection
         3   24 hours after very active infection

It seems that the toxic agent brings the living cell and thus (besides other things) also the water metabolism to a state of passivity at the very early beginning of the disease process and active infection. As a result of this, the natural vitality of the plant is also decreased, i.e. the natural variability is lowered in so far as it affects variations in water deficiency. That is why, in the case of the wounded variants (where wound reactions cannot be expected to be so harmful as the effect of toxins in infection) a much greater variability of water deficiency was found up to 3%. In the control groups where a normal course of physiological functions is expected, the fluctuations in water deficiency among the individual variants was marked. The variability reached, approximately, 4% and more.

Although it is not possible to generalise the above, nevertheless, together with the results concerning water movement, the data obtained give a general orientation of the process of pathological wilting. It was again confirmed that immediately after wounding important processes and metabolic changes occur in the apricot plant.

In the case of water deficiency and water movement it may be observed that after a certain time under the influence of certain agents some of the functional processes favourable to the plant become muffled. On the other hand, biochemical reaction unfavourable to the plant acquire a dominating character. This fact is most marked in the case of infected variants (Majerník, Frič 1963).

Discussion

The influence of toxic substances on the absorption of phosphorus by plant roots has not been investigated.

Baldacci et al. (1959) reported a direct relation between the absorption of ^{32}P and the transpiration rate in the diseased plant. Russel and Shorrocks (1958) found that the absorption of ions from nutrient solution by the healthy plant may depend on increased transpiration, if there is a high salt content in the plant and in the solution.

On the question of the opening of the stomata after wounding, our results support the opinions expressed by Willis, Yemm and Balasubramaniam (1963).

If we were to ascribe the increased absorption of radioactive phosphorus in our experiments to influences other than increased transpiration, there still remains an outstanding amount above the level of the control plants that can be explained in no other way than by transpiration. In the case of wounding and, especially, of infection, information obtained so far (Gäumann 1958) points to an increased permeability of plasma to water rather than to phosphorus only.

We are seeking an approach to the cause of the rapid loss of water and wilting. It seems that by discovering the principle of the dysfunction of stomata we have achieved a first step towards this goal.

Summary

On the basis of previous work where dysfunction of stomata was found after wounding or infection as well as after the application of toxic agents, studies on water movement have been continued using ^{32}P.

It was again found that a short time after wounding water movement was increased due to dysfunction of the stomata transpiration and the uptake of ^{32}P being also increased.

Owing to increased water movement water deficiency of these plants was held within the same limits as in the healthy plants. After regeneration of the tissues transpiration and ^{32}P uptake became normal.

In the infected variants water movement first increased, as in the case of wounded plants. Later, due to permanent dysfunction of the stomata, transpiration continued to increase but water uptake was already disturbed.

It is better to work with branches of apricot trees than with leaves of tomato plants because of less variability.

References

Baldacci, E., Betto, E., Foa, R., Volpi, A.: Absorption and transfer of phosphorus (P^{32}) in healthy and diseased plants. — Comitato nazionale per le ricerche nucleari. Roma 1959.

Čatský, J.: The role played by growth in the determination of water deficit in plants. — Biol. Plant. *1* : 277—286, 1959.

Dimond, E.: Pathogenesis in the wilt diseases. — Ann. Rev. Plant Physiol. *6* : 329—350, 1955.

Gäumann, E.: Über die Wirkungsmechanismen der Fusarinsäure. — Phytopath. Z. *32* : 359—398, 1958.

Majerník, O.: [Periodicity of growth and development of apricots in relation to their precautious death.] — (In Slovak.) — Biológia *16* : 110—111, 1961.

—, Nižňanský, A.: Transpirationsintensität der infizierten und verwundeten Aprikose in Bezug auf ihren Wasserhaushalt. — Biológia *17* : 321—331, 1962.

—, Frič, F.: [Pathological wilting of apricot from aspect of metabolism.] — Biológia *18* : 5—14, 1963.

—, Janitor, A.: Influence of toxic matter in plant tissue followed with the help of radioactive phosphorus P^{32}. — Biológia *18* : 489—497, 1963.

Russel, R. S., Shorrocks, V. M.: The effect of transpiration on the absorption of inorganic ions by intact plant. — Radioisotop. in sci. Res. *5* : 286—303, 1958.

[Seedlings of apricot and their pathological wilting.] — Slovak Acad. Sci., Bratislava 1958.

Willis, A. J., Yemm, E. W., Balasubramaniam, S.: Transpiration phenomena in detached leaves. — Nature *199* : 265—267, 1963.

Discussion

H. Polster: The investigation of transpiration in the leaves of poplar infected with rust (*Puccinia larix populina*) showed:

Phase 1 — incubation phase with slight decrease in transpiration,
Phase 2 — growth phase of fungus with greatly increased cuticular and stomatal transpiration,
Phase 3 — necrotic phase with a decrease in transpiration to zero.
Greatly infected leaves dry up more quickly than healthy leaves and also show visual evidence of greater water loss.

W. R. Müller-Stoll: The question is asked for the estimation method of ^{32}P—activity in the plant tissues and the author's opinion on the physiological background of the increased phosphorus uptake as found by the author. It seems possible that injurious influences, such as wounding and fungal infection, are followed by a rise in phosphorus incorporation into nucleotides, e.g. ATP, in connection with metabolic restitution processes.

O. Majernik: ^{32}P uptake was measured according to counts per min. on disks, cut off from fresh material. In relation to the connection between ^{32}P uptake and its incorporation into nucleotides, we rely on the data in the literature cited in the discussion. We did not carry out such experiments ourselves. Our results only concern ^{32}P uptake by way of transpiration stream.

THE DEMONSTRATION OF A POROMETER FOR PHYSIOLOGICAL AND ECOLOGICAL INVESTIGATIONS

P. STREBEYKO

Institute of Plant Physiology, University of Warsaw, Warszawa 64, Poland

In connection with the importance of stomatal movement in water stress in plants we must determine stomatal movement. We have some methods for stomatal closure determination but it seems to be necessary to look for new more simple and quick methods. Such methods would be of great importance for ecological and agricultural investigations. I would like to present a very simple porometer (Fig. 1). It consists of medical pincers provided with two

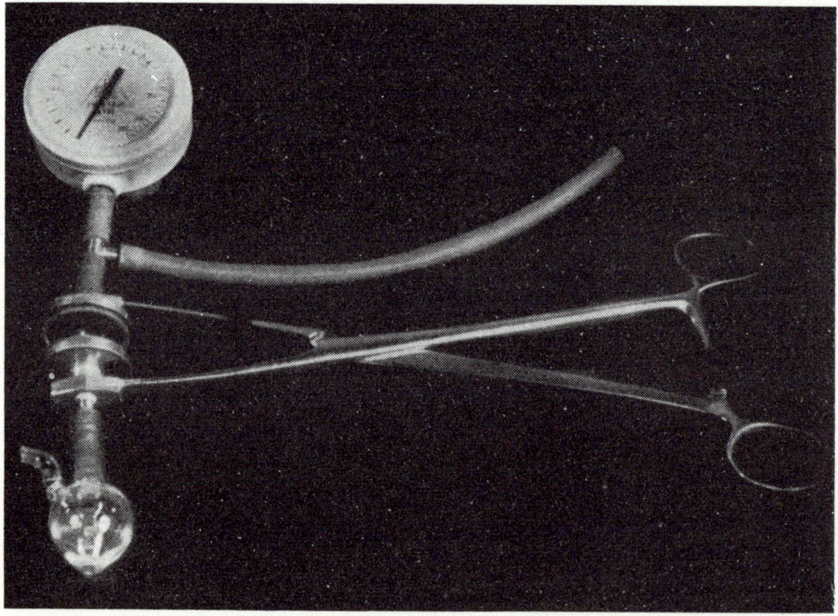

Fig. 1. General view of the porometer.

little boxes. We close a leaf blade between these boxes. A stream of air under slight pressure comes into one of the boxes and passes through the leaf to the other box if the stomata are opened. This stream of air can be observed in the little vessel containing some water. If the stomata are closed the air needs relatively high pressure to pass the leaf blade. I think that this pressure could be a measure of stomatal closure.

The investigations are limited to the leaves provided with stomata on both sides of the blade, but many species have such leaves. Many plants of great agricultural importance belong to this group.

Discussion

V. M. Sveshnikova: Many authors studying water relationships among individual species of plants have shown that some species are characterized by quite definite stomatal regulation of water loss, while other species do not have this kind of control and water loss is regulated by other systems.

G. F. Makkink: The apparatus of Alvim (Peru) is suitable for field work because pressure is produced by compressing air in the rubber cuff of a sphygmomanometer.

J. Čatský: Allow me one remark on the construction of the apparatus: Stomata react very sensitively and quickly to darkness. Therefore I think it is not good to close the leaf in a dark chamber during the measurement.

G. F. Makkink: The stomata are not in the dark if you make the cups of plexiglass.

G. V. Lebedev: If we start from the construction of the apparatus and the method of work, to what extent can we assume the existence of a correction (= correlation factor) to decreasing pressure by escape of air from the leaf tissue in the horizontal direction ? If such a coefficient exists what is its value (in mm.Hg and in percentage of measured pressure) ?

P. Strebeyko: We are only testing the apparatus and for the time being have not made any similar correction. The escape of the air from the leaf tissue in the horizontal direction seems to be impossible because the margins of the porometer close tightly the leaf blade and therefore prevent the gas exchange between the investigated part and other ones.

THE WATER RELATIONS OF TREE SEEDLINGS. V. GROWTH AND ROOT RESPIRATION IN RELATION TO OSMOTIC POTENTIAL OF THE ROOT MEDIUM

P. G. JARVIS and MARGARET S. JARVIS

Institute of Physiological Botany, Uppsala, Sweden*

It has earlier been suggested that the water potential at the roots of rapidly transpiring plants will be much lower (negatively) than that measured in the soil mass (Jarvis and Jarvis 1963b and c). The difference depends on the rate of transpiration, the conductivity of the soil to water, and the concentration of the effective roots. To study plant response in relation to a known water potential at the root surface, culture solutions containing osmotic substrates have been used. In an earlier paper, response of transpiration in short term experiments was examined (Jarvis and Jarvis 1963b).

To complement studies on growth in response to soil water potential (Jarvis and Jarvis 1963a), it was thought desirable to compare the growth of the same four species [*Betula verrucosa* Ehrh., birch; *Populus tremula* L., aspen; *Pinus sylvestris* L., Scots pine; and *Picea abies* (L.) Karst., Norway spruce] in relation to a known water potential (i.e. osmotic potential) at the root surface. Before beginning the work on tree seedlings, the effects of various osmotic substrates on the growth and water relations of *Lupinus albus* in sterile root cultures were tested (Jarvis and Jarvis 1963c). These experiments provided the essential information on which the following is based.

It is generally assumed that the development of low water potentials in plants reduces growth as a result of loss of turgor causing stomatal closure. However, the additional possibility exists that water potentials directly affect metabolism. As a preliminary step, root respiration of the four species has been measured in relation to the osmotic potential of the solution. Studies of root respiration in relation to the concentration of salts or sucrose in the medium were made earlier (e.g. Takaoki 1957). In our investigation, and in the growth studies, polyethylene glycol (Carbowax 1540) was used as the osmotic substrate.

Fuller discussion of the problem and methods is given in our two papers cited above.

---

* *Present address*: C.S.I.R.O., Division of Land Research and Regional Survey, P.O. Box 109, Canberra, A.C.T., Australia.

Methods

Growth experiments

The three treatments were the same as in the preliminary investigation with lupins, viz., Control: the root medium was nutrient solution alone (optimum solution for birch, Ingestad 1957), osmotic potential (π) —5 joules/ /hectogram (ca. 0·5 atm.), Treatment I: osmotic potential due to the polyethylene glycol (PEG) added to the culture solution (π_{add}) was —12 j/hg., Treatment II: π_{add} was —27 j/hg.

Pine and spruce: Sterile seedlings from locally obtained seed were grown in light on nutrient agar in test-tubes at 20° C. When the cotyledons were expanded and the radicle about´ 3 cm. long, the seedlings were transferred to aerated sterile solutions in 500 ml. widenecked conical flasks, three seedlings per flask (for a description of the method, see Lundeberg 1960, Jarvis and Jarvis 1963c). They were grown for 35 days in a growth room with the following conditions: air temperature 20° C, light intensity 9×10^4 erg/cm.²sec. from a ramp of fluorescent tubes (General Electric, power groove, de luxe cool white), photoperiod 18 hours in 24. As net assimilation rate was not estimated, relative growth rate was calculated according to Fisher (1921), as $(\log_e W_2 -$ $- \log_e W_1)/t$, where W_1 is the average seedling dry weight at time of transference to the flasks, determined from a harvested sub-sample, W_2 the individual seedling dry weight at the final harvest, and t the time between harvests. This formula does not presuppose any particular relationship between weight and time. The number of seedlings harvested is shown in Table 1.

Birch and aspen: Seedlings were raised in sand culture, birch from local seed and aspen from seed from Mykinge experimental station. When they had grown to a 4—5 leaf stage they were transferred to aerated unsterile culture solutions, in black, 1·5 litre, square plastic containers. The solutions were renewed every 5 days to minimize infection and to ensure adequate nutrient supply; the volume was maintained by daily additions of distilled water. There were two containers in each treatment, each with 4 or 5 plants. The results presented are for two experiments with birch, each lasting 18 days, lumped in the analysis, and one experiment with aspen lasting 12 days. Prior to and during the experiments the seedlings were grown in a growth room with the following conditions: air temperature $20° C \pm 1° C$, light intensity ca. 4×10^4 ergs/cm.²sec. (98% in the range 300—700 mμ) from a selection of Phillips fluorescent tubes (see Jarvis and Jarvis 1963a), photoperiod 18 hours in 24, relative humidity in the light 40—50% and in the dark 60—70%.

Leaf area (by blue-printing and planimetering) and dry weights of leaves, stem and roots were determined for a sample of plants at the beginning of the growth period and for each seedling at harvest at the end of the experiment. At the first harvest the areas of the leaves on the unharvested plants were determined from the relationship between area and the product length $\times$ $\times$ breadth. In the same way, leaf areas were determined during the experiments at intervals so that it was subsequently possible to calculate the average leaf area during the growth period for each plant separately. The method of growth analysis used was the same as that in previous experiments with the same species growing in soil (Jarvis and Jarvis 1963a). Net assimilation rate $\left(\dfrac{1}{LA} \cdot \dfrac{dW}{dt}\right)$ was calculated as: $\dfrac{W_2 - W_1}{\overline{LA} \cdot t}$ where W_1 and W_2 are plant dry weights at the first and second harvests of the experiment, respectively. $\overline{LA}$ the mean leaf area, and t the time between harvests. Relative growth rate was calculated as the product of net assimilation rate and mean leaf area ratio (leaf area/plant dry weight). Mean leaf area ration was estimated as:

$$LAR_1 + \frac{(LAR_2 - LAR_1)(\overline{LA} - LA_1)}{LA_2 - LA_1}$$

where LAR is the leaf area ratio, LA the leaf area and the suffixes 1 and 2 denote the first and second harvest, respectively. This assumes that changes in leaf area ratio follow a similar course to changes in leaf area and that the mean leaf area ratio coincides with the mean leaf area. It is emphasized that the purpose of the first harvest is only to provide estimates of leaf area ratio of the unharvested plants. Since this is not very variable, large samples are not required. From this estimate and the leaf areas of the unharvested plants, estimates of the initial weights of these plants can be obtained and hence individual estimates of net assimilation rate and relative growth rate made. Other derived quantities have been calculated, e.g. root/shoot ratio and specific leaf area (leaf area/leaf dry weight).

Internal water relations: The methods used for the determinations of relative turgidity, the relation between leaf water potential and relative turgidity, and the relation between stomatal aperture and relative turgidity have been described previously (Jarvis and Jarvis 1963c, pp. 487—489).

Measurement of root respiration

Measurement of root respiration in relation to osmotic potential of the medium was carried out by Warburg constant volume manometry on the apical 1—2 cm. of actively growing roots. Wheat and maize root tips were used in preliminary experiments and the method below was finally adopted for comparative measurements.

Plants were grown in a growth chamber, see 'Birch and aspen' above, in nutrient solution culture ($\pi = -5$ j/hg.). Root tips were taken from seed-

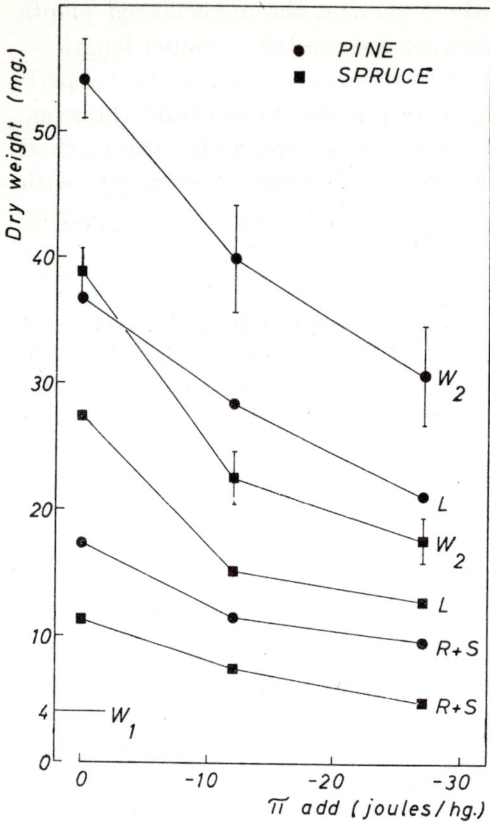

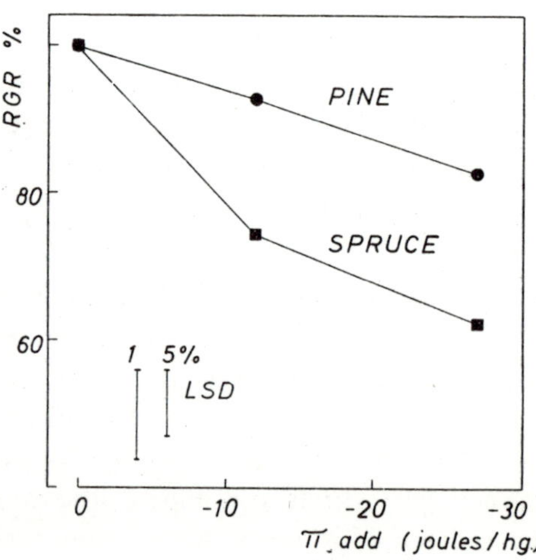

Fig. 1. The relation between dry weight of seedlings of pine and spruce and the osmotic potential of the root medium. The osmotic potential (π) of the control solution was —5 joules/hg. π_{add} is additional osmotic potential, due to added polyethylene glycol. Standard errors are shown by the vertical lines. Mean dry weight of the seedlings at the beginning of the experiment (W_1) was 4 mg. The experimental period was 5 weeks. W_2 = total plant dry weight. L = leaves. R + S = roots + stem.

lings when they were about the size of the growth experiment plants at the end of the experiments (see Table 3 and Fig. 1).

As the root tips were cut off they were stored in distilled water for up to $^3/_4$ hour until sufficient had been obtained and divided up into equal samples of about 30 per flask. Phosphate buffer (Sörensen's), M/15, was diluted with distilled water and/or PEG solution to give a solution of M/60 buffer, pH 5·5, at different osmotic concentrations determined by the amount of PEG. Each flask contained 2·8 ml. of this solution. The central well contained a filter paper wick and 0·2 ml. of 10% KOH.

The root tips were lightly dried on filter paper and put into the flasks. Half an hour was allowed for equilibration, so that the tips had been in the solution for about $^3/_4$ hour before recording began. The water bath temperature was 25° C. The manometers were read at half

Fig. 2. The relation between relative growth rate of seedlings of pine and spruce and the osmotic potential of the root medium. Least significant differences (LSD) at P = 0·05 and P = 0·01 are shown by the vertical lines. Experimental details as for Fig. 1.

hourly intervals for 3 to 4 hours, during which period the rate of pressure change was constant. Hence the slope of the straight line relating pressure with time was used in calculating the rate of oxygen uptake. After the manometric measurements, the roots were washed on filter paper and dried so

Table 1.

The number of pine and spruce seedlings harvested and the leaf weight ratios (and standard errors) at harvest. C = control; in I, π_{add} = —12 j/hg.; in II, π_{add} = —27 j/hg.

	Treatment		
	C	I	II
Number of plants harvested			
Pine	24	14	15
Spruce	12	10	13
Leaf weight/total weight, mg./g.			
Pine	678 ± 10	713 ± 13	687 ± 12
Spruce	705 ± 16	670 ± 18	725 ± 16

that the oxygen uptake could be expressed per unit of dry weight. There was about 5 mg. per flask.

Respiration was measured at seven different osmotic potentials in the range 0 to —140 j/hg., between 4—8 determinations being made at each one. Aver-

Table 2.

Some properties of the birch and aspen seedlings at the first harvest; means with standard errors.

Measurement	Birch	Aspen
Harvested plants		
Leaf area ratio, cm.²/g.	248 ± 19	318 ± 10
Leaf weight ratio, mg./g.	620 ± 14	690 ± 7
Specific leaf area, cm.²/g.	454 ± 25	460 ± 12
Shoot/root ratio	4·21 ± 0·32	3·94 ± 0·25
Number of plants	9	6
Unharvested plants, control treatment		
Mean dry weight, mg.	32	122
Mean leaf area, cm.²	8·6	38·8

age rates of oxygen uptake at each osmotic potential have been plotted (Fig. 7) but the regression lines have been calculated using all the separate determinations.

Results

Growth of pine and spruce:

The average dry weights at harvest (total, stem + roots and leaves) are shown in Fig. 1. All dry weights decreased with decreasing solution osmotic potential. To compare the response of the two species relative growth rates have been calculated and expressed as percentages of the control rates (Fig. 2). The relative growth rates of the controls, with standard errors, were: pine, 498 ± 14, spruce 450 ± 21 mg./g. week. It is immediately apparent that spruce is more sensitive than pine in response to decrease in osmotic potential in the root medium. In both the greater sensitivity of spruce and its lower relative growth rates, these results are in agreement with those obtained earlier in soil in the same growing conditions (Jarvis and Jarvis 1963a). There were no significant differences between treatments or species in the ratio of leaf dry weight/total dry weight, i.e. leaf weight ratio (Table 1).

Growth of birch and aspen:

The size, dry matter distribution and number of plants at the first harvest are given in Table 2. A large part of the comparatively large variation for birch arises from the lumping in the final analysis of two separate experiments with plants at slightly different stages. In each experiment the variation was not greater than that given for the aspen plants and the appropriate leaf area ratio, not the overall average in Table 2, was used in estimating the initial weights of the unharvested plants.

Data from the second harvest are given in Table 3. The growth responses,

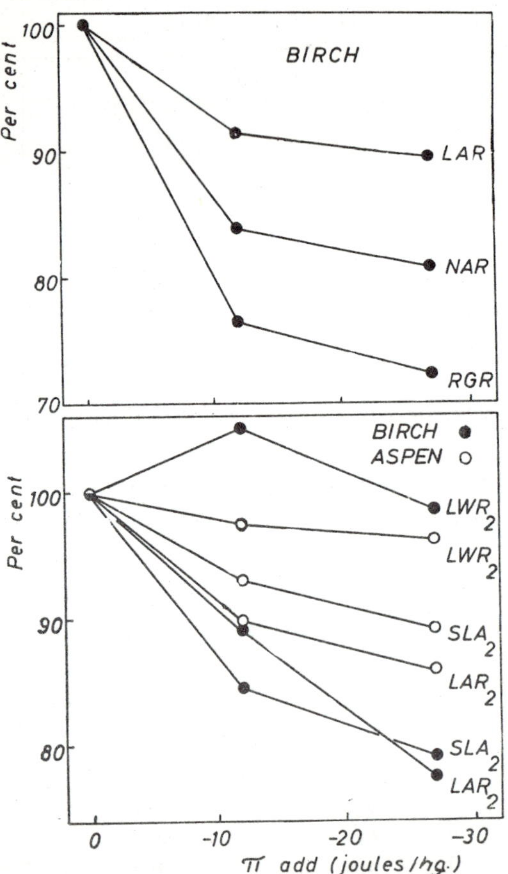

Fig. 3. (a) Net assimilation rate (NAR), mean leaf area ratio (LAR) and relative growth rate (RGR) of birch in PEG treatments as a percentage of the control.
(b) Leaf weight ratio (LWR), final leaf area ratio (LAR$_2$) and specific leaf area (SLA$_2$) of birch and aspen in PEG treatments as a percentage of the control.

Table 3.

The average net assimilation and relative growth rates, together with some morphological features of the birch and aspen seedlings at the second harvest; means with standard errors. C = control; in I, π_{add} = —12 j/hg.; in II, π_{add} = —27 j/hg.

Measurement and species	Treatment			L.S.D.
	C	I	II	$P = 0.05$
Number of plants harvested				
Birch	17	14	15	
Aspen	8	9	8	
Leaf area per plant, cm.²				
Birch	100·3	47·5	33·2	
Aspen	148·9	132·2	102·1	
Dry weight per plant, mg.				
Birch	453·4	232·1	185·1	
Aspen	493·1	478·8	388·2	
Net assimilation rate, g./m.² wk.				
Birch	32.1 ± 1.2	26·9 ± 1.3	25·9 ± 1·2	3·5
Aspen	24·6 ± 0·9	22·9 ± 0·9	25·5 ± 0·9	n.s.
Relative growth rate, mg./g. wk.				
Birch	851 ± 34	650 ± 37	615 ± 36	102
Aspen	764 ± 28	678 ± 28	744 ± 30	n.s.
Final leaf area ratio, cm.²/g.				
Birch	230 ± 8	205 ± 9	178 ± 8	24
Aspen	301 ± 7	270 ± 7	258 ± 8	22
Final leaf weight ratio, mg./g.				
Birch	632 ± 10	665 ± 11	623 ± 11	30
Aspen	678: ± 12	661 ± 12	652 ± 13	n.s.
Final specific leaf area, cm.²/g.				
Birch	363 ± 10	307 ± 11	287 ± 11	31
Aspen	444 ± 13	411 ± 13	395 ± 14	40
Final shoot/root ratio				
Birch	5·81 ± 0·25	5·68 ± 0·27	3·89 ± 0·26	0·74
Aspen	4·54 ± 0·31	3·75 ± 0·31	3·46 ± 0·33	0·95

as for pine and spruce, are very similar to those obtained earlier in soil. Net assimilation rate and relative growth rate of birch decreased with decreasing water potential whereas the aspen rates were not significantly affected. Another similarity is that the aspen net assimilation rates were lower than the birch rates but were compensated for by higher leaf area ratios so that the relative growth rates were very similar. For both species, decreasing water potential also caused decreased leaf area and leaf area ratio. The smaller leaf areas in PEG treatments I and II are the result of lower rate of leaf production, earlier senescence and abcission, and smaller size of the leaves (Table 4). The relative contributions of the reductions in leaf area ratio and

T a b l e 4.

The effects of the applied moisture stresses in reducing the leaf area of birch and aspen seedlings. C = control; in I, π_{add} = —12j/hg.; in II, π_{add} = —27 j/hg.

Treatment	Aspen			Birch		
	C	I	II	C	I	II
No. of leaves per plant produced during the experiment	5·3	5·2	4·9	4·7	3·9	3·2
No. of leaves per plant abscissed during the experiment	0·3	0·1	0·2	0·6	1·0	1·9
Area per leaf, cm.², at final harvest	13·1	10·3	8·2	10·5	6·4	5·7

net assimilation rate to the reduction in relative growth rate of birch are shown in Fig. 3A. From Fig. 3B it is clear that birch leaf area ratio was more sensitive to decreasing water potential than aspen. Leaf area ratio may be

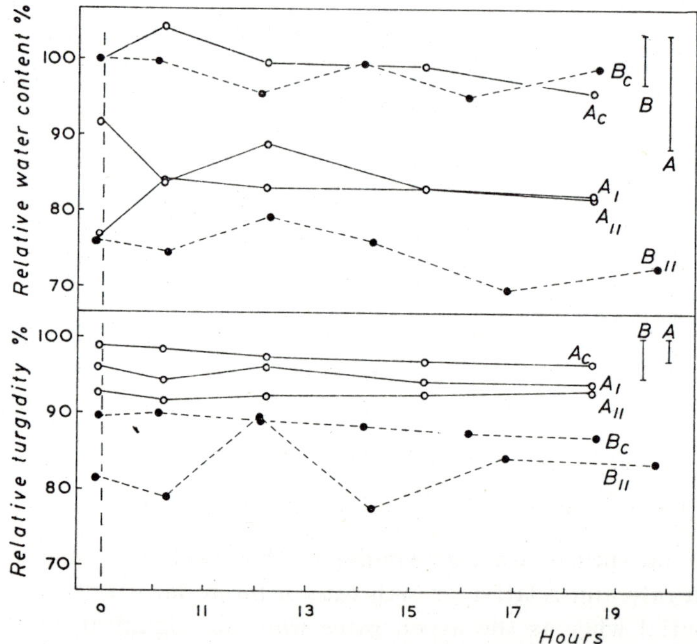

Fig. 4. Relative water content and relative turgidity through the day of leaves from birch and aspen PEG treatments and control. Relative water content is water content, (fresh wt. — dry wt.)/dry wt., as % of the control water content at the end of the 6 hour dark period. LSD is least significant difference at P = 0·05.

considered as made up of two components: the leaf weight ratio (LW/W) × × specific leaf area (LA/LW). In both species the leaf weight ratio was largely unaffected by the treatments (Table 3, Fig. 3B) and hence the reductions in leaf area ratio were accompanied by equivalent reductions in specific leaf area (Fig. 3B), i.e. the leaves had greater dry weight per unit area. This leaf modification was also less pronounced in aspen + than in birch leaves.

The shoot/root ratio also decreased with decreasing water potential in both species, i.e. root growth was stimulated. Birch root response was greater than aspen, although birch root development was still less in all treatments than in corresponding aspen treatments.

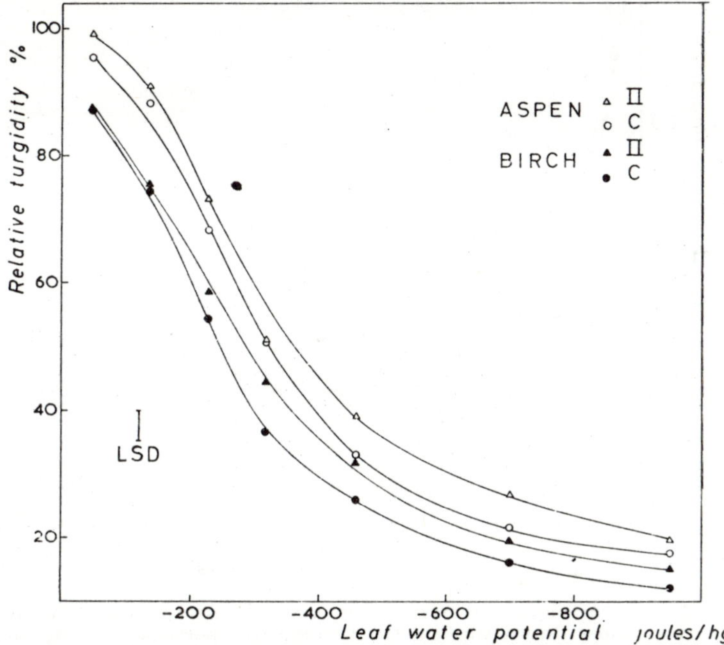

Fig. 5. The relation between relative turgidity and water potential for birch and aspen leaves from PEG treatment II (π_{add} = —27 j/hg.) and control. The dotted line indicates the approximate critical level for cell damage. LSD is least significant difference at P = 0·05.

In the responses of leaf area ratio, specific leaf area and root/shoot ratio, these results obtained in solution culture agree with those in soil only over the wettest part of the range, at very high, sub-optimal, water potentials. In soil birch also showed larger responses than aspen.

Internal water relations:

Leaf relative turgidities during the day are shown in Fig. 4B. Those of birch were lower than aspen, and those of both birch and aspen from the PEG treatments were lower than the controls both before and after illumination began. The relative water contents (Fig. 4A) were more variable but suffice to demonstrate that the relative turgidities in the PEG treatments were not lower because of excessive solute uptake, with a concomitant large uptake of water when the leaf discs were put to saturate, but were lower because the actual water content was lower (cf. Jarvis and Jarvis 1963c). Curves of the relation between relative turgidity and leaf water potential (Fig. 5) show that at the same potentials the relative turgidities of leaves from the PEG treatments were higher than the controls. This may be because of a decrease in tissue osmotic potential or in tissue matric potential, i.e. in

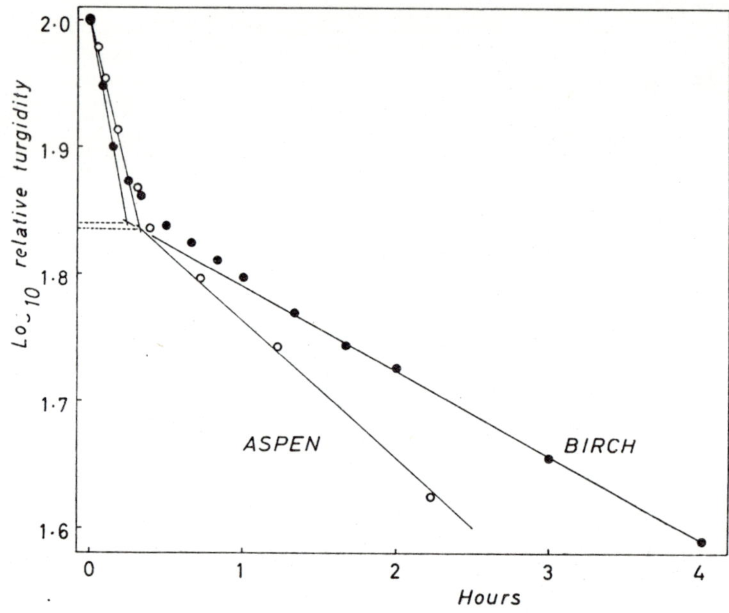

Fig. 6. Examples of the change in relative turgidity of detached leaves (logarithmic scale) with time. Lines drawn through the two straight portions of each curve have been extrapolated to intersect, and relative turgidities at the point of intersection are compared in Table 5.

the hydration properties of the tissue. Of possible significance to the latter are the increases in leaf dry weight per unit area noted above. At the same leaf water potentials, the relative turgidities of aspen leaves were higher than those of birch leaves.

Sample curves of the rate of decrease in relative turgidity of detached leaves under standard conditions are shown in Fig. 6. The data obtained from such curves are collected in Table 5. The relative turgidities at the point of stomatal closure given by the intersections is about 70% for both species and both treatments. The rate of reduction in relative turgidity by cuticular water loss was not less in the PEG treatments thus showing that the increase in leaf mass per unit area is not adaptive in this connection. The cuticular rate of reduction was significantly greatly in aspen than in birch and the stomatal rate was the reverse.

T a b l e 5.

The relative turgidity at a defined point of stomatal closure, determined from the intersection of the extrapolated straight line portions of the curve of reduction in relative turgidity (log. scale) with time (see Fig. 6). The rate of reduction in relative turgidity was obtained from the first 10 min. of water loss. Full details and the conditions in which the determinations were made are given by Jarvis and Jarvis (1963c).

	Aspen		Birch		L.S.D. P = 0·05
	Control	PEG II	Control	PEG II	
Relative turgidity of stomatal closure and equivalent water potential in j/hg.	65·8 ± 2·3 —238	72·3 ± 2·3 —228	70·0 ± 2·7 —128	71·0 ± 2·6 —150	n.s.
Rate of reduction in RT (%/min.) by:					
a) Cuticular water loss	0·23 ± 0·03	0·37 ± 0·03	0·18 ± 0·03	0·17 ± 0·03	0·09
b) Stomatal water loss	1·33 ± 0·12	1·01 ± 0·12	1·69 ± 0·14	1·74 ± 0·14	0·40
Stomatal/cuticular	5·8	2·7	9·4	10·2	

R o o t r e s p i r a t i o n i n r e l a t i o n t o o s m o t i c p o t e n t i a l o f t h e m e d i u m :

Oxygen uptake was reduced by decreasing osmotic potential between 0 to —140 j/hg. in root tips of all the four species. The reductions were proportional to the osmotic potential and highly significant linear regressions were fitted (Fig. 7 and Table 6). While the response of the four tree species was similar, it differs appreciably from the curvilinear response of wheat roots (Fig. 8) with a

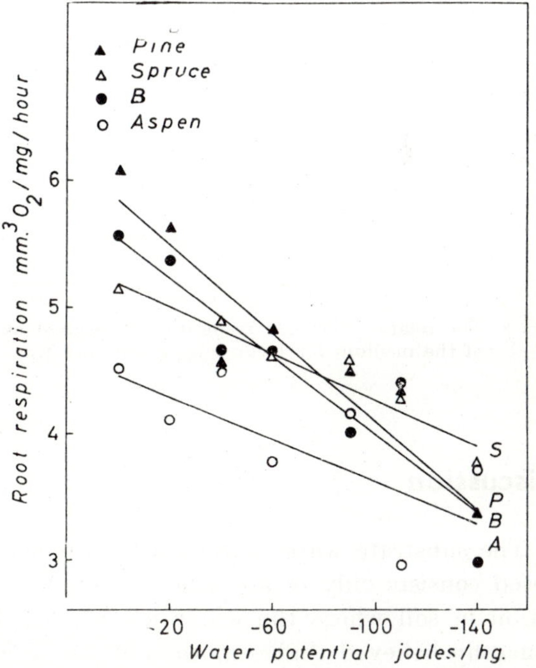

Fig. 7. The relation between respiration, measured as oxygen uptake, and osmotic potential of the medium for apical root segments 1—2 cm. long of pine, spruce, birch and aspen. LSD is least significant difference at P = 0·05. The standard errors of the slopes of the regressions are given in Table 6.

distinct maximum rate at about —40 j/hg. Pine root respiration was significantly more sensitive than spruce root respiration, and birch significantly more sensitive than aspen.

Table 6.

The relation between root respiration and the water potential of the root medium.

	Regression of y on x (y = a + bx)	Standard error of b
Aspen	y = 4·458 — 0·0841 x	± 0·0237
Birch	y = 5·544 — 0·1547 x	± 0·0263
Pine	y = 5·856 — 0·1755 x	± 0·0278
Spruce	y = 5·190 — 0·0923 x	± 0·0137

Aspen and birch; and pine and spruce, are significantly different (P = 0·01).

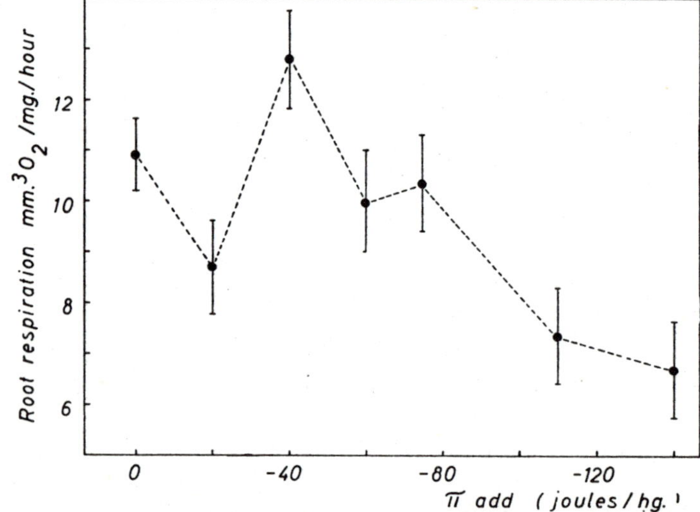

Fig. 8. The relation between respiration, measured as oxygen uptake, and osmotic potential of the medium for 1 cm. long wheat root tips.

Discussion

The substrate water potential to which plants rooted in solutions are exposed consists only of an osmotic potential and hence differs from the condition in soil where the water potential is largely a matric potential. It is generally believed that osmotic and matric potentials have similar effects on growh (see for example, Wadleigh and Ayers 1945 and Richards and Wadleigth 1952). However, Collis-George and Sands (1962), showed that the germination rate of some seeds was retarded much more by low matric potentials than by similar osmotic potentials. It is apparent from Slatyer's experimental data (1961) that for an osmotic potential to produce a result similar to that of a matric potential, the root must act as an efficient osmometer to the solute, which is not able to enter the plant to a significant degree. In the experiments with lupins, cited above, this condition seemed to be satisfied by polethylene glycol.

The order of increasing sensitivity of the relative growth rate of the four species to decreasing osmotic potential in the root medium is aspen < pine < < birch < spruce. This differs from that in the soil (Jarvis and Jarvis 1963a), where birch was less sensitive than pine. Comparison between the pine spruce and birch-aspen experiments is perhaps not justified because the conditions were not the same, but, if the comparison is allowed, this result is not unexpected. A detailed discussion of the possible bases for differences in species response has been given in the earlier papers in this series. In soil the water potential to which the roots are exposed depends on the conductivity of the soil, the transpiration rate and the concentration of the effective roots, whereas in solution it is the same for all roots. It may be suggested that the results in these experiments depend on differences in sensitivity of the leaves (i.e., in the relationships between water potential and turgidity, and turgidity and stomatal aperture, and in the direct metabolic response to leaf water potential) to the same applied potentials at the roots, whereas in soil the results also partly depend on differences in the potentials developed at the roots. In soil the larger root concentration of birch than pine would result in higher water potentials at the roots: in solution this relative advantage would not occur. Response to moisture stress induced in solutions is more directly physiological; it is interesting that there were not greater differences in growth rate response or in the pattern of response of birch and aspen growth between the soil and solution culture experiments.

The relative turgidities through the day (Fig. 4) together with the relative turgidities for stomatal closure (Table 5) show that the applied moisture stress was not sufficient to cause stomatal closure in aspen and was probably not sufficient to cause increased stomatal diffusion resistance in birch. Results of a supplementary experiment in which transpiration rates of birch and aspen in control and PEG solutions were measured by weighing, support this deduction. In PEG treatment II both relative turgidity and water content were much lower in birch than in aspen.

From the relative turgidities during the day and the relation between relative turgidity and leaf water potential (Fig. 5) the leaf water potentials occuring during the day can be estimated (Table 7). It is noticable that leaf water potentials of aspen were somewhat lower than those of birch although growth was reduced less. Reduction in growth of birch, as compared with aspen, cannot then be attributed to lower leaf water potentials, but must be due to a greater sensitivity to those occurring. This greater sensitivity might

Table 7.

The range of leaf water potentials occurring through the day. Data from Fig. 4 and 5. Values at the end of a 6 hour dark period are in brackets. The starred value is exceptionally high, see Fig. 4.

Species	Leaf water potentials j/hg.	
	Control	II (π_{add} — 27 j/hg.)
Aspen	—10 to —34 (—10)	—118 to —128 (—120)
Birch	—34 to —52 (—38)	—38* to —118 (—88)

be in the control of stomatal aperture, since at the same water potential relative turgidity is lower in birch, or in response of metabolism. In support of the latter, it is interesting to note firstly, that root respiration of birch was much more reduced by decreasing osmotic potential than aspen root respiration (Fig. 7); and, secondly, that decrease of transpiration rate of birch in

Table 8.

The leaf water potentials at the end a 6 hour dark period, and the difference in potential between leaves and solution in joules/hg. The corrected leaf water potentials are for leaves of the same water content as the control leaves.

Treatment	Actual			Corrected	
	π_{sub}	ψ_{leaves}	$\psi_{lvs} - \pi_{sub}$	ψ_{leaves}	$\psi_{lvs} - \pi_{sub}$
Aspen: Control	— 5	— 10	— 5	—10	— 5
$\pi_{add} =$ —27 j/hg.	—32	—120	—88	—92	—60
Birch: Control	— 5	— 38	—33	—38	—33
$\pi_{add} =$ —27 j/hg.	—32	— 88	—56	—67	—35

response to sudden decrease of osmotic potential of the root medium was similar to that of aspen (Jarvis and Jarvis 1963b).

The early morning leaf water potentials (Table 8) have been derived from the data in Fig. 4 and 5. The leaf water potentials, ψ_{leaf}, of the plants growing in the osmotic substrate are much lower than in similar experiments with lupins (Jarvis and Jarvis 1963c, Table 8). The leaf water potentials and the difference in water potential between leaves and solution ($\psi_{\text{leaves}} - \pi_{\text{sub}}$) for the PEG treatment plants are much larger than for the controls. These figures do not, therefore, support the suggestion (Slatyer 1961, Jarvis and Jarvis 1963c) that the leaf water potential of π_{add} treatment leaves is generally lower than that of control leaves by an amount equal to π_{add}. Part of the lower water potential of the PEG treatment leaves was due to a lower water content (see Fig. 4). This was corrected for by recalculating ψ_{leaf} to the water content of the controls on a simple osmotic concentration basis. The remaining difference in water potential between control and PEG treatment leaves must depend on a lower osmotic or matric potential. The former might have arisen metabolically, or possibly by PEG uptake.

The decrease in root respiration of all four species and wheat in response to decreasing osmotic potential of the root medium differs from many earlier results (see Takaoki 1957) showing increases when salts or sucrose were used as the osmotic substrate. Increased respiration of wilting leaves has often been demonstrated (Iljin 1957). The response patterns with the tree seedlings and wheat are similar to certain of Takaoki's results for halophytes with NaCl as osmotic substrate, but our species cannot be classifield as halophytes. There is good evidence for doubting the usefulness of salts as osmotic substrates (see Slatyer 1961).

References

Collis-George, N., Sands, J. E.: Comparison of the effects of the physical and chemical components of soil water energy on seed germination. — Austral. J. agric. Res. *13* : 575—584, 1962.

Fisher, R. A.: Some remarks on the methods formulated in a recent article on the quantitative analysis of plant growth. — Ann. appl. Biol. *7* : 367—372, 1921.

Iljin, W. S.: Drought resistance in plants and physiological processes. — Ann. Rev. Plant Physiol. *8* : 257—272, 1957.

Ingestad, T.: Studies on the nutrition of forest tree seedlings I. Mineral nutrition of birch. — Physiol. Plant. *10* : 418—439, 1957.

Jarvis, P. G., Jarvis, M. S.: The water relations of tree seedlings I. Growth and water use in relation to soil water potential. — Physiol. Plant. *16* : 215—235, 1963a.

—, —: The water relations of tree seedlings III. Transpiration in relation to osmotic potential of the root medium. — Physiol. Plant. *16* : 269—275, 1963b.

—, —: Effects of several osmotic substrates on the growth of *Lupinus albus* L. seedlings. — Physiol. Plant. *16* : 485—500, 1963c.

Lundeberg, G.: The relationship between pine seedlings *(Pinus silvestris* L.*)* and soil fungi. — Svensk bot. Tidskr. *54* : 346—359, 1960.

Richards, L. A., Wadleigh, C. H.: Soil water and plant growth. — *In*: Soil Physical Conditions and Plant Growth, ed. Shaw, B. T., Academic Press, N. Y., 1952.

Slatyer, R. O.: Effects of several osmotic substrates on the water relationships of tomato. —
    Austral. J. biol. Sci. *14* : 519—540, 1961.
Takaoki, T.: Relationships between plant hydrature and respiration II. Respiration in relation
    to the concentration and the nature of external solutions. — J. Sci. Hiroshima Univ.,
    Series B, Div. 2. *8* : 73—79, 1957.
Wadleigh, C. H., Ayers, A. D.: Growth and biochemical composition of bean plants as condi-
    tioned by salt moisture tension and salt concentration. — Plant Physiol. *20* : 106—132,
    1945.

Discussion

P. E. Weatherley: I would like to ask Dr. Jarvis whether the penetration by the polyethylene glycol into the plants was a problem. We have been unable to condition leaf disks in polyethylene glycol solutions because of penetration. Of course leaf disks are not exactly whole plants but I wonder whether you had evidence for or against penetration.

P. G. Jarvis: Before starting on these comparative studies we carried out some preliminary experiments with *Lupinus albus* in which we compared the effects of several osmotic substrates on growth and leaf water relations. From studying the leaf water relations (relative turgidity: relative water content, relative turgidity for stomatal closure, relation between leaf water potential and relative turgidity) we found good evidence that much NaCl had entered the leaves and altered the osmotic potential status. Plants grown on polyethylene glycol, however, were similar in these properties to the control plants. The slight variations shown may have been due to the uptake of a little of the substrate but might also have been due to osmotic adaptation. There were large differences between the leaf water relations of the plants grown on NaCl and on polyethylene glycol solutions. We cannot be certain that no polyethylene glycol is taken up but we think it unlikely that very much is. We should like to label it to confirm this but it is not available commercially and it is expensive to have it prepared specially. (See also B. E. Jones, Plant Physiol. *39* (supplement): lviii. 1964.)

B. Slavík: How great were the changes in water potential of plants after cultivation in osmotic medium ?

P. G. Jarvis: The plants grown on the PEG solutions showed a shift in the relationship between relative turgidity and leaf water potential such that PEG treatment plants have a lower leaf water potential for the loss of a given amount of water. This helps to maintain a favourable gradient for water uptake from solution to leaf without requiring such a large reduction in turgidity (See Table 8). Relative turgidities through the day were fairly constant (Fig. 4B). The corresponding range of leaf water potentials together with the early morning values are given in Table 7. It can be seen in Table 8 that the lower leaf water potentials of the PEG treatment plants were mostly due to loss of water and concentration of the cell sap in birch, but not in aspen.

THE INFLUENCE OF WATER STRESS ON THE RELATIONSHIP BETWEEN CO₂ UPTAKE AND TRANSPIRATION

W. LARCHER

Institute of Botany, University of Innsbruck, Austria

Water stress soon forces the plant to reduce its water consumption. The leaf manages this rapidly and efficiently by constricting the width of the stomatal opening. Depending upon species and morphology, transpiration may fall to $1/3$ to $1/30$ of the full value (Kamp 1930, Pisek and Winkler 1953, Stålfelt 1956). Stomatal regulation also restricts the CO_2-uptake by the leaves, however, and when the stomata are completely closed, uptake essentially ceases. Thus the two gas exchange processes are linked, and to a certain degree transpiration and photosynthesis and finally, water consumption and dry-matter production also are linked.

These facts were known to the agronomists by the end of the last century. Thus in 1883 Hellriegel calculated the "transpiration coefficient" (equation [1]) for certain crops as an expression of economical water use in dry-matter production.

$$k_T = \frac{Wc \ [kg.]}{Dmp \ [kg.]} \qquad [1]$$

where k_T = transpiration coefficient, Wc = water consumption [kg.], Dmp = dry-matter production [kg.].

Instead of the transpiration-coefficient Maximov and Alexandrov (1917) introduced the "efficiency of transpiration" (equation [2]).

$$e_T = \frac{Dmp \ [g.]}{Wc \ [kg.]} \qquad [2]$$

e_T = efficiency of transpiration.

At the beginning of this century Briggs and Shantz (1913), and Maximov and Alexandrov (1917) investigated the relationship between water consumption and dry-matter production of a large number of plants native to dry regions as well as of certain crop plants. They hoped to find a usable constant expressing adaptation of these plants to dry habitats. It appeared that this hope would not be realized, since among the plants of dry regions were some which on the average transpired actively during the entire growing season but were unable to produce much more than other plants which transpired less. Among the plants investigated, only the ephemerous annuals were exceptional by their high productivity per unit transpiration; i.e. they yielded considerable dry-matter with relatively low transpiration. The efficiency values of plants in the Tiflis region and on the Great Plains of North America were all in the range of 1—4 grams of dry-matter per kilogram of water consumed (equivalent to a transpiration-coefficient of 1,000 to 250 kilograms of water consumption per kilogram of dry-matter production, values that are well within those found in the temperate zone (see Stocker 1929).

Yet more promising results were obtained by Maximov (1923) in experiments in which plants were raised in pots with soil water added to be equivalent to either 40% or 60% of field capacity (FC). All of the plants grown in the drier soils yielded much less dry-matter than control plants. Nevertheless, 7 of the total of 12 plants in the dry soils reduced their transpiration more than their dry-matter production, so that these plants transpired more profitably than the controls which received an optimal amount of water. Tulaikov (1922) obtained analogous results in field trials with agricultural plants, carried out in southeastern Russia, and Tumanow (1927) in laboratory experiments with the common bean *(Phaseolus vulgaris)* carried out in Leningrad.

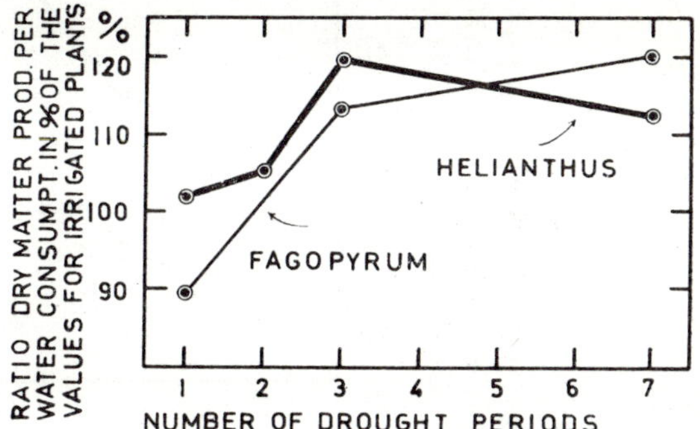

Fig. 1. Efficiency of transpiration (dry-matter production in *mg.* per *g.* water transpired) of *Helianthus annuus* and *Fagopyrum esculentum* as a function of the number of drought periods which occurred during the growing season. The efficiency of the control plants which were watered regularly is considered to be 100 %. According to data of Tumanow (1927).

In addition Tumanow measured water consumption and dry matter production of sunflowers *(Helianthus annuus)* and buckwheat *(Fagopyrum esculentum)* throughout development from seedlings to harvest and calculated the efficiency of transpiration for each stage of development (Fig. 1). In these experiments irrigated controls accompanied the test plants growing under field conditions. Seven short periods of drought occurred during the growing season. Tumanow then calculated the efficiency of transpiration of the sunflower and the buckwheat plants growing under field conditions, after each of the seven dry periods, comparing these values with those for irrigated controls harvested at the same time. Some of Tumanow's data are shown graphically in Fig. 1. It is apparent that the efficiency of transpiration is not only not reduced by drought, but that it is somewhat increased, providing that wilting does not occur too frequently or last too long. Furthermore,

Helianthus seems to be more sensitive to frequent wilting than *Fagopyrum*.

Not only annual herbs are able to utilize water more rationally when soil moisture is rather low, but woody plants, at least their seedlings, also exhibit this phenomenon, as recent experiments of Eidmann (1962) have shown. Using potted plants Eidmann determined dry-matter production and transpiration of 3 to 5 year-old conifer seedlings as a function of various external

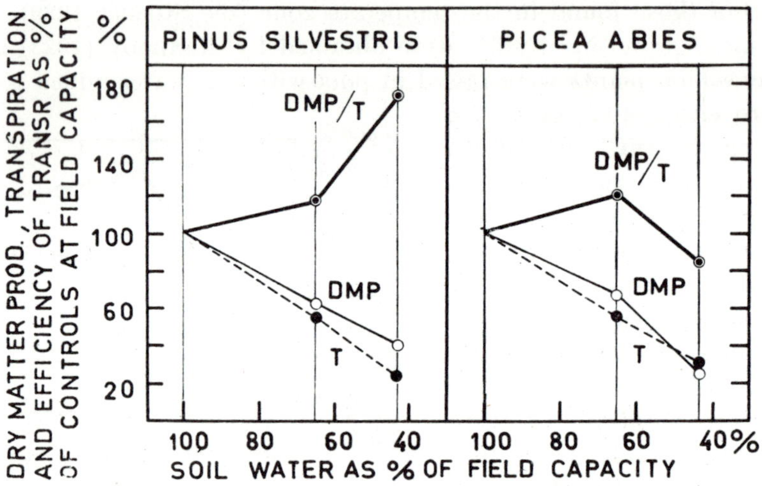

Fig. 2. Water consumption ("T", dots, dashed lines), dry matter production ("DMP", circles, thin lines) and efficiency of transpiration (ratio "DMP/T", circles with points, heavily drawn lines) of 3 to 5 year old seedlings of *Pinus silvestris* and *Picea alba*. Plants were grown in a sandy humus soil with moisture contents of 100, 65 or 42 percent of field capacity. The permanent wilting point of the soil was about 35% field capacity. With *Pinus silvestris*, transpiration is more strongly restricted compared to dry matter production the drier the soil. With the more drought sensitive *Picea abies*, water is more profitably used only with moderately dry soils; with extreme soil dryness the dry-matter production and concurrently the efficiency of transpiration rapidly collapse. According to data of Eidmann 1962.

environmental factors. One part of this program was established to clarify the influence of water stress. In order to begin the experiment with uniform material, all plants were provided with optimal water conditions for one year. At the beginning of the growing season of the following year, the water content of the soil in the pots was established at 42%, 65% or 100% field capacity (20% moisture content expressed as percentage of soil dry weight). The lowest value was still above the permanent wilting point (35% FC) of the sandy humus soil. Eidmann's data from Scotch pine *(Pinus silvestris)* and spruce *(Picea abies)* are summarized and expressed graphically in Fig. 2.

Dry-matter production, transpiration and efficiency of transpiration of the plants provided with abundant water are shown in Fig. 2 as 100%. As in the previous example the drier the soil the less mass is produced by the plants:

At 65% of FC, the seedlings produce not more than 64—67% and at 42% of FC only 26—39% as much dry-matter as in soil at field capacity. Again, however, at least at the first level of drying (65% of FC), the water consumption of both species is reduced more than the yield. In the driest soil (42% of FC) on the other hand, very specific differences begin to appear, from which one may indeed draw certain conclusions relating to water stress tolerance. While *Picea abies*, as well as *Thuja plicata* and *Pseudotsuga taxifolia* which Eidmann has also studied, definitely transpires less efficiently under strong water stress than when they are provided with ample water, *Pinus silvestris* and *Tsuga heterophylla* are actually able to restrict their transpiration more strongly under these conditions. Both species loose only $^1/_5$ as much water in soil at 42% of FC as the controls with sufficient water. Dry-matter production, on the other hand, falls much more slowly with increasing soil moisture stress in the case of the apparently tolerant conifers than in the representatives of the spruce type.

The course of the curves expressing efficiency of transpiration proves to be an excellent aid to recognition and characterisation of the different responses of the various species when they are subjected to water stress. Beyond this, all of these results seem to imply that the highest dry-matter production is to be purchased with extravagant water consumption but that water will be more profitably spent after a partial restriction in the intensity of gas exchange has taken place.

We are indebted to Polster (1950) for the first extensive and direct comparison of intensities in gas exchange, i.e. comparison of CO_2-uptake with transpiration. Later Koch (1957) succeeded in directly measuring the photosynthesis/transpiration ratio, (P/T quotient, equation [3]), not only for single leaves but also for an entire plant community.

$$P/T = \frac{P \, [mg. \, CO_2]}{T \, [g. \, H_2O]} \tag{3}$$

where P = photosynthesis as CO_2-uptake per unit area (weight) and unit time, T = transpiration as H_2O-output per unit area (weight) and unit time.

In both of these classical studies the question of possible effects of the drying soil remained in the background. The transpiration and assimilation curves made by Loustalot (1945) with *Carya* seedlings on gradually drying soils, and results of recent investigations from Polster and his team (Polster, Weise and Neuwirth 1960, Neuwirth and Polster 1960) allow us to expect not only that the efficiency of transpiration will increase with increasing soil dryness on the average over an entire growing season but that, during the initial phases of hydroactive stomatal closing, stomatal transpiration will under certain conditions be also more strongly restricted than the CO_2-uptake.

Theoretical considerations also agree. Thus Stålfelt (1935 p. 753) investigated the influence of stomatal width upon the diffusion of CO_2 and water vapor and determined that in still air the effect of stomatal width was essentially the same for both transpiration and CO_2-uptake. In turbulent air, on the other hand, changes in stomatal aperture, especially at extreme widths, have a greater influence upon transpiration than upon CO_2-uptake. The difference becomes greater as the light intensity decreases and the diffusion pressure deficit of the air increases. In other words: For diffusion processes into and out of the leaf, the Mitscherlich law holds in the sense that the limiting factor is that which is present at a minimum (Stålfelt 1956, p. 390). With narrow stomatal openings, the stomatal resistance (r_s) becomes limiting; with wide stomatal openings, the external environmental factors are more effective and, as Gaastra (1959, 1962) has recently demonstrated in detail, the diffusion resistance of the external air layer (r_a) and of the membranes of the mesophyll cells (r_m) are also limiting. Under conditions of natural CO_2-supply the influence of open stomata on the CO_2-diffusion is relatively small at high values of mesophyll resistance. With the vegetables studied by Gaastra the mesophyll resistance for CO_2-diffusion was 2 to 4 times that of the stomatal resistance for CO_2 when the stomata were completely open. For the H_2O-vapor diffusion, however, the mesophyll resistance, in leaves with ample water, is much smaller than the stomatal resistance. Gaastra comes to the conclusion that "diffusion takes place through a series of resistances, and the concept that the rate is determined by the sum of these resistances ($r_a + r_s$ for transpiration and $r_a + r_s + r_m$ for photosynthesis) implies that transpiration depends upon the stomatal opening more than photosynthesis" (Gaastra 1959, p. 61).

With wide open stomata the two gas exchange processes are promoted by environmental factors at completely different levels—transpiration is determined especially by the leaf temperature and by the vapour pressure deficit of the air, and CO_2 penetration primarly by the CO_2 partial pressure gradient between leaf interior and air, which develops due to active photosynthesis (high light intensity). We should also consider the CO_2 released by respiration being a metabolic process which is highly dependent upon temperature. So, for example, under a given light intensity, in one case (perhaps in the morning at fairly low temperature and saturation deficit of the air), the CO_2-uptake may be intense, but the transpiration through wide open stomata may be only moderate, yielding a high P/T quotient. In another case (noon, with high saturation deficit of the air and relatively high temperature), at the same light intensity and perhaps stomata that are still wide open, transpiration may have increased significantly while photosynthesis remained at the same level or perhaps even dropped as the result of increasing respiratory CO_2-output, now the P/T quotient is low. There are sufficient examples to be found in the papers of Koch and Polster.

In the natural fluctuation of climatic environment factors it is extremely difficult to study the effect of water supply on the P/T quotient. Here, at least so far as we are interested in understanding basic trends, the laboratory experiment is superior to the field experiment. Thus the assignment of the laboratory experiment can obviously only be to determine the trends in the course of the P/T relationship. The absolute values of the P/T quotient are exclusively valid only for the specific conditions present during the experiment and frequently they are too high, since as a rule transpiration is reduced in the cuvette (see also Koch 1957)

Actually it can be shown under laboratory conditions, that as water begins to become limiting, water vapour output is more strongly restricted than CO_2-uptake. This was shown by continual recording of transpiration and apparent assimilation of gradually drying

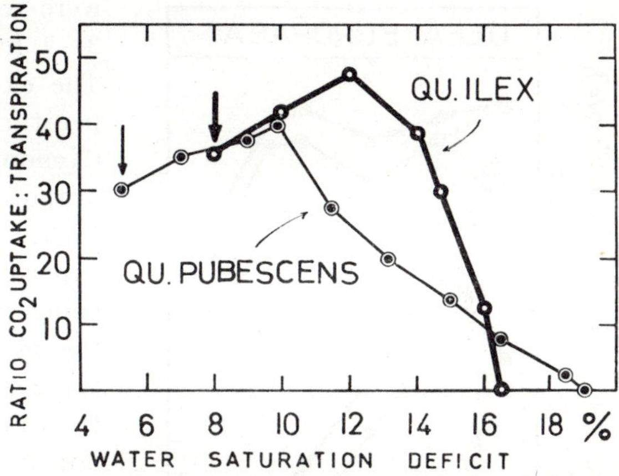

Fig. 3. The course of the P/T quotient (ratio photosynthesis per transpiration expressed in mg. CO_2 per g. H_2O) for drying, excised twigs of *Quercus ilex* and *Quercus pubescens* under standard conditions (see text). At the twig water deficit marked by the arrows, transpiration and photosynthesis reached their highest absolute values. From Larcher 1960.

excised twigs of the evergreen *Quercus ilex* and the deciduous *Quercus pubescens* (see Fig. 3. The twigs were held at 22° C and exposed to an illumination level of 10,000 lux (Xenon high pressure lamp). Evaporation in the cuvette was very slight, about 0·24 cm.³ per hr. [Piche].

Characteristically, the P/T quotient was not highest when the twigs were assimilating and transpiring most rapidly, i.e. at a water saturation deficit (W.S.D., given in % of the saturation weight of these twigs) of ca. 4% *(Quercus pubescens)* to 8% *(Quercus ilex)*, but only later, at a W.S.D. from 10 to 12%. Only at still higher W.S.D. does the CO_2-uptake fall off faster than transpiration and the P/T quotient drop steeply towards zero (complete stomatal closure).

These results required further verification by experiments on branches that have not been separated from the parent plant, for we cannot eliminate the possibility that the excision of the twig has some way influenced the gas exchange processes, or at least that the artificially rapid drying — in 6 — 8 hr.

after excision the oak twigs suffered a W.S.D. at 16—18% — invalidated the measurements (Larcher 1963a). Therefore I allowed 4-year-old olive trees *(Olea europaea* ssp. *sativa)*, which had been transplanted to pots a year before the beginning of the experiment and brought to leaf water saturation just before the experiment, to gradually dry out in the pots. During the week in which the increasing water stress became effective in soil and plants, transpiration and CO_2-uptake of one year old twigs were registered daily, each morning, for 3—4 hr. with the infrared gas analyzer. The experimental conditions were: Temperature 25—26° C, Illumination 30,000 lux (Xenon high pressure lamp), CO_2 350 p.p.m., evaporation in the cuvette ca. 0·35 cm³/hr. Piche. Since these were only preliminary experiments and the same branches were being used each day, I decided to forgo the measurement of W.S.D. build-up in the leaves. The results from W.S.D. controls taken from parallel branches could not be directly transferred to the test branches.

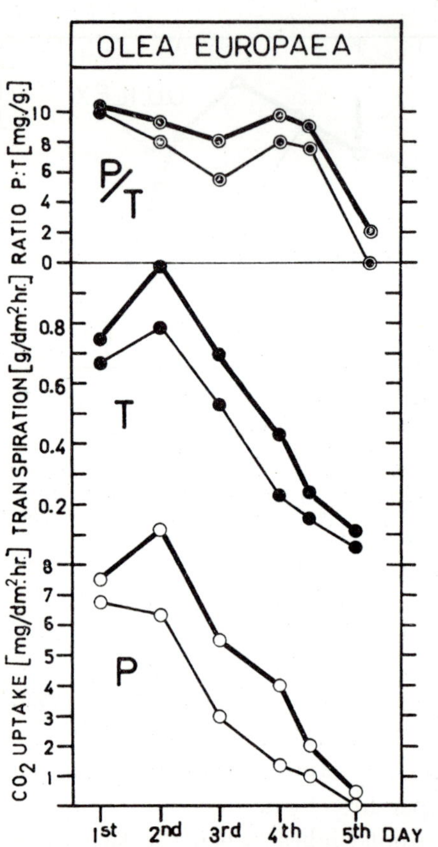

Fig. 4. The course of CO_2-uptake (P), transpiration (T) and P/T ratio of attached one year old branches of *Olea europaea* ssp. *sativa* during increasing soil dryness under standard conditions (see text). The trees, ca. 50 cm. high, were watered for the last time just before the first measurement. Photosynthesis is quoted as mg. CO_2-uptake per dm.² leaf area (both sides of the leaf⁻) and hr., transpiration as g. H_2O-output per dm². leaf area (both sides) and hr. From Larcher 1963b.

In the morning as soon as the light was turned on, transpiration and CO_2-uptake climbed rather rapidly to their peak values, where they remained for some hours in the first two runs, in the third only for some 30 to 40 minutes, and in the fourth only for 20 minutes. In evaluating the data I have used only the highest values of transpiration and CO_2-uptake, obtained at each day. These yielded the picture illustrated by Fig. 4.

Photosynthesis remained uniformly fairly high during the first two days in 4 of a total of 5 test runs. In one instance the peak value was not observed with turgid material but first on the following day after a slight drop in water saturation, a phenomenon which has been described several times in the literature (Schneider and Childers 1941, Loustalot 1945, Negisi and Satoo 1955). From the third day on, the CO_2-uptake of all test plants began to sink, and on the fifth day it had come to a stand still in most cases. Apparently by then the stomata would no longer be open. The olive leaves then had a W.S.D. of 15 to 18% of saturation weight.

With turgid material transpiration was noticably slow in starting and reached its peak value in all test plants only on the following day. From then it dropped steadily.

The P/T quotient reached a peak at two times: at the beginning of the experiment with excess water in the soil, and again as the soil reached the medium dry condition. The data are too tenous to permit detailed conclusions from the curves, but the results of the experiments with olive trees and with oaks seem to indicate that in the course of plant drying there may be a stage at which the most CO_2 can be assimilated with the most frugal water consumption.

There remain the problems of checking the observations, extending them to a larger number of plant species, and above all of causally explaining this phenomenon, which may be important for plant life in dry regions.

I am indebted to Prof. Frank Salisbury for his help in translating the manuscript.

Summary

As the soil begins to dry out various herbs and seedlings of woody plants are able to reduce their water consumption more than their dry-matter production, the degree being dependent upon and characteristic of the species. The circumstance that the relation between the CO_2-uptake and transpiration which is most favourable for both water and energy balance of the plant requires a certain degree of stomatal aperture but not necessarily maximum apertures, may contribute, at least partially to this phenomenon, which is important for the success of plants in habitats which are occasionally or continually dry.

It was in fact demonstrated in three cases, the evergreens *Olea europaea* and *Quercus ilex* and the deciduous *Quercus pubescens*, that at the most frugal water consumption these plants assimilate the most when the intensity of their gas exchange is partially restricted, depending on experimental conditions, to 30—80% of their maximum value.

References

Briggs, L. J., Schantz, H. L.: Relative water requirement of plants. — U.S. Dep. Agr. Bureau Plant Ind. Bull. No. 284—285, Washington 1913.

Eidmann, F.: Der Wasserverbrauch der Holzarten im Durchschnitt der Vegetationsperiode. — Int. Symp. Baumphysiologen Innsbruck, 50—54, 1962.

Gaastra, P.: Photosynthesis of crop plants as influenced by light, carbon dioxide, temperature and stomatal diffusion resistance. — Med. Landbouwhogeschool Wageningen *59* : 1—68, 1959.

—: Photosynthesis of leaves and field crops. — Neth. J. agric. Sci. *10* : 311—324, 1962.

Hellriegel, F.: Beiträge zu den naturwissenschaftlichen Grundlagen des Ackerbaues. — Vieweg u. Sohn, Braunschweig, 1883.

Kamp, H.: Untersuchungen über Kutikularbau und kutikuläre Transpiration von Blättern. — J. wiss. Bot. *72* : 403—465, 1930.

Koch, W.: Der Tagesgang der „Produktivität der Transpiration". — Planta *48* : 418—452, 1957.

Kramer, P. J.: Photosynthesis of trees as affected by their environment. — *In*: Thimann, J. V.: The Physiology of Forest Trees. Ronald Press, New York, 1958.

Larcher, W.: Transpiration and photosynthesis of detached leaves and shoots of Quercus pubescens and Quercus ilex during desiccation under standard conditions. — Bull. Res. Counc. Israel *8D* : 213—224, 1960.

—: Die Eignung abgeschnittener Zweige und Blätter zur Bestimmung des Assimilationsvermögens. — Planta *60* : 1—18, 1963a.

—: Orientierende Untersuchung über das Verhältnis von CO_2-Aufnahme zu Transpiration bei fortschreitender Bodenaustrocknung. — Planta, *60* : 339—343, 1963b.

Loustalot, A. J.: Influence of soil moisture conditions on apparent photosynthesis and transpiration of pecan leaves. — J. Agr. Res. *71* : 519—523, 1945, ref. Kramer 1958.

Maximov, N. A.: Physiologisch-ökologische Untersuchungen über die Dürreresistenz der Xerophyten. — J. wiss. Bot. *62* : 128—144, 1923.

—. Alexandrov, W. G.: [Efficiency of transpiration and drought resistence] (In Russian). — Trav. Jard. Bot. Tiflis *19* : 1917, ref. Maximov 1923.

Negisi, K., Satoo, T.: Soil moisture in relation to apparent photosynthesis and respiration of Akamatu and Sugi seedlings. — J. Jap. For. Soc. *37* : 100—103, 1955.

Neuwirth, G., Polster, H.: Wasserverbrauch und Stoffproduktion der Schwarzpappel und Aspe unter Dürrebelastung. — Arch. Forstwesen *9* : 789—810, 1960.

Pisek, A., Winkler, E.: Die Schliessbewegungen der Stomata bei ökologisch verschiedenen Pflanzentypen in Abhängigkeit vom Wassersättigungszustand der Blätter und vom Licht. — Planta *42* : 253—287, 1953.

Polster, H.: Die physiologischen Grundlagen der Stofferzeugung im Walde. — München, Bayer. Landwirtschaftsverlag 1950.

—, Weise, G., Neuwirth, G.: Ökologische Untersuchungen über den CO_2-Stoffwechsel und Wasserhaushalt einiger Holzarten auf ungarischen Sand- und Alkali-(„Szik") Böden. — Arch. Forstwesen *9* : 949—1014, 1960.

Schneider, G. W., Childers, N. F.: Influence of soil moisture on photosynthesis, respiration and transpiration of apple leaves. — Plant Physiol. *16* : 565—583, 1941.

Stålfelt, M. G.: Spaltöffnungsweite als Assimilationsfaktor. — Planta *23* : 715—759, 1935.

—: Die stomatäre Transpiration und die Physiologie der Spaltöffnungen. — *In*: Ruhland, W.: Encyclopedia of Plant Physiology, Vol. III, p. 351—426, Berlin-Göttingen-Heidelberg 1956.

Stocker, O.: Experimentelle Ökologie der Pflanzen. — Tabulae Biologicae V, p. 510, Junk, Berlin 1929.

Tulaikov, N.: [The water requirements of crop plants in the southerstern Russia as results of the data obtained from field trials.] In Russ. — Bull. Agron. exper. Sta. Saratow *3* : 1—11, 1922, ref. Tumanov 1927.

Tumanow, J. J.: Ungenügende Wasserversorgung und das Welken der Pflanzen als Mittel zur Erhöhung ihrer Dürreresistenz. — Planta *3* : 391—480, 1927.

Discussion

G. F. Makkink: When dealing with production efficiency or the reciprocal of it, transpiration requirements, we must distinguish two points. First, production as a whole over a longer period, secondly, the production rate (photosynthesis) over a short period. The results of Tumanov, Briggs and Schanz, Kiesselbach and many others, as to the transpiration requirement (coefficient) for the harvest as a whole, can be fully explained by combining the rectilinear relationship between CO_2 absorption and radiation, which is a saturation curve. Böhning and Burnside showed that this last curve is almost identical for the important agricultural crops. De Wit has explained that within the lower range of radiation values production efficiency is statistically constant, and within a higher range of radiation it is also true when transpiration is first divided by evaporation. On the other hand, for short periods, it is necessary, as Larcher has shown, to consider gradients of moisture stress, resistances and structural properties of leaves and the layer above it.

B. Slavík: On Fig. 2 the curves of photosynthesis and transpiration run parallel and in comparable numerical quantities close to one another. The reliability of the ratio depends, therefore, very strongly on usually large middle errors of the averages.

W. Larcher: In this case we are concerned with a trend and not with the magnitude of the ratio.

P. G. Jarvis: It has been suggested that the differences in the ratio assimilation: transpiration for *Pinus* and *Picea* which Dr. Larcher has shown, taken from the data of Eidman, are due to chance or error. I would just like to say that we have found similar differences in this ratio for *Pinus* and *Picea* (Jarvis and Jarvis, Physiol. Plant. *16* : 215—235, 1963 Figure II).

THE INFLUENCE OF DECREASING HYDRATION LEVEL ON PHOTOSYNTHETIC RATE IN THE THALLI OF THE HEPATIC *Conocephallum conicum*

B. SLAVÍK

Institute of Experimental Botany, Czechoslovak Academy of Science,
Praha, Czechoslovakia

The effects of the hydration level of the photosynthetizing tissue on the rate of CO_2 assimilation can, in principle, be three:

1. Water stress causing, at supraoptimal W.S.D. [according to Stålfelt's terminology (1955)], hydroactive closure of the stomata, thereby decreasing their diffusibility and as a result of this causing a deterioration in the CO_2 supply of the leaf mesophyll. The relative effect of this closure is greatest when the apertures of the stomata are small.

2. Hydration of the cytoplasmatic ultrastructure in which photosynthetic processes take place, has a direct effect on the enzymatic activity of photosynthesis.

3. Hydration of the cuticle, the epidermal cells and the cell membrane of the intercellular cells decreases their permeability for CO_2, resp. HCO_2 ions and thus again the supply of green cells with CO_2.

Although some authors underestimate (e.g. Pisek and Winkler 1956) or even deny the extrastomatal influencing of the photosynthetic rate by water stress, all three modes of action of water deficit on the assimilation of CO_2 given above, are probable. The question remains as to their proportional contribution to the empirically known and measured decrease in photosynthetic rate with increasing water stress.

In view of the photoactivity of the stomata it is practically impossible to exclude the influence of the state of the stomata when measuring photosynthetic rate. Mathematical elimination by calculating the diffusibility of the stomata from data of their apertures is uncertain. I attempted to circumvent these difficulties in previous work when I demonstrated that different sites of the leaf blade of tobacco even at full water saturation in situ (with zero diffusion pressure deficit) have permanent differences in osmotic potential. It was actually demonstrated that in the basal part of the leaf blade where osmotic potential is higher (lower osmotic pressure of the cell sap), the photosynthetic rate is significantly higher (by 22%) than in the apical part with lower osmotic potential. With full water saturation in situ it would have been expected that the stomata would be fully open at both sites. Although the density and size of the stomata were different at the two sites the relative area of stomata pores was surprisingly similar (it differed by only about 2·5%). This means that although the stomata factor was not obviously different and the water saturation deficit was quite minimal and almost equal, the base and apex of the blade had significant differences in osmotic potential and photosynthetic rate calculated per unit area. The chlorophyll content of the apical part is in fact higher. Thus a higher osmotic potential was statistically significantly associated with a higher photosynthetic rate (Slavík 1963a, 1963b).

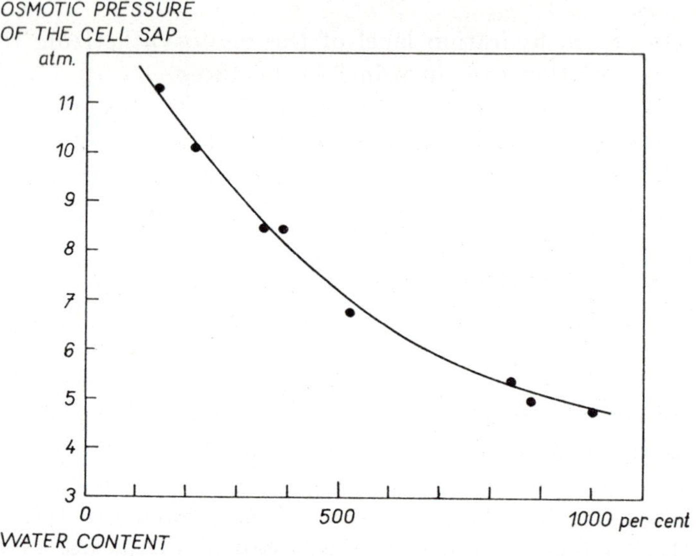

Fig. 1. Relation between water content (abscissa: in per cent of dry weight) and osmotic pressure of the cell sap (ordinate: in atmospheres) in material used.

The influence of the stomatal factor can be reliably eliminated in two further ways: Either by complete hydroactive closure which can only be attained with high values of W.S.D., or by the selection of material without working stomata.

We choose the second alternative in order to investigate changes in photosynthesis rate at low values of W.S.D.

Material and Methods

The material used was hepatic thallus *Conocephallum conicum* (L.) Dum. grown originally in sand cultures (cf. Ensgraber 1954).

The P.R. of parts of young thalli in small plexiglass chambers was measured by means of a differentially measuring infrared CO_2-analyzer (Slavík and Čatský 1963) at $25 \pm 0\cdot5°$ C, $55 \pm 5\%$ relative air humidity, 300 p.p.m. CO_2, illumination $6\cdot8 \times 10^4$ erg.cm.$^{-2}$sec.$^{-1}$ (Tungsraphot B 500 W, 220 V bulbs), air stream of 5 litres per hour per 300 mg. average thallus fresh weight. The determination began with fully water saturated thalli which gradually dehydrated due to transpiration during determination. The transpiration loss was measured gravimetrically every hour during the 8-hour measuring period, followed by 16 hours dark period during which the thalli were placed in an

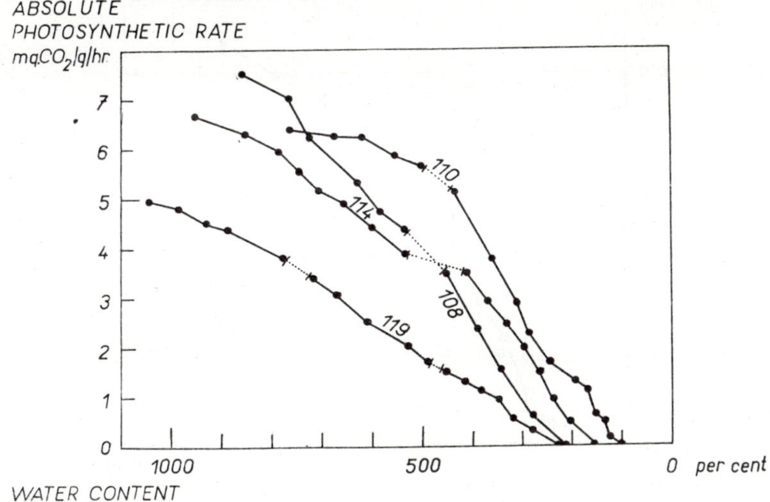

Fig. 2. Examples of the time course of the dependence of absolute photosynthetic rate (ordinate: in mg. CO_2 per g. dry weight per hour) on the water content (abscissa: in per cent of dry weight).

atmosphere almost saturated with water vapour. After 2 or 3 measuring periods the P.R. fell to zero. In similar material the relation between water content and osmotic potential of the cell sap expressed from thalli killed at $100°$ C measured cryoscopically was determined (Fig. 1).

Results and Discussion

Fig. 2 shows some examples of the actual measurements, i.e. the course of the dependence of absolute photosynthetic rate (ordinate) during decreasing water content (abscissa). The interruptions in the lines correspond to periods when the thallus was placed in a closed atmosphere in the dark. On Fig. 3 the statistically elaborated average values of the results of the whole experimental series are given together with the middle errors ($\pm s_{\bar{x}}$). On the abscissa is the water content in percent dry weight, on the ordinate the relative photosynthetic rate expressed in percent of the photosynthetic rate of each sample at 700% water content.

Already from these results we can arrive at several conclusions: First, starting with maximal water saturation of the thallus (about 700% water content) entailed a decrease in photosynthetic rate for each decrease in water content. This dependence is quite steep so that at 350% water content, which

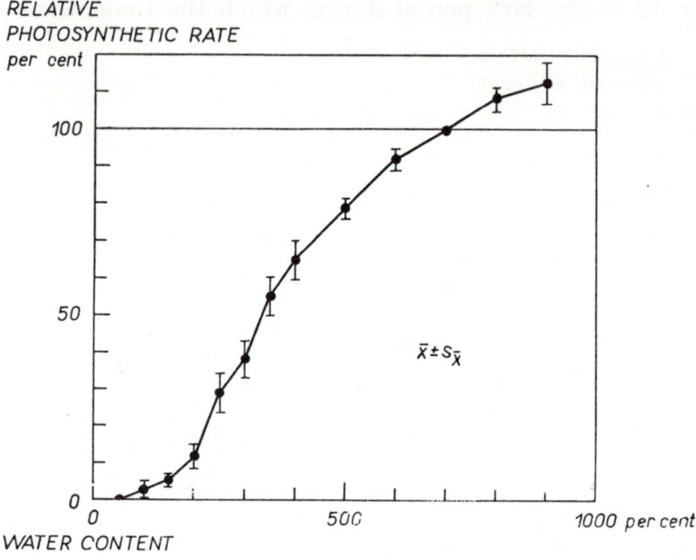

Fig. 3. Dependence of relative photosynthetic rate (i.e. expressed in per cent of the absolute photosynthetic rate [per dry weight] of each sample at 700 per cent water content: ordinate) on the water content (abscissa, in per cent of dry weight). Each point is expressed as mean $\pm$ middle error ($\bar{x} \pm s_{\bar{x}}$).

corresponds to a water saturation deficit of about 50%, it falls to a half.

Of even greater interest are the results of converting water content to osmotic potential according to the relationship found empirically and shown in the Fig. 1. Fig. 4 represents this dependence of relative photosynthetic rate (ordinate) on the osmotic potential of thallus (abscissa). It is also clear here that with the slightest decrease in osmotic potential (i.e. increase in osmotic pressure in cell sap) there is a definite fall in photosynthetic rate. It is interesting that in the range between full saturation, which corresponds to an average osmotic potential of —5 atm., and the osmotic potential of —8·5 atm. a definite linear dependence was found between osmotic potential and photosynthetic rate. Zero values of photosynthetic rate were attained at average osmotic pressure of 12·6 atm., corresponding to a water saturation deficit of about 90 percent.

From the above it follows (1) that in the material used without working stomata, the hydration of the tissue limited the intensity of CO_2 assimilation directly, i.e. without the interaction of stomata (2). This decrease occurred even at very low values of water saturation deficit so that no optimal water saturation deficit was found (3). A linear dependence of photosynthetic rate and osmotic potential of the cell sap was found within the range between full water saturation and a water saturation deficit of about 50 per cent,

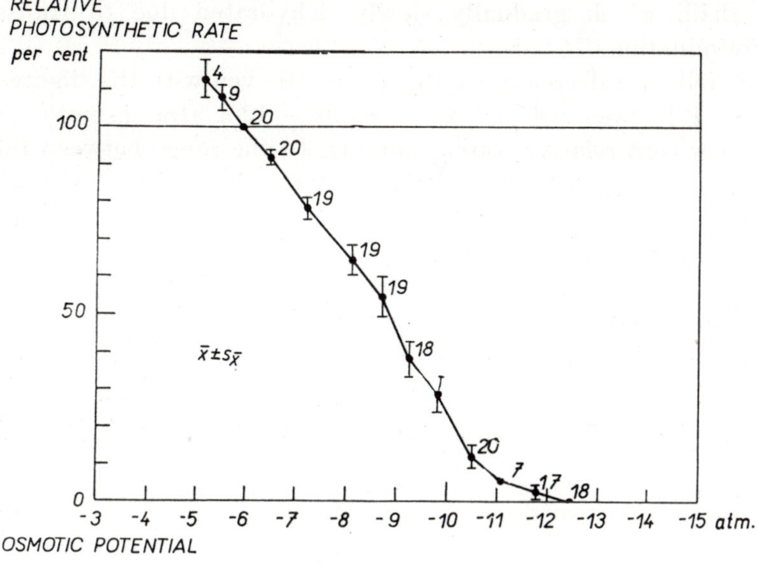

Fig. 4. Dependence of relative photosynthetic rate (ordinate: expressed in the same manner as in Fig. 3) on the osmotic potential of the cell sap (abscissa: in atmospheres). Figures at each mean value represent number of individual measurements.

the regression value being a decrease of approximately 10 per cent of maximal photosynthetic rate for a decrease of 10 per cent in osmotic potential (4). From the points (1) and (2) it also follows that the occurrence of an optimal water deficit with highest photosynthetic rate values is due only to a hydropassive closure of the stomata with full saturation (cf. Brilliant 1924, Alekseev and Gusev 1937, Stålfelt 1935, 1955, 1960, Stocker 1960, Larcher 1960 and others).

So it was confirmed experimentally that the decrease of photosynthetic rate in developing water saturation deficit of the photosynthesizing tissue is influenced essentially not only by the diffusibility of the stomata, i.e. by their hydroactive or hydropassive changes, but also by hydration itself and that, not only at high values of water saturation deficit, but also with low values of W.S.D., as demonstrated in material without active stomata. This influence, as follows from the above regression, is considerable.

Summary

In stomata-free thallus of the hepatic *Conocephallum conicum* the relation between relative water content and osmotic potential of the cell sap respectively and the photosynthetic rate (P.R.) was measured.

The P.R. of parts of young thalli was measured by means of a differentially measuring infrared CO_2-analyzer. The determination began with fully water saturated thalli which gradually slowly dehydrated due to transpiration during determination.

The P.R. fell with decreasing water potential even with the slightest water deficit. Zero P.R. was reached with about —12·6 atm. osmotic potential (about 100 per cent relative water content). In the range between full water saturation (average osmotic potential —5 atm.) and 8·5 atm. (about 350 per cent water content) a clearly linear relationship between osmotic potential and P.R. was found.

From these results it follows that the hydration of the photosynthesizing tissue limited the rate of the CO_2 assimilation directly even without the interaction of stomata, beginning already with lowest water saturation deficits

References

Alekseev, A. M., Gusev, N. A.: [Influence of environmental moisture conditions on water relations, hydration of protoplasma and carbon assimilation.] In Russ. — Uch. Zap. Kazan. Univ. *95* : 7, 1937.

Brilliant, W. A.: La teneur en eau dans les feuilles et l'énergie assimilatrice. — C.R. Acad. Sci. (Paris) *178* : 2122—2124, 1924.

Ensgraber, A.: Über den Einfluss der Austrocknung auf die Assimilation und Atmung von Moosen und Flechten. — Flora *141* : 432—475, 1954.

Larcher, W.: Transpiration and photosynthesis of detached leaves and shoots of *Quercus pubescens* and *Quercus ilex* during desiccation under standard conditions. — Bull. Res. Counc. Israel *8D*: 715—759, 1960.

Pisek, A., Winkler, E.: Wassersättigungsdefizit, Spaltenbewegung und Photosynthese. — Protoplasma *46* : 597—611, 1956.

Slavík, B.: On the problem of the relationship between hydration of leaf tissue and intensity of photosynthesis and respiration. — *In*: Rutter, A. J. and Whitehead, F. H. (ed.). The Water Relations of Plants (Symp. Brit. ecol. Soc. London 1961, ed. Blackwell Sci. Publ., 225—234, Oxford 1963a.

—: The distribution pattern of transpiration rate, water saturation deficit, stomata number and size, photosynthetic and respiration rate in the area of the tobacco leaf blade. — Biol. Plant. *5* : 143—153, 1963b.

—, Čatský, J.: Differentially measuring infrared analyzer with an air-conditioned exposure chamber for photosynthetic rate measurements. — Biol. Plant. *5* : 135—142, 1963.

Stålfelt, M. G.: Die Spaltöffnungsweite als Assimilationsfaktor. — Planta *23* : 715—748, 1935.

—: The stomata as a hydrophotic regulator of the water deficit of plants. — Physiol. Plant.
8 : 572—593, 1955.

—: Das Kohlendioxyd. (Allgemeine Physiologie und Ökologie der Photosynthese. b) Die
Abhängigkeit von äusseren Faktoren. *In*: Pirson, W., ed.: Die CO_2 Assimilation, in: Handb.
d. Pflanzenphysiologie *5/2* : 81—99, 1960.

Stocker, O.: Physiological and morphological changes in plants due to water deficiency. —
Arid Zone Res.*15* : 63—94, UNESCO, Paris, 1960.

Discussion

P. G. Jarvis: Would Dr. Slavík explain what he means by osmotic potential and how is it measured ? Is it determined on the expressed sap ?

B. Slavík: The osmotic potential was measured cryoscopically in the sap expressed from killed (at 100° C) tissue.

J. Úlehla: Have the elastic deformations of the tissues connected with the increasing water saturation deficit been considered ?

B. Slavík: No, they were not measured. Of course, they can only play a role with higher water deficit values.

J. Úlehla: The volume changes accompanying the increasing W.S.D. could perhaps be connected with decrease in light absorption by reduction of surface exposed to light.

B. Slavík: May be. In small W.S.D. values these changes would not be very great.

P. E. Weatherley: Were the changes in osmotic potential entirely attributable to the changes in water content ?

B. Slavík: I suppose so, for in similar experiments no changes in relative contribution of osmotic active substances dissolved in cell sap to the total osmotic potential were observed, during wilting. So presumably no active osmotic regulation took place.

W. E. Müller-Stoll: Presumably the course of the curves presented by the author is somewhat dependent on the calculation base of photosynthetic rate since with prior desiccation the light absorbing surface of the moss samples decreases. In a stage of advanced water loss and shrinking of thalli, in particular, greater differences between the trend of photosynthesis values calculated to surface and calculated to dry matter content are to be expected.

B. Slavík: Prof. Müller-Stoll is right when drawing attention to the possible influence of volume as well as surface changes due to shrinking of the thalli with high water deficits, as mentioned already by Dr. Úlehla. The photosynthetic rates with these extreme water saturation deficits are not taken into account for the argumentation which concerns chiefly the influence of small water deficits.

WATER SATURATION DEFICIT AND PHOTOSYNTHETIC RATE AS RELATED TO LEAF AGE IN THE WILTING PLANT

J. ČATSKÝ

Czechoslovak Academy of Sciences, Institute of Experimental Botany,
Praha, Czechoslovakia

The changes of photosynthetic rate in leaves of different ages in the course of wilting of whole plants have not been hitherto investigated experimentally. The papers demonstrating the dependence of photosynthetic rate on water saturation deficit in leaf tissue (e.g. Scarth and Shaw 1951, Pisek and Winkler 1956) deal with this problem indirectly only. The aim of this work was to determine — in the framework of the broader study of the physiological heterogeneity of plants — the course of water saturation deficit and its corresponding photosynthetic rate in leaves of different ages in the course of permanent wilting of whole plants, induced by increasing soil moisture stress. In addition, the findings can, on the basis of not hitherto used methodical treatment, contribute to an explanation of the dependence of photosynthetic rate on water saturation deficit in higher plants.

Material and Methods

The material used for the experiments was approximately hundred-day-old fodder cabbage plants [*Brassica oleracea* L., convar. *acephala* (DC.) Alef., var. *medullosa* Thell.] grown in pots and slowly wilting due to decrease in soil moisture from water loss by transpiration. Water saturation deficit was determined by the leaf-disk extrapolation method (Čatský 1962, 1965). The photosynthetic rate was measured by serial gasometric determination using a six-unit colorimetric apparatus (Čatský and Slavík 1960, Čatský 1960) under near natural conditions (300 to 350 p.p.m. CO_2, 25 to 28° C). Every leaf measured was irradiated with a light bulb in a water cooled mirror channel. The density of irradiation was $1 \cdot 1 \times 10^5$ erg.cm.$^{-2}$ sec.$^{-1}$ (corresponding to about 20 klx.).

Results and Discussion

Two homogenous groups of ten plants were used for one determination. In one group the photosynthetic rate was determined during wilting in leaves of different age. In the other group the average water saturation deficit was measured in the corresponding leaves.

The results are shown in the graphs. The results of one experiment are given as an example; two further similar experiments gave practically the same results.

As was found previously by the direct determination of water saturation deficit (Čatský 1962), in the presence of a total deficit of water in the plant, the leaves of different ages wilt unevenly. First the oldest leaves wilt and die, then the adult leaves and lastly the youngest. This course of wilting was confirmed also in this experiment (Fig. 1). Here the preference given to young leaves in total water stress in the plant, is clearly evident and also the uneven development of water saturation deficit in leaves of different age.

The differences in the decrease of photosynthetic rate also corresponds to the uneven development of water saturation deficit in leaves of different age (Fig. 2). From both graphs it is clear that during wilting of the whole plant, young leaves preserve photosynthetic activity longer than old leaves, due to their better water supply.

This preference given to young leaves is shown even better in Fig. 3, where the photosynthetic rate is given in relation to the age of the leaves at different stages of wilting. As is known from the literature and confirmed in our experiments (Šesták and Čatský 1962), the photosynthetic rate calculated per unit leaf area changes during the ontogenesis of the leaf: at first it increases and regularly reaches a maximum

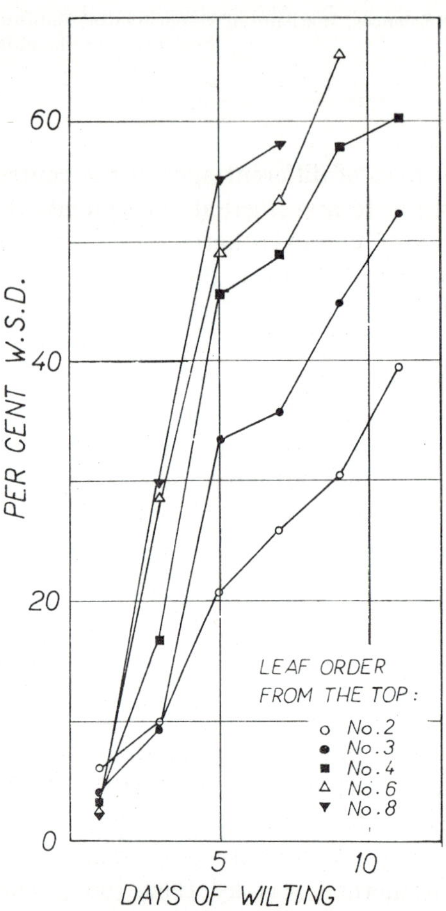

Fig. 1. Development of water saturation deficit in fodder cabbage plants induced by increasing soil moisture stress. Abscissa: days of wilting, corresponding to increasing soil moisture stress, ordinate: water saturation deficit in per cent.

before the attaining of maximal leaf size, then on ageing of the leaf it falls to zero. These changes in photosynthetic rate in ontogenesis are naturally expressed in the simultaneous determination of photosynthetic rate in leaves of different ages from the same plant, where maximum photosynthesis is found in maturing leaves. This is valid only when the plant is well supplied with water. Fig. 3 shows such curves, in a saturated plant, and in a plant

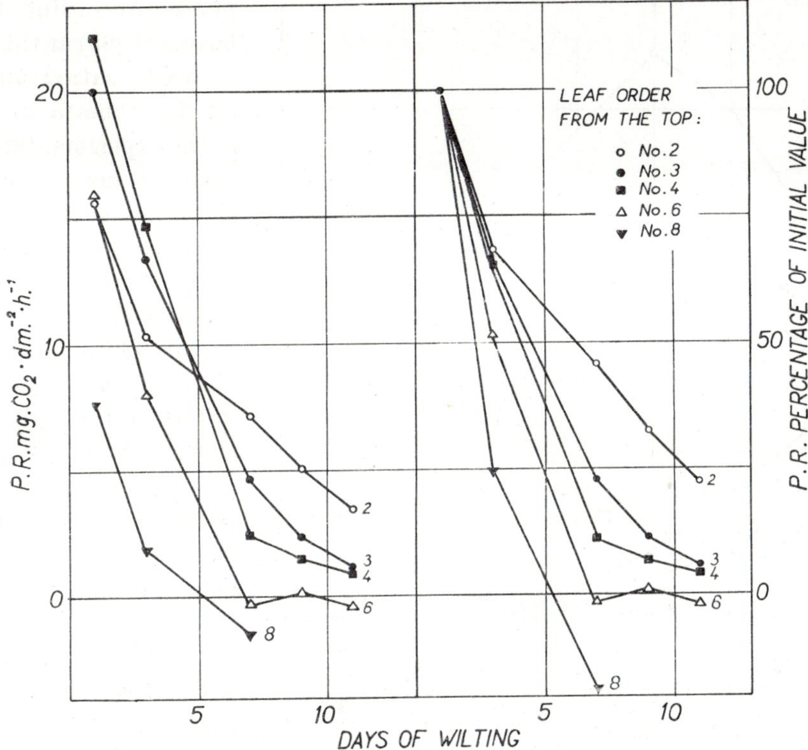

Fig. 2. Decrease in photosynthetic rate during wilting of potted fodder cabbage plants. Abscissa: days of wilting, corresponding to increasing soil moisture stress. Ordinate: photosynthetic rate in mg CO_2 . dm.$^{-2}$ h.$^{-1}$ (for left graph), and in relative values in per cent of initial values (for right graph).

with an average water saturation deficit of about fifteen per cent. On further wilting of the plant the form of the curve clearly changes and again the preference given to the youngest leaves is clearly evident.

The graphic expression of the dependence of photosynthetic rate on water saturation deficit may be of help in explaining the question of the extent to which stomata control contributes to the fall in photosynthetic rate in wilting and the extent to which hydration of the protoplasm and other factors resp. are involved. Fig. 4 shows these relationships. As compared with Solarová's

(1965) data, the first, rising part of the curve corresponds to hydropassive closure of the stomata. Maximal photosynthetic rate was found at about 5 per cent water saturation deficit. Then, the rapid fall in photosynthetic rate with increasing water stress is caused primarily by the hydroactive closure of the stomata. This fall in photosynthetic rate caused by the stomata, ends roughly at 20 per cent water saturation deficit, where the curve passes to a practically linear phase. According to Solarová (1965) at this value of water saturation deficit the stomata are closed in the experimental plants used by us. In view of this, we must consider the residual photosynthetic rate to be cuticular photosynthesis. Its rate at this point amounts to about 25 per cent of the initial total photosynthetic rate which, according to the data of Solarová, also corresponds to the ratio of total and cuticular transpiration in this material. The further fall in photosynthetic rate with increasing water saturation deficit is no longer dependent on the stomata and is caused by decreasing hydration of the assimilating tissue of the leaves only (comp.

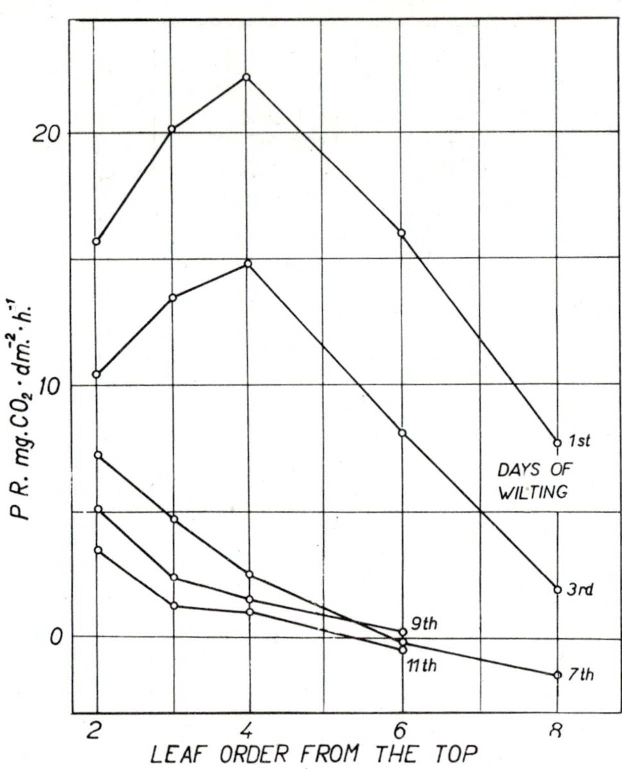

Fig. 3. Photosynthetic rate as related to leaf age during wilting of fodder cabbage plants. Abscissa: leaf order from the top. Ordinate: photosynthetic rate in mg. $CO_2 . dm.^{-2} h.^{-1}$.

Slavík 1965), and other factors respectively. Since the experiments presented here were made for a different purpose, the factors playing an important role in the dependence of photosynthetic rate on water saturation deficit (stomata aperture, hydration) were not investigated. Nor' was the diffusion resistance of mesophyll for carbon dioxide according to Gaastra (1959, 1962) taken into account.

It is an interesting fact that differences among leaves of different age cannot be observed in the general course of the curve. Only points denoting the oldest leaves, to a considerable extent due to an obvious decrease in chlorophyll content (Šesták and Čatský 1962).

Maximal photosynthetic rate irrespective of the age of the leaves was observed in a very narrow range of water saturation deficit so that in this

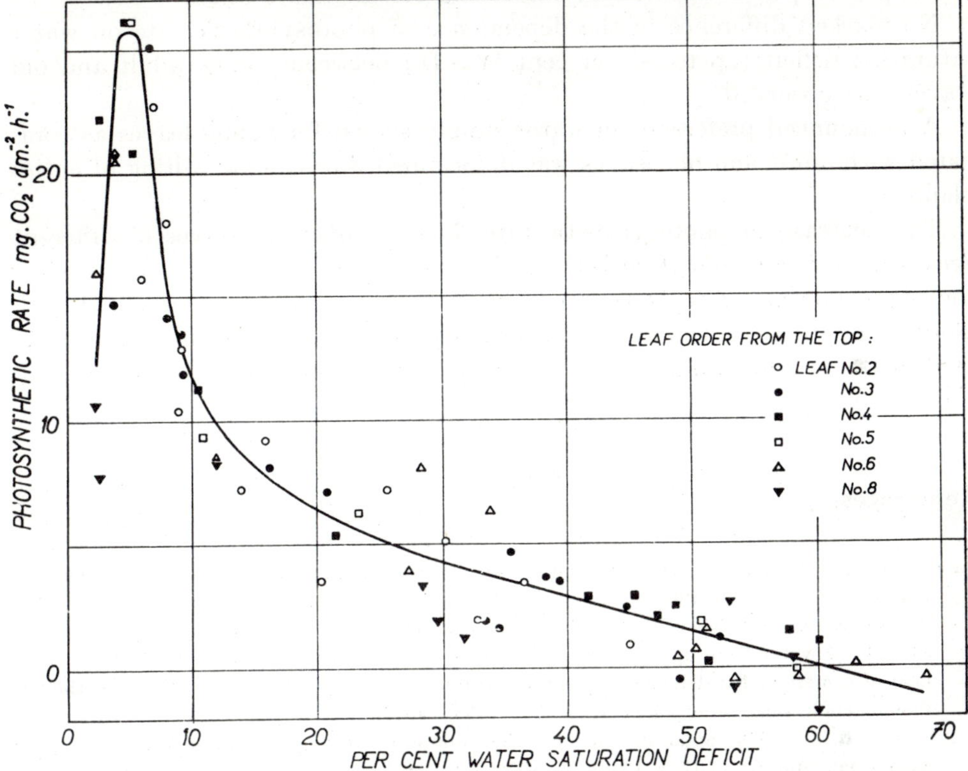

Fig. 4. The dependence of photosynthetic rate on water saturation deficit of leaf tissue in fodder cabbage. Abscissa: water saturation deficit in per cent, ordinate: photosynthetic rate in mg. CO_2 . dm.$^{-2}$ h.$^{-1}$.

range it was not possible to detect significant differences in the sensitivity of leaves of different age to water deficit. On the other hand, these differences were found in the same material by Solarová (1965) in the reaction of stomata to water stress. It is possible that this discrepancy arises because different factors contribute to the above dependence of photosynthetic rate on water saturation deficit, these factors being in different way dependent on the leaf age and changes in water balance.

Summary

The changes in water saturation deficit and photosynthetic rate in fodder cabbage leaves of different ages were measured during permanent wilting of whole potted plants induced by increasing soil moisture stress.

No marked difference in the dependence of photosynthetic rate on water saturation deficit (up to 60 per cent W.S.D.) between young, adult and old leaves was observed.

A pronounced preference in water supply shown for young leaves as compared with adult and old leaves was determined during slow wilting of entire plants.

The decrease in photosynthetic rate during wilting in leaves of different ages is in basic agreement with the increase of water saturation deficit values: the photosynthetic rate of young leaves falls more slowly as compared with adult and old leaves, the maximum photosynthetic rate being in water saturated plants in adult (middle) leaves, and in wilted plants in the youngest leaf.

References

Čatský, J.: Zur Frage der pH-Bestimmung bei colorimetrischen Assimilationsmessungen. — Planta *55* : 381—389, 1960.

—: Water saturation deficit in the wilting plant. The preference of young leaves and the translocation of water from old into young leaves. — Biol. Plant. *4* : 306—314, 1962.

—: Leaf-disc method for determining water saturation deficit. — Internatl. Symp. Methodol. Plant Ecophysiol., Montpellier 1962 : 351—358, 1965.

—, Slavík, B.: [A field apparatus for the determination of photosynthetic rate.] In Russ. with Engl. Summ. — Biol. Plant. *2* : 107—112, 1960.

Gaastra, P.: Photosynthesis of crop plants as influenced by light, carbon dioxide, temperature, and stomatal diffusion resistance. — Med. Landbouwhogesch. Wageningen *59* : 1—68, 1959.

—: Photosynthesis of leaves and field crops. — Neth. J. agric. Sci. *10* : 311—324, 1962.

Pisek, A., Winkler, E.: Wassersättigungsdefizit, Spaltenbewegung und Photosynthese. — Protoplasma *46* : 597—611, 1956.

Scarth, G. W., Shaw, M.: Stomatal movement and photosynthesis in *Pelargonium*. II. Effects of water deficit and of chloroform: photosynthesis in guard cells. — Plant Physiol. *26* : 581—597, 1951.

Šesták, Z., Čatský, J.: Intensity of photosynthesis and chlorophyll content as related to leaf age in *Nicotiana sanderae* hort. — Biol. Plant. *4* : 131—140, 1962.

Slavík, B.: The influence of decreasing hydration level on photosynthetic rate in the thalli of the hepatic *Conocephallum conicum*. — In: B. Slavík (ed.): Water Stress in Plants. Proc. Symp. Praha *1963* : 195—202, Praha 1965.

Solarová, J.: Stomata reactivity in the leaves of different insertion level during the wilting. — In: B. Slavík (ed.): Water Stress in Plants. Proc. Symp. Praha *1963* : 147—154, Praha 1965.

Discussion

P. G. Jarvis: Dr. Čatský's results are the only ones I know of which show net photosynthesis occurring by carbon dioxide uptake through the cuticle when the stomata are closed. This must result from a very low cuticular diffusion resistance in fodder cabbage and this resistance might be expected to be different in leaves of different ages. Has Dr. Čatský any information on how the cuticular diffusion resistance in fodder cabbage compares with other species or varies with leaf age ?

J. Čatský: No, I have not.

H. Polster: The results show that in comparing the investigated species and varieties it is necessary to respect the character of leaf.insertion. How do you explain the great increase in photosynthetic rate in the experimental plants on the increase in water saturation deficit from 0 to 5 percent ?

J. Čatský: In our experiments the great increase in photosynthetic rate at the lowest values of W.S.D. corresponded basically with the hydropassive opening of the stomata at these W.S.D. values, as determined in the same material by Solarová. The above increase could be explained by these changes in the stomata opening. However, in view of Gaastra's findings on the significance of diffusibility of the mesophyll with limitation of photosynthetic rate, this explanation cannot be considered to be quite sufficient.

G. V. Lebedev: I wish to draw the attention of workers doing research on the dependence of the photosynthesis of plants on water relations, to the results of Prof. Belikov who showed that, during the wilting of plants, photosynthesis is maintained for a certain time at a relatively equal level and only later does its value rapidly fall. The watering of wilting plants leads to a rapid increase in photosynthetic rate even when the water balance is still not regained.

J. Čatský: We know the work of Prof. Belikov well. In evaluating these results, obtained with whole plants, it is necessary to remember, that 1. leaves of different age wilt unevenly in the presence of total water deficit, starting from the oldest, 2. when suddenly increasing the soil moisture content, saturation usually occurs very rapidly, particularly in young leaves and again unequally in leaves of different age, and 3. the fall in photosynthetic rate with increasing W.S.D. usually takes place in three phases (increase, rapid fall, slow fall) as I demonstrated in the last figure in my contribution. Changes in photosynthetic rate of the whole plant with increasing soil moisture stress are thus the result of the interaction of all these factors. The results of Prof. Belikov can easily be explained on this basis and the conclusions are not in conflict with our findings.

RELATIONSHIP BETWEEN CHLOROPHYLL CONTENT AND PHOTOSYNTHETIC RATE DURING THE VEGETATION SEASON IN MAIZE GROWN AT DIFFERENT CONSTANT SOIL WATER LEVELS

Z. ŠESTÁK and J. VÁCLAVÍK

Institute of Experimental Botany, Czechoslovak Academy of Sciences,
Praha, Czechoslovakia

From the sparse literature on the subject, the isolated facts are known that the soil water level is reflected in the photosynthetic rate (Ashton 1956, Belikov 1960, Hagan et al. 1957, Kursanov et al. 1933, Mikhaleva 1956, Shchukina 1959, Simonis 1941, Terentev and Rakovskiy 1961) or in the chlorophyll content (Buzover 1959, Hübner 1949, Onishchenko 1960) of the leaves. Changes in the relationship between the chlorophyll content and the photosynthetic rate in the ontogenesis of the leaves and plants grown in soil with different constant water levels have, however, evidently not been studied hitherto. This was, therefore, investigated in our experiments with maize grown in soil, whose water level was maintained throughout the experiment at different constant soil water levels.

Material and Methods

Maize (*Zea mays* L., var. Stupická raná) plants were grown in Mitscherlich vegetation pots in clay soil from humous horizon of eluated chernozem. The soil moisture was maintained constant at 90, 60 or 30% maximum capillary capacity throughout the whole vegetation season. By means of weighing (3 times in a week) the calculated quantities of water were injected irregularly into the soil horizon. The soil surface was covered with a disk of gray PVC tilt.

On the 38th day after sowing (June 10, 1963) and then twice at 14 day intervals and once after the interval of 28 days, the chlorophyll content and the photosynthetic rate at each leaf insertion level of all three variants were determined.

The p h o t o s y n t h e t i c r a t e was measured by the increase in dry weight in isolated disks of leaf tissue (area 0·5 cm.²) exposed for 6 hr. under constant conditions of illumination, i.e. about 30,000 lx, at temperature $25° \pm 0\cdot2°$ C, with normal air CO_2 concentration (0·03 vol.%) and at full water saturation. The principle of the method was described by Bartoš et al. (1960), the apparatus being used in our experiments is demonstrated in Šesták et al. (1965).

C h l o r o p h y l l $(a + b)$ content was determined colorimetrically in acetone extracts and was calculated from calibration curves according to crystallized chlorophyll standards (Šesták and Ullmann 1964).

All data were calculated per unit leaf area. Every value of photosynthetic rate represents the mean from 3 parallel samples, each of them being formed by 12 disks taken from 4 leaves of the same age. Values for chlorophyll content were determined in an adequate quantity of plant tissue.

Results and Discussion

Interesting differences between the 90 and 60% maximum capillary capacity variants representing, according to the literature, increased and optimum soil water level, were found (Fig. 1, 2 and 3). The results of the 30% variant were not included in this comparison. The effect of soil moisture stress in the variant with 30% soil moisture was, in fact, clearly evident not only in photosynthetic characteristics but also in the inhibition of growth. The development of these plants was retarded throughout the greater part of ontogenesis and it is, therefore, not possible to make a simple comparison between the results attained with the 30% variant and those of the 90 and 60% variants.

Fig. 1 gives the relationship between the chlorophyll content and photosynthetic rate in single leaves, determined at the given time intervals. The chlorophyll content and photosynthetic rate decreased with ageing of the plant. In the phase of ear formation (4th set of determinations) respiration already exceeded photosynthesis. In the first two sets of determinations, the photosynthetic rate was obviously dependent on the amount of chlorophylls. The chlorophylls in young leaves were photosynthetically more active, in older leaves the proportion of nonactive forms of chlorophylls evidently increases. This significant change in the activity of the chlorophylls occurs before the maximum chlorophyll concentration per unit leaf area is attained and before the leaf blade reaches maximum area. Before the onset of the steep fall in the activity of the chlorophylls, the leaf attains its maximum photosynthetic production per unit area, i.e. reaches the stage of "photosynthetic maturity". With ageing of the whole plant, the maximum attainable chlorophyll concentration and maximum photosynthetic rate reached under the same conditions, decline. The degree of dependence of photosynthetic rate on the amount of chlorophylls also gradually falls and in the third set of determinations, at the beginning of the formation of the male inflorescence, is already very low.

Fig. 2 shows the photosynthetic rate, the amount of chlorophylls and the thickness of the leaf blade (expressed as dry weight per unit area of leaf) in leaves of different age, determined at each of the given time intervals. The samples show an increase in these indices up to maximum values attained in

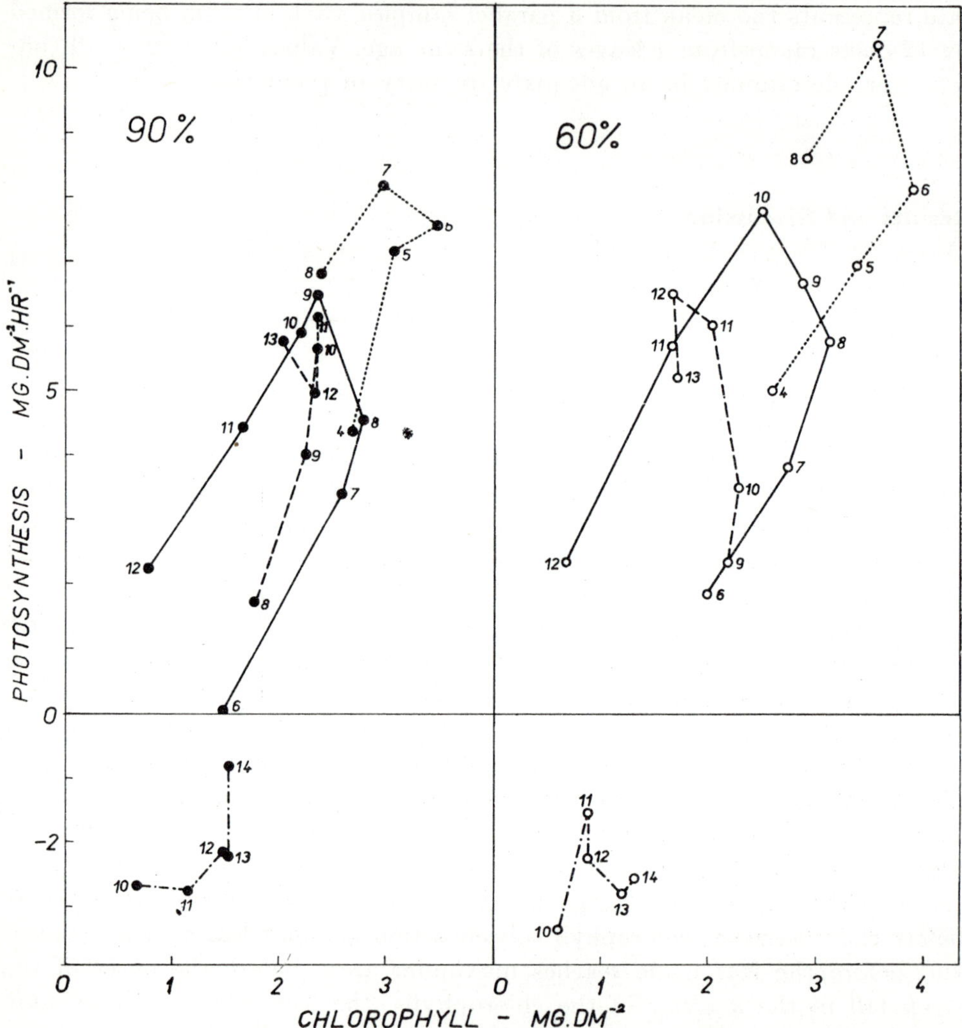

Fig. 1. Relationship between photosynthetic rate (ordinate) and chlorophyll ($a + b$) content (abscissa) in leaves of different age (Nos. 4 to 14 — numbered from the base of the stem) during the vegetation season ($\cdots$ 38th day after sowing, ———— 52nd day, — — — 66th day, — · — · — 94th day). Plants were grown at 90% (●) and 60% ○) soil moisture.

relatively young leaves, with a subsequent fall. In the first two sets of determinations, in plants from variants with 60% soil water level these young leaves are thicker, have more chlorophyll and more intense assimilation than

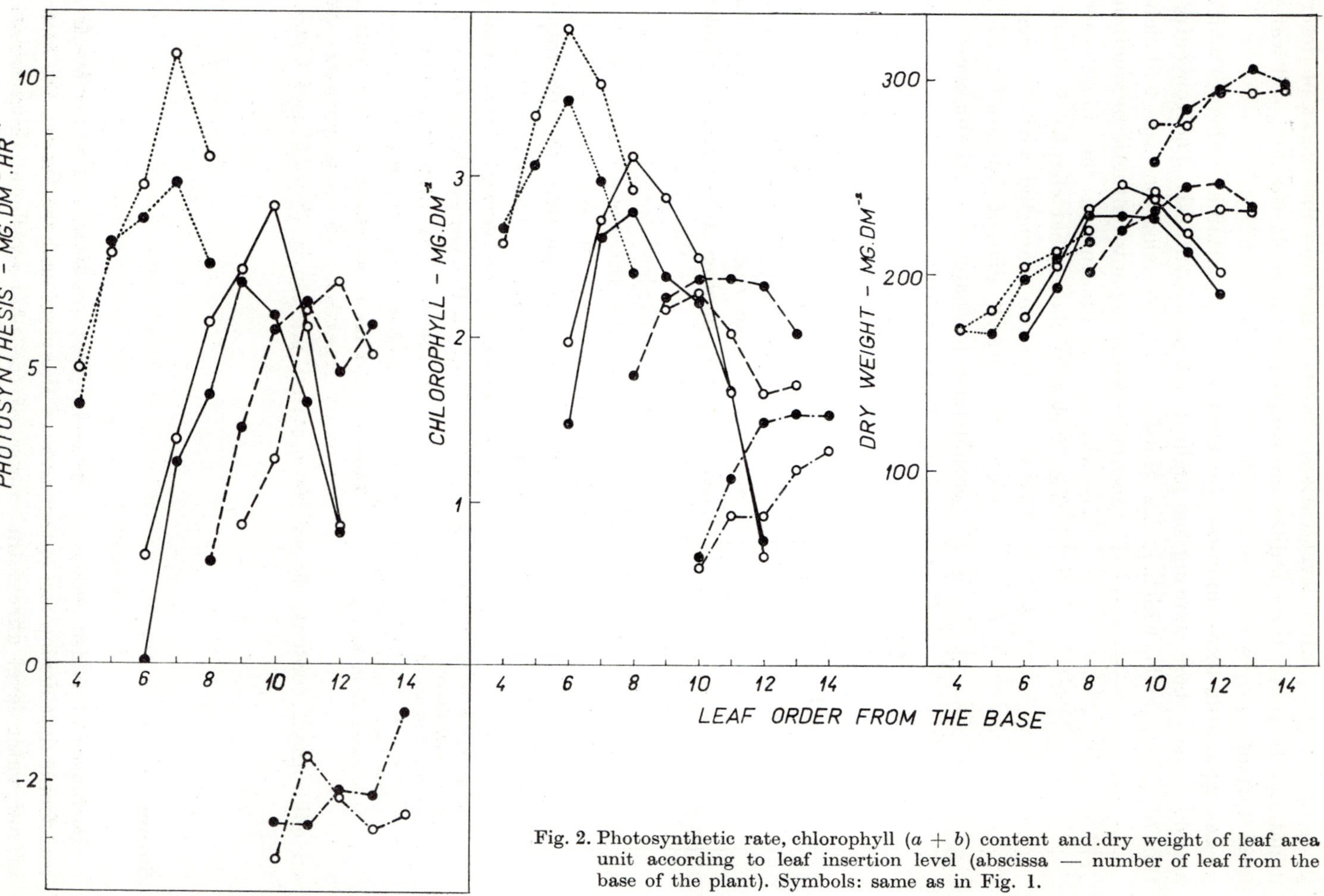

Fig. 2. Photosynthetic rate, chlorophyll (a + b) content and dry weight of leaf area unit according to leaf insertion level (abscissa — number of leaf from the base of the plant). Symbols: same as in Fig. 1.

those of the 90% water level variant. In the third and fourth set of determinations, however, these indices are somewhat higher in the 90% soil water level variant.

The maximum total chlorophyll content is always attained in leaves which are older than those showing the highest net photosynthesis. The concentration of chlorophylls, which in the third set of determinations was sufficient to ensure positive photosynthetic production, is already insufficient in the fourth set. Fig. 2 clearly shows that changes in photosynthetic rate are connected with the changes of chlorophyll amount, but are inversely proportional to changes in leaf thickness.

Fig. 3 demonstrates the above described relationship with the aid of the so-called "assimilation number", which is the photosynthetic rate divided by the chlorophyll content. We have already pointed out in previous work (Šesták 1963) that this assimilation number cannot be an index of the dependence of photosynthetic rate on chlorophyll content, since this linear relationship begins to be evident, especially in older leaves, only when there is a certain positive amount of chlorophylls per unit leaf area. Fig. 3 clearly shows that the photosynthetically most active chlorophyll forms are always present in the younger leaves and in all cases only in the variant with 60% soil water level. The smallest range of assimilation numbers was found in the first set of determinations. This is evidence of the highest degree of dependence of photosynthetic rate on chlorophyll content, as we have stated previously (Šesták and Čatský 1962).

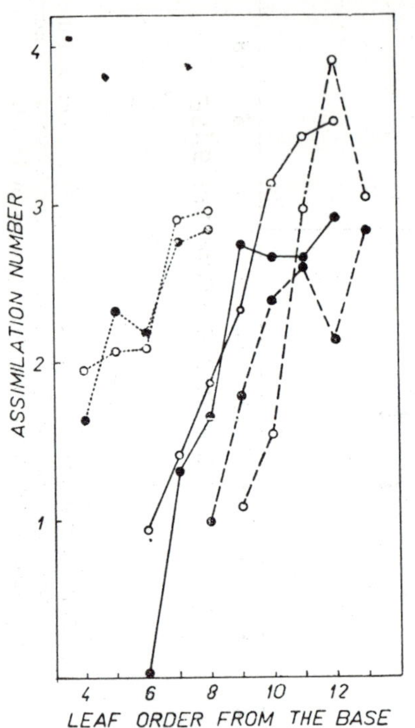

Fig. 3. Assimilation numbers (ordinate — calculated from the dry weight increase of experimental disks) according to leaf insertion level (abscissa). Symbols: same as in Fig. 1.

Summary

In plants of maize grown at three different constant soil water levels it was found that during the ontogenesis of the plant and the leaves, the factor of soil water level affects the course of the photosynthetic characteristics (photosynthetic rate measured in water-saturated disks of leaf tissue), chlorophyll content and thickness of the leaves. The inadequate supply of the plant with water (30% soil moisture) makes itself felt markedly throughout the whole course of the vegetation period. Obvious differences between the variants with 90 and 60% soil water level were found particularly in chlorophyll content and the photosynthetic rate of young leaves: in plants aged 38 and 52 days, 60% soil water level appeared to be more favourable, while later 90% soil water level was best. The photosynthetic rate changes in relation to chlorophyll content; leaves which are formed later attain a lower maximum chlorophyll concentration and a lower maximum photosynthetic rate. The chlorophylls are relatively most active in young leaves, and always in plants grown at 60% soil water level. A decrease in the photosynthetic activity of the chlorophylls occurs in the period after the leaf has attained "photosynthetic maturity" and before it reaches maximum chlorophyll concentration per unit area of the leaf blade. The closest relationship between the chlorophyll content and the photosynthetic rate was found in 38-day-old plants. The photosynthetic rate changed inversely with changes in the thickness of the leaf blade.

References

Ashton, F. M.: Effects of a series of cycles of alternating low and high soil water contents on the rate of apparent photosynthesis in sugar cane. — Plant Physiol. *31* : 266—274, 1956.

Bartoš, J., Kubín, Š., Šetlík, I.: Dry weight increase of leaf disks as a measure of photosynthesis. — Biol. Plant. *2* : 201—215, 1960.

Bclikov, P. S.: [Regulation of photosynthetic rate by the plant organism.] In Russ. — Dokl. Mosk. selsko-khoz. Akad. im. Timiryazeva *57* : 5—26, 1960.

Buzover, F. Ya.: [Chlorophyll amount and photosynthetic rate in potato.] In Russ. — *In*: Problemy fotosinteza, pp. 646—652. Izdat. Akad. Nauk SSSR, Moskva 1959.

Hagan, R. M., Peterson, M. L., Upchurch, R. P., Jones, L. G.: Relationships of soil moisture stress to different aspects of growth in Ladino clover. — Soil Sci. Soc. Amer. Proc. *21* : 360—365, 1957.

Hübner, R.: Untersuchungen über Chlorophyllgehalt und Werteigenschaften bei Futterpflanzen in Abhängigkeit von der Wasserversorgung und Belichtung und der Nährstoffversorgung im Feldversuch. — Ztschr. Acker- u. Pflanzenbau *91* : 374—414, 1949.

Kursanov, A., Blagoveshchenskiy, V., Kazakova, M.: [Effect of soil moisture on physiological processes and chemical composition of sugar beet.] In Russ. — Bull. Soc. Nat. Moscou, S. Biol. *42* : 171—198, 1933.

Mikhaleva, E. N.: [Characteristics of relationship between gas exchange and dry matter accumulation in plants grown in different conditions.] In Russ. — Eksper. Bot. *11* : 116—164, 1956.

Onishchenko, L. I.: [Influence of soil moisture on the number of chloroplasts and chlorophyll accumulation in sugar beet leaves.] In Russ. — Belocerk. selsko-khoz. Inst. Nauch. Zapis. *10* : 43—49, 1960.

Šesták, Z.: On the question of the quantitative relation between the amount of chlorophyll, its forms, and the photosynthetic rate. — Colloque Internat. C.N.R.S. No. 119 — La Photosynthèse, pp. 343—356. Paris 1963.

—, Čatský, J.: Intensity of photosynthesis and chlorophyll content as related to leaf age in Nicotiana sanderae hort. — Biol. Plant. *4* : 131—140, 1962.

—, Ullmann, J.: Srovnání metod stanovení chlorofylů. I. Spektrofotometrické a kolorimetrické metody. [A comparison of methods of chlorophyll determination. I. Spectrophotometric and colorimetric methods.] — Rostlinná Výroba *10* : 1197—1206, 1964.

—, Čatský, J. et coll.: Metody studia fotosyntetické produkce rostlin. [Methods of the study of plant photosynthetic production.] — Publ. House Czechoslovak Acad. Sci., Praha 1965. (In press.)

Shchukina, A. I.: [Mineral nutrition and water relations as factors determining photosynthetic productivity of maize leaves.] In Russ. — Uchen. Zap. Kuybishevsk. Gos. Pedag. Inst., Estestvozn. *22* : 93—104, 1959.

Simonis, W.: Über den Einfluss des Bodenwassergehaltes auf die Kohlensäureassimilation und den Ertrag bei Einjährigen. — Ber. dtsch. bot. Ges. *59* : 52—68, 1941.

Terentev, V. M., Rakovskiy, V. D.: [On the effect of peat soil on carbon dioxide assimilation by maize plants.] In Russ. — Dokl. Akad. Nauk BSSR *5* : 401—404, 1961.

Discussion

P. G. Jarvis: Transpiration continuously removes water from the soil. When this water is replaced by the usual methods part of the soil becomes wet while the remainder remains dry. The water content of the soil is then no longer homogenous. I would like to ask how Dr. Šesták manages to grow plants at constant soil water content ?

Z. Šesták: Water loss was determined by regular weighing of cultivation pots. It was not made up by pouring water onto the soil surface, but by injecting water into various soil levels. In this way the moisture level was maintained throughout the experiment with only very slight variations. Evaporation from the soil surface was restricted by covering it with a layer of PVC.

W. C. Visser: With respect to the shape of the relation between soil moisture stress ψ and plant response, it may be of importance to mention that a theoretical consideration shows that the simplest relation is obtained by plotting the plant response against $1/\psi$. If plotted against the moisture content V the same shape is obtained by plotting against W^m with m of the order of 3. By plotting against ψ or V, the quantitative relation remains the same but the shape of the curve changes in such a way that the relation is far more difficult to value or to explain.

W. R. Müller-Stoll: Numerous investigations have proved that the chlorophyll content in leaves in general is not a limiting factor to photosynthetic rate with normal CO_2-content of the air. Aurea-varieties of wood species showed, inspite of a greatly reduced chlorophyll content, no or only slight decrease in photosynthetic activity as compared with normal green forms. The example of sun and shadow leaves also shows that in conditions of light saturation the chlorophyll content does not limit the photosynthetic rate. Shadow leaves with a higher amount of chlorophyll exhibit reduced photosynthetic rates, sun leaves behave reversely. The results presented by Dr. Šesták need not necessarily force the conclusion of a direct connection between chlorophyll content and photosynthetic rate. The decrease in both quantities may, inspite of the simultaneous trend, have its cause within quite different ranges of metabolism.

Z. Šesták: We do not think that it is possible to refer photosynthetic rate to the t o t a l content of chlorophylls in the material used. In recent years the experiments of American workers at the Carnegie Institution in Washington and the Soviet school of Prof. Krasnovskiy, have shown that there are from two to four forms of chlorophylls, particularly of chlorophyll *a*. These forms are created by different links of chlorophyll with lipo-protein complexes and differ in vivo in the spectral properties. On extraction of the chlorophylls with organic solvent a uniform solution is obtained, however, in which the differences between the forms disappear. Forms of chlorophyll in vivo differ in their photosynthetic activity: perhaps they are active and non-active forms, perhaps active and reserve. It is more probable, however, that the individual forms take part differently in various stages of the photosynthetic reaction, which today are not usually divided into light and dark stages (v. theory of two photochemical systems in photosynthesis — Duysens et al., Nature *190* : 510—511, 1961). The contribution of the different forms of chlorophyll to the total balance of photosynthesis expressed as dry matter formation or gas exchange, is thus obviously unequal. For simplification we can say that the photosynthetic rate is directly dependent on the content of one, the active form, of chlorophyll *a*. It is obviously not correct, however, that there is no relationship between chlorophyll content and photosynthetic rate. The opinion is conditioned by our hitherto poor possibilities of determining the relative content of the different forms of chlorophyll in vivo. In American experiments, where they used a derivative spectrophotometer specially constructed for the purpose, noteworthy results were obtained which indicate the direction for further research. At the present time we do not have access to such an apparatus. However, in all or most of our simple experiments, when we have worked with plants of different age or with leaves of different age or by affecting plants under different conditions (Šesták and Čatský, Biol. Plant. *4* : 131—140, 1962; Šesták, Photochem. and Photobiol. *2* : 101—110, 1963; Šesták, Coll. Intern. C.N.R.S., No. 119, p. 343—356, 1963), we have found a linear relation between chlorophyll concentration and photosynthesis, concordant, of course, only for a given experimental series. It can be seen on Fig. 1. of this paper, where we show two relationships, one for young and the other for old leaves. Since, however, we determine net photosynthesis only from a certain positive amount of chlorophylls $(a + b)$ we must be careful in expressing the relationship especially insofar as it concerns assimilation number. There is, of course, always a danger in comparing normal plants with aurea-varieties, since genetically determined mutations can differ basically in the ratio of individual forms of chlorophylls. This is indicated, for example, by the very specific reactivity of chlorophylls in aurea-varieties to light (v. Sagromsky, Z. Naturforsch. *11b* : 548—560, 1956). There are obviously similar differences in the ratios of the forms of chlorophyll in sun and shadow leaves; the discordance between chlorophyll content and photosynthetic rate in sun and shadow leaves in some publications is due to not respecting the differences between young and old leaves and to measuring at unsuitable intensities of illumination.

W. R. Müller-Stoll: Evidence of purely statistical correlation between two measured quantities does not permit the presumption of a direct relation within the metabolic processes, without further assurance. This ought to be supported by additional proofs. Chlorophyll content is merely a part of a very complicated internal mechanism. It seems to be more likely that the depression in photosynthetic activity with increasing age of leaves as found by Dr. Šesták might have its cause in the range of photosynthetic dark-reaction, and that at the some time chlorophyll content is reduced without this fact being responsible for the rate of photosynthesis (two phenomena of physiological ageing probably independent of each other).

Z. Šesták: I would like to emphasize that in our experiments we worked with leaves and tissues saturated with water and with quite constant conditions of exposure in the apparatus. It is thus natural that we attribute the difference in photosynthetic rate that was found to differences in the content of chlorophylls, with which they have a concordant character. The correlation of photosynthetic rate with other characteristics, e.g. leaf thickness, was always lower. It cannot be excluded that the photosynthetic rate may be connected with some other factor which we have not established and which bears a constant relation to the determined chlorophyll content, e.g. with the concentration of some enzyme (e.g. Smillie, Plant Physiol. *37* : 716—721, 1962) — this we do not know. Since, however, chlorophyll is the absorber and transmitter of solar energy it has without doubt an important role in the mechanism of photosynthesis, as clearly shown in numerous works in recent years and it is, therefore, not illogical that we place it in connection with the total balance of photosynthesis.

WATER SATURATION DEFICIT OF INTENSELY PHOTOSYNTHESIZING LEAF TISSUE WELL SUPPLIED WITH WATER

L. NÁTR

Cereals Research Institute, Kroměříž, Czechoslovakia

Bartoš, Kubín and Šetlík (1960) worked out a method for determining the photosynthesis rate of leaf tissue using disks and Čatský. (1960) used this method to determine the water saturation deficit of leaves. We modified the method for the leaves of cereals, in the same way as Rychnovská and Bartoš (1962) for wild grasses, by using leaf segments in place of disks and placing them on a special frame (Nátr and Špidla 1961). Two segments 2 cm. long are cut out from the leaves. One is placed on the special frame so that its cut edge is in contact with a moistened spongy polyurethane strip. The frame with segments in position is placed in an assimilation chamber with constant illumination, air temperature and CO_2 content. The other segment is used to determine dry weight at the beginning of the experiment. The photosynthetic rate can then be calculated from the increase in dry weight of the first segment. Since conditions can be regulated in the assimilation chamber to near the optimum for photosynthesis, values of increase in dry weight of up to 18 mg.dm.$^{-2}$hour^{-1} are obtained, which means that the dry weight of the segment increases up to 50% in 10—12 hours of exposure. These values are obtained with an intensity of illumination of about $1 \cdot 5 \times 10^5$ erg.cm.$^{-2}$ sec.$^{-1}$, $0 \cdot 1$ vol.$\%$ CO_2, air temperature $25°$ C, relative air humidity 55%, air flow 200 l.min.$^{-1}$, total area of segments maximally 2 dm.^{2}.

Since the length of the segment is only about 2 cm. it can be assumed that there is a good degree of water saturation of the leaf tissue. On the other hand, however, conditions for intense transpiration are present in the assimilation chamber: intense illumination, relatively high temperature and rapid air flow. In addition, the position of the segment on the stand permits water loss by transpiration from both the upper and under surface of the leaf segment. For these reasons it cannot be excluded that the leaf segments may have a definite, although not very high, water saturation deficit in these favourable conditions for photosynthesis and transpiration.

In the first series of experiments to measure water saturation deficit in the leaf segment, the leaf tissues of mature plants grown under field conditions, were used. In this series it was found that the fresh weight of the segments during their 8-hour exposure at an illumination of 10,000—40,000 lux was greater than after a subsequent 18-hour exposure of the segments in the dark and in an atmosphere saturated with water vapour. In evaluating the results, changes in dry weight due to the accumulation of assimilates in the light and to respiration in the dark, were obviously taken into consideration.

In view of these results a series of measurements was later carried out on the leaves of young plants.

Material

Spring barley plants were grown in sand cultures with fluorescent tube illumination. Measurements were made on 10 to 20 day-old plants. After cutting the leaves they were saturated with water for two hours in the dark. Segments were then made from them, placed on the stand and put into the assimilation chamber. The fresh weight of the segments was then determined on a torsion balance at one- or 2-hour intervals for 5—8 hours of exposure. After exposure the segments in place on the stand were put into the dark for 20 hours to determine water deficit in the usual way. In some series the segments were made from slightly wilting leaves with a water saturation deficit of about 10%. The course of saturation was then determined in two parallel series of these segments under given intensity of illumination and in the dark.

Results and Discussion

In no determinations did we confirm the conclusions from the measurements of the first series with leaf segments from adult plants growing in field conditions. The results showed that the water deficit of leaf segments of young plants growing under constant conditions was nearly always less than 1% and often reached zero values. No effect of different intensities of illumination could even be demonstrated. If the segments showed a certain deficit at the beginning of the experiment complete saturation of the leaf tissue was reached already after two hours, both on exposure of leaf tissue in the dark and with the highest intensity of illumination.

Unfortunately the method of weighing the segments used did not permit the determination of water saturation deficit of 0—1% with greater precision.

It was thus shown that this method of measuring photosynthetic rate ensures almost immediate replacement of water loss with high transpiration. It is not possible to determine the degree to which this complete water saturation of the tissue might negatively influence the photosynthetic rate. It is, however, stated (Pisek and Winkler 1956 and others) that the highest photosynthetic rate is reached in the presence of a definite small degree of water saturation deficit. It can certainly be assumed that a possible decrease in photosynthesis rate in these leaf segments would be very small since the increase in dry weight per unit area of leaf surface was fairly high.

Summary

Observations were made on the water deficit of assimilating tissue of leaf segments of spring barley. The segments were placed in a special stand which ensured their water saturation and the stand was placed in an assimilation chamber under controlled conditions. The measured values of water saturation deficit of the segments under conditions favourable for very intense photosynthesis and transpiration showed that the given method ensures the practically immediate replacement of transpired water. The water saturation deficit of the segments was in the range of 0—1 %. The method of determining water saturation deficit used does not permit the determination of these small values with greater precision.

References

Bartoš, J., Kubín, Š., Šetlík, I.: Dry weight increase of leaf disks as a measure of photosynthesis. — Biol. Plant. *2* : 201—215, 1960.

Čatský, J.: Determination of water deficit in disks cut out from leaf blades. — Biol. Plant. *2* : 76—78, 1960.

Nátr, L., Špidla, J.: Application of the leaf-disk method to the determination of photosynthesis in cereals. — Biol. Plant. *3* : 245—251, 1961.

Pisek, A., Winkler, E.: Wassersättigungsdefizit, Spaltenbewegung und Photosynthese. — Protoplasma *46* : 597—611, 1956.

Rychnovská, M., Bartoš, J.: Measurement of photosynthesis by dry weight increment of samples composed of leaf segments. — Biol. Plant. *4* : 91—97, 1962.

Discussion

W. Scheumann: We would be very interested in these methods if the measured values correspond absolutely or at least relatively to the assimilation rate of "leaves in situ". Were these relationships investigated?

L. Nátr: No. Such experiments were not carried out.

Z. Šesták: In our previous experiments (Šesták and Čatský, Biol. Plant. *4* : 131—140, 1962) also no negative influence of full water saturation of experimental disks on photosynthetic rate has been found. This was affirmed by comparing the results obtained by leaf disk method with the results of infra-red analyzer determinations on intact leaves: no substantial difference was found (naturally, after multiplication of the values obtained by the former method and expressed originally in units of dry weight, by the coefficient of 1·4).

THE GROWTH AND WATER SUPPLY OF INDIVIDUAL ORGANS OF THE POTATO

G. MEINL

Institute for Plant Breeding, German Academy of Agricultural Science,
Berlin, Gross-Lüsewitz, near Rostock, German Democratic Republic

In investigating growth, development and the production of substance in agricultural crops we used time-lapse cinematography in addition to other methods. By this means photos were obtained at 20-minute intervals in plants grown in Mitschlerlich pots throughout the entire vegetation period. Changes

Fig. 1. Mitscherlich pots with specially installed 16 mm. camera, installations for flash-light and switch-clock (in background).

in volume or length of the individual organs were obtained by evaluatin single photos. In potatoes one camera was used to film growth of the overground parts, a second to film root growth and a third to film tuber growth. Fig. 1 shows the apparatus used.

The evaluation of these three films gave interesting data which are the subject of this paper.

Elongation of the shoot attained maximal values in the afternoon hours (16 hr.), it then greatly decreased and in the morning hours even attained negative values. It is obvious that these were not true negative values but were evidently produced by changes in turgor due to the raising of the leaves from the sleep position which occurs at this time (Fig. 2).

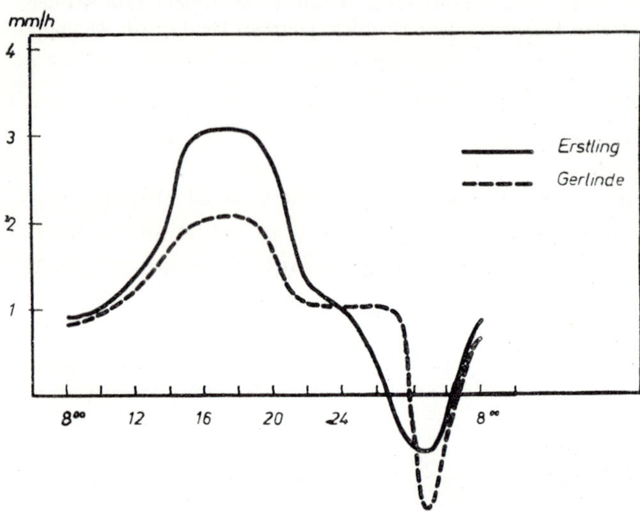

Fig. 2. Mean increment of shoot of Erstling and Gerlinde varieties during day.

The roots did not show this diurnal growth rhythm. It is necessary to note, however, that the technique of filming was much more difficult than in the overground parts. In addition, they had much smaller increments due to the relatively greater error in evaluation and therefore did not give a clear picture.

The tubers again behaved differently. A definite rhythm of volume increment was maintained, particularly in young plants (Fig. 3).

In the time between 16 and 7 hr. the tubers grew most intensely, in the remainder of the day the increments were slight.

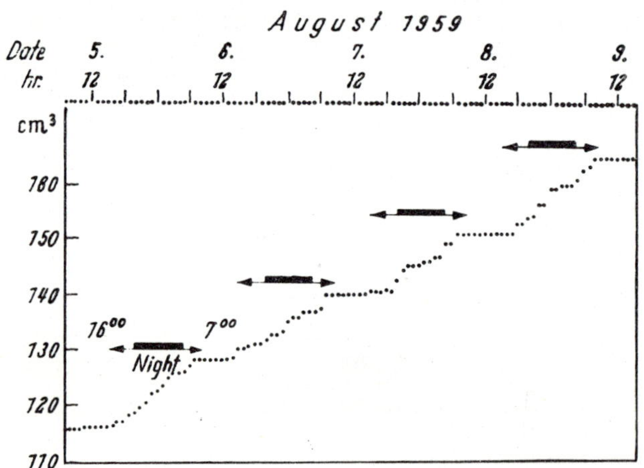

Fig. 3. Hour-increments in potato tubers.

In parallel with these rhytmic increments in volume there was a raising and lowering of the small tuber, shown in Fig. 4.

This tuber weighing about 1·5 g. was placed at the end of an approximately 4 cm. long stolon freely protruding into a space. At about 8 hr. the tuber started to fall slowly and reached the lowest point between 14 and 16 hr. Then between 18 hr. and 1 hr. it was again raised to the initial position. The path of movement was about 12 mm. long. This rhythm, conditioned by changes in turgor, was in accord with the colume increments of the other tubers.

These findings already suggest that the water supply of the individual organs has a considerable effect on rhythmic processes.

A further fact which cannot be overlooked is that the measurable changes in the volume of plant organs are most often the result of elongation growth of cells. The dry weight of these tissues shows that $4/5$ of the volume changes were produced by water uptake and only $1/5$ by increase in dry weight. It would be expected, therefore, that volume increment would be maximal at the time of highest water supply.

In our case the taking up of water by the underground supply organs due to transpiration occurred in daylight. As against this, the apical growth centre obtained water from the ascending water flux and elongation processes could take place. At this time these parts were well supplied with dissolved sugars and as a result of increased power of suction, could take up further water.

Fig. 4. Still from film (tuber showing movements described in text).

The evening closure of the stomata reduced water consumption so much that the tubers could now use the water taken up. At this time sugar reserves are laid down in the tubers and in this way the osmotic gradient is increased.

We are, therefore, of the opinion that water economy, photosynthesis and the transport of sugars contribute to the different rhythm of growth of the individual organs. It would now be desirable to study the effect of changing the external conditions, mainly the water supply, on this rhythm. In the experiment described, care was taken to maintain soil moisture capacity constant (95—100%).

(A 16 mm. film was then shown of the growth of the potato tuber made by time-lapse cinematography.)

References

Engel, K. H., Engelhardt, K., Raeuber, A.: Über den Einsatz der Filmkamera bei phänometrischen Untersuchungen an Kartoffeln. — Naturwiss. *47* : 333, 1960.

Engel, K. H., Meinl, G., Raeuber, A.: Das tagesperiodische Wachstum der Kartoffelknollen und der Verlauf der Temperatur. — Z. angew. Meteorol. *4* : 364—369, 1964.

Meinl, G.: Über den Tagesgang der Assimilation und des Kohlenhydratgehaltes von Kartoffelblättern. — Z. Bot. *51* : 388—398, 1963.

Discussion

J. Úlehla: Changes in the weight of tubers during the day have also been observed by Soviet authors.

G. Meinl: I know this work. It was not concerned, however, with alternating changes in overground and underground mass.

Z. Šesták: Did you try to change this rhythm of tuber formation by artificial modification of light and dark periods ?

G. Meinl: No, there was no modification. The experiment was started in a glasshouse without side-walls.

THE INFLUENCE OF SOIL DROUGHT ON THE VITALITY OF FOREST PLANTS

H. POLSTER

Institute for Forest Plants Selection, German Academy of Agricultural Sciences,
Graupa bei Pirna/Sa., German Democratic Republic

The vitality of the plant, that means its ability to perform actively all biochemical and physiological processes conditioning life, is impaired by water deficiency. The increase in reaction resistances by means of depriving the plant of its free intramicellar water important for the chemical transpositions paralyses "active vitality" with increasing dehydration of the organs (Stocker 1947). The active vitality of the plant which declines with the increasing water deficits becomes quite evident in gaseous exchange studies in intermittent light (alternating periods of light and dark) under climatic conditions.

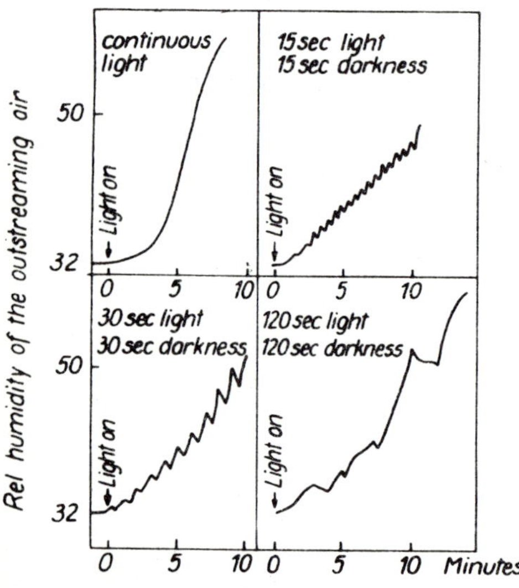

Fig. 1. Light induced stomata movements (Virgin 1956).

Results and Discussion

Virgin (1956) demonstrated in his experiments on turgescent wheat leaves with the Corona hygrometer that already intervals of 15 seconds in the change of light and dark led to visible transpiration regulations, so-called "fluctuations" (Fig. 1). Polster and Reichenbach (1957) found that in poplars these transpiration regulations are correlated with the stomata regulation capacity in dependence upon hydration (i.e. water condition) of the plant. In alternating periods of light and dark there was always in the light phase — parallel and proportional to stomata width — a higher intensity of transpiration than in the dark phase, and the transpiration regulations declined according to the increase of plant desiccation (Fig. 2). While the "fluctuations" of the full-vital poplar plant reached almost half of the momentary transpiration intensity they are hardly recognizable in deturgescent shoots.

When the number of light reactions as well as their duration decrease, there is also a continuous decline in the water deficit the plant can withstand, or in other words: the more unfavourable the hydration of the plant and the weaker its resistance to water losses due to constitution or special physiological properties, the less also its transpiration reaction in the change of light and dark (Table 1). Thus, we may conclude that transpiration regulations — we call them so because of their comparibility to stomata regulations — represent a measure for plant vitality.

According to Pisek and Winkler (1956) the stomata reactions during alternating water balances do not only control transpiration but also photosynthesis. With increasing water deficit in the plant organs there is also a decrease in assimilation intensity parallel to the reducing stomata width. Consequently, a decline of vitality must be reflected in CO_2 exchange as well.

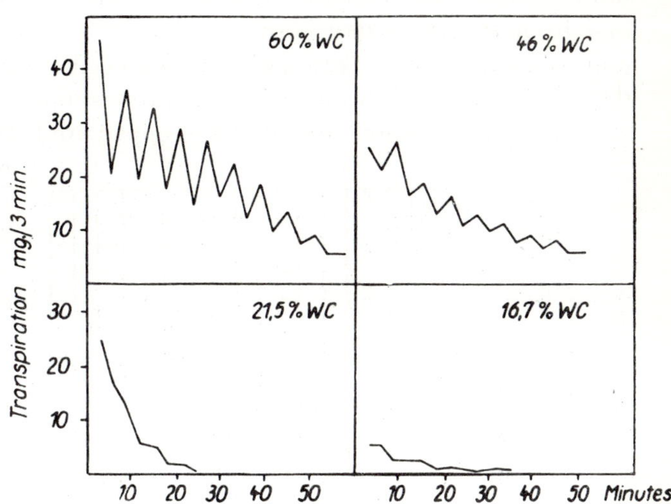

Fig. 2. Fluctuations of transpiration intensity related to desiccation of poplar leaves cultured in soil of different water capacities (W.C.) (Polster and Reichenbach 1957).

T a b l e 1.

Light reactions and water deficits (Polster and Reichenbach 1957)

Reaction limiting water deficit %	25·7	23·0	23·0	22·4	21·0	19·1	17·6	17·0	13·8	7·2	6·6
Light reactions number	9	8	7	6	6	5	4	4	3	2	1
Duration (min.)	57	51	45	39	39	33	27	27	21	15	9

Polster and Fuchs (1960) have carried out corresponding experiments in the change of light and dark with young potted conifers under constant conditions (7,000 lux, 22—22° C, 63—67% rel. humidity). As can be seen in Fig. 3

in full turgescence and vitality of the test plants, the needles of *Larix decidua* show maximum of transpiration and assimilation as well as the greatest amplitudes between dark transpiration on the first day of measuring. With falling turgescence first the transpiration decreases (second day of measuring) and from the third day both processes of gaseous exchange — assimilation and transpiration — are radically reduced with signs of distinct vitality decrease. On the fifth day without water supply the net assimilation as well as dark transpiration are finally stopped and only small water losses are recognizable in the light phases. As already found by Pisek and Winkler, respiration, however, is hardly changed in the dark: in the moment when net assimilation is being stopped, respiration is just intensive enough to balance the CO_2 exchange. Already a few hours after rewatering of the soil and plant both assimilation and transpiration set in again (Fig. 4): the amplitudes of the gaseous exchange reactions are clearly recognizable: vitality begins to be restituted. After eight days full vitality is regained, since the initial intensities of gaseous exchange are reached again. This is also evident with the coefficient $\dfrac{A + R}{R}$ during de-

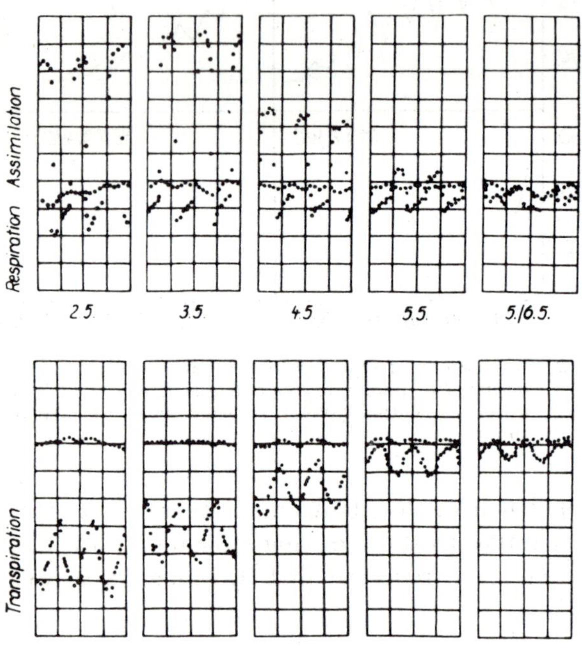

Fig. 3. Regulations of transpiration, assimilation and respiration in *Larix decidua* in intermitting light under the influence of progressing soil drought (Polster and Fuchs 1960).

hydration and after resaturation (Table 2). At approximately 60 per cent of water capacity (W.C.) of loamy sand soil this coefficient amounts to 5·1 on the second day, then drops to 1·0 when net assimilation is stopped and rises to 5·7 after full rewatering of soil (Table 2).

From the total course of the investigation it can be concluded that the dehydration process of the test plant still occurred in the sphere of active vitality. The observation that the larch tree had lost approximately 7 per cent of its needles when resaturation sets in, proves, moreover, that dehydration was just interrupted when the "sublethal water deficit" was reached. In the sense of Pisek and Winkler (1953) this means a water loss causing about 5 per cent irreversible injuries. Consequently, the cessation of net assimilation and dark transpiration we found may be characteristics for reaching the sublethal water deficit as vitality limit in regard to premortal plant damages. Oppenheimer (1963), however, is no longer of the opinion — in contrary to his former point of view — that initial injuries in leaves are a criterion for reaching the sublethal water deficit. Therefore, he replaces the term "sublethal water deficit" by that of "permanent turgor loss point"; that means the critical water loss no longer admitting complete resaturation — measured by the fresh weight of excised leaves.

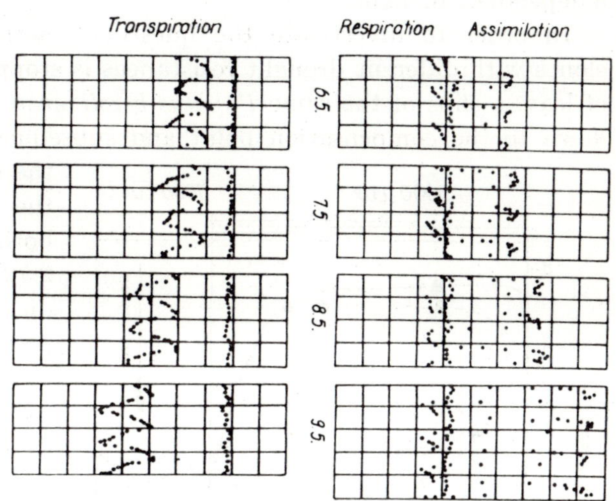

Fig. 4. The same experiment as in Fig. 3 — after water resaturation up to full hydration (Polster and Fuchs 1960).

When using one-year old slips of poplar clones *(Populus berolinensis)* grown in soils of 70% and 30% water holding capacity, it was found that all plants — those well supplied with water and also kept rather dry — stop

Table 2.

The economic coefficient of CO_2 gaseous exchange under the influence of progressing soil drought and rewatering (Polster and Fuchs 1960).

Day	2. 5.	3. 5.	4. 5.	5. 5.	5./6. 5.	6. 5.	7. 5.	8. 5.	9. 5.
W.C. %	60			→	10	100	100	100	100
$\dfrac{A + R}{R}$	4·7	5·1	3·5	2·1	1·0	3·7	4·0	4·7	5·7

their net assimilation after 120 hours (5 days). The influence of dehydration on photosynthesis is so strong that increasing the light intensity from 3,000 lux to 7,000 lux does not even produce any net assimilation. However, photo-

synthesis rises immediately after the rewatering of the soil and plant (Fig. 5).
These findings not only prove that during the dehydration process the test
plants remained actively vital but also that the assimilation reactions were
independent of light.

In order to investigate the question whether photosynthesis in poplar
clones with different drought resistances is stopped at different water deficits
of leaves, slips of the same *Populus berolinensis* clones were repeatedly dried
down to the compensation point and subsequently resaturated (Fig. 6). At
the beginning of the experiment the low water deficits (W.D.) of both clones amounting only to 2·72 and 2·77, respectively, indicate full turgescence of the plants. Corresponding to the preceding figures, the curves of net assimilation of both clones drop to near thé compensation point within four to five days and then rise approximately to the first assimilation maximum after rewatering of the soil. Respiration shows only faint reactions. According to Stocker, the plants remained again "actively vital" notwithstanding the temporary paralysing of vitality due to dehydration. As could be expected this process which is quite similar for both clones, always occurs

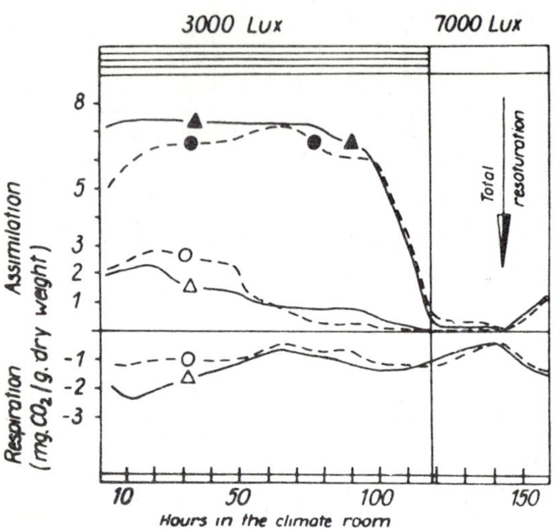

Fig. 5. Effect of light on CO₂ gaseous exchange under
the influence of progressing soil drought and
water resaturation. *Populus berolinensis*:
Clone 83 — ▲ 70 % W.C., △ 30 % W.C.;
Clone 56 — ● 70 % W.C., ○ 30 % W.C.
(Weise and Fuchs 1964).

when water is added. On the other hand, it is characteristic for the drought-
resistant clone 83 (Kamischin, district Volgograd, U.S.S.R.) that the maximum assimilation intensity is significantly superior to that of the drought-
sensitive clone 56 (Dresden-Klotzsche) and that the compensation point is
only reached when there is a greater water deficit of the leaves (W.D.
18·10 vs W.D. 14·65). Thus, the clone with higher drought-resistance is not
only able to withstand stronger dehydration but also to stop its assimilation
when there are greater water losses.

Studies on the biochemical reactions during the dehydration process after
resaturation of the plant carried out with needle press saps of *Picea abies*,
led to the following results (Fig. 7): Net assimilation and transpiration drop
with decreasing leaf turgescence whereas sugar concentration and dry matter
percentage rise. Paper chromatography reveals a marked increase of the

saccharose level as against the percentage of glucose and fructose. Soil drought is reflected by increasing suction forces of more that 50,000 Ω. After resaturation the situation is inverse, recognizable in the increase of gaseous exchange and decrease of sugar concentration and dry matter. Consequently, the capacity for synthetizing sugars and particularly that of saccharose, is an indispensible condition for the restitution capacity after dehydration.

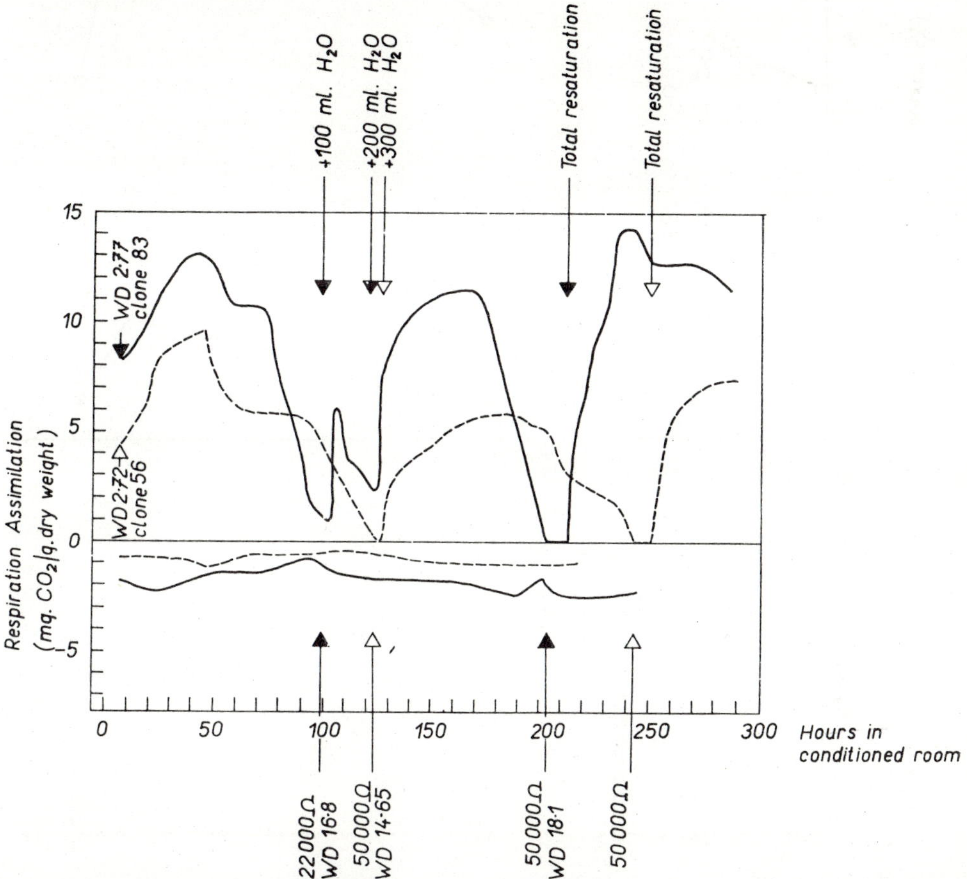

Fig. 6. Photosynthesis and respiration of different drought resistant poplar clones in reiterated soil drought and resaturation experiments (Weise and Fuchs 1964).

Summary

The dependence of plant vitality upon the soil moisture level can be determined by a water deficit limiting the dark/light reaction, indicating the minimum water level for plant reactions to light by stomata regulation (Polster and Reichenbach 1957). The restriction of active vitality during increasing soil drought under constant experimental conditions leads to the stop-

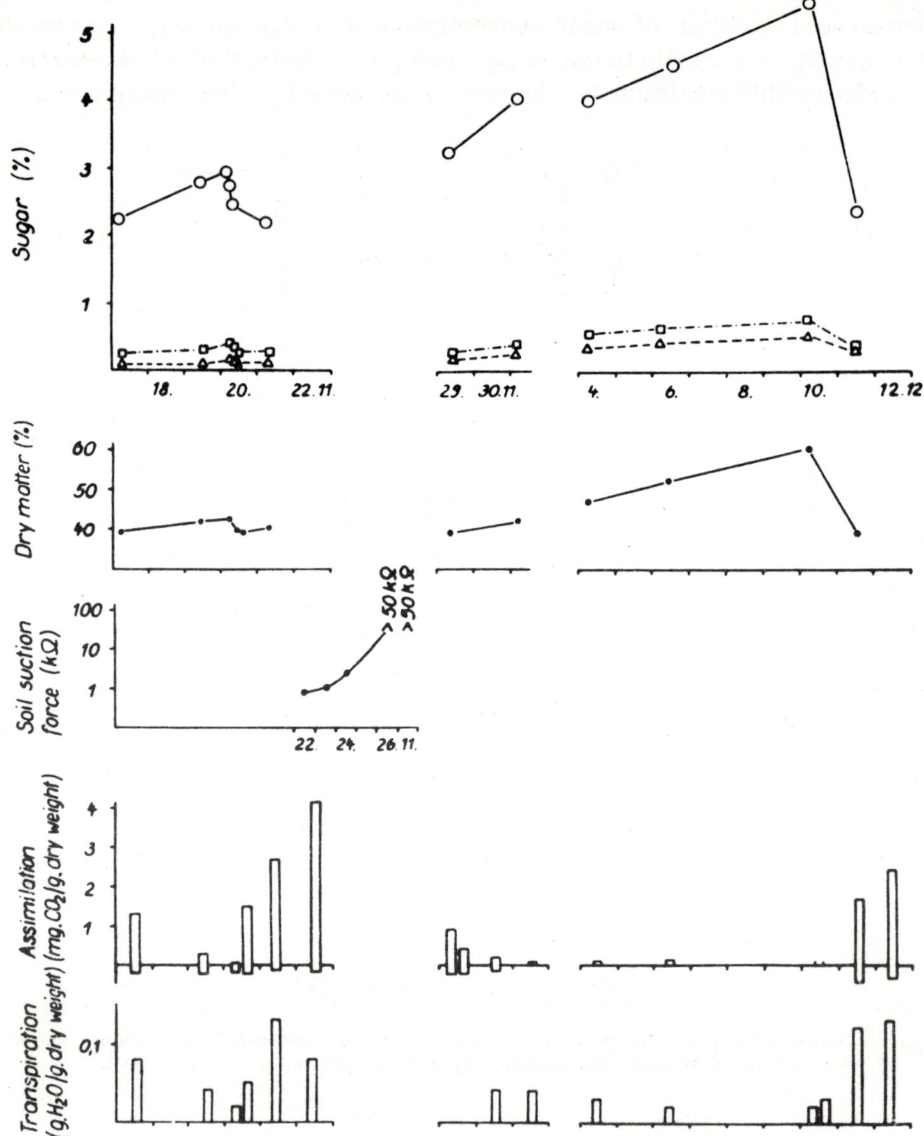

Fig. 7. Sugar concentration (○ sucrose, □ glucose, △ fructose) in relation to dry matter, assimilation and transpiration in reiterated soil drought and resaturation experiments (Weise and Börtitz 1964).

ping of net assimilation and dark transpiration when sublethal water deficits are reached as a limit for premortal damages. Soil rewatering prior to visible damage restitutes vitality in young conifers. This is recognizable in the regaining of gaseous exchange amplitudes preceding dehydration after optimum water saturation of the leaf organs (Polster and Fuchs 1960). The assimilation reactions of poplar plants under drought influence are quite independent of light. When there are two poplar clones of different drought-resistance the more resistant one stops net assimilation at higher water deficits in the leaves than the other (Weise and Fuchs 1964). The biochemical reactions correspond to the gaseous exchange reactions during dehydration, since with increasing soil drought the carbohydrate level of young *Picea abies* rises — particularly the saccharose concentration — whereas simultaneously CO_2 exchange activity and transpiration decrease. The regain of full vitality depends primarily on the persistent biosynthesis capacity of saccharose (Weise and Börtitz 1964).

References

Oppenheimer, H. R.: Zur Kenntnis kritischer Wasser-Sättigungsdefizite in Blättern und ihrer Bestimmung. — Planta *60* : 51—69, 1963.

Pisek, A., Winkler, E.: Die Schliessbewegung der Stomata bei ökologisch verschiedenen Pflanzentypen in Abhängigkeit vom Wassersättigungszustand der Blätter und vom Licht. — Planta *42* : 253—278, 1953.

—, —: Wassersättigungsdefizit, Spaltenbewegung und Photosynthese. — Protoplasma *46* : 597—611, 1956.

Polster, H., Reichenbach, H.: Ein Verfahren zur Prognose der vitalen Dürreresistenz durch Ermittlung des Stomataregulationsvermögens abgeschnittener Pflanzensprosse. — Biol. Zbl. *76* : 700—721, 1957.

—, Fuchs, S.: Der Einfluss intermittierender Belichtung auf die Transpiration und Assimilation von Fichte und Lärche bei Dürrebelastung. — Biol. Zbl. *79* : 465—480, 1960.

Stocker, O.: Probleme der pflanzlichen Dürreresistenz. — Naturwiss. *34* : 362—371, 1947.

Virgin, H. I.: Light-induced stomatal movements in wheat leaves recorded as transpiration. — Physiol. Plant. *9* : 280—303, 1956.

Weise, G., Börtitz, S.: Biochemische und gasstoffwechselphysiologische Untersuchungen an Fichte *(Picea abies* [L.] Karst) und Omorikafichte *(Picea omorica* Pančič) nach Dürrebelastungen unter standardisierten Bedingungen. — Biol. Zbl. *83* : 19—25, 1964.

—, Fuchs, S.: Die Sistierung der Nettoassimilation und ihre Beziehung zur Dürreresistenz. — Untersuchungen an zwei *Populus berolinensis*-Klonen unter standardisierten Bedingungen. — Biol. Zbl. *83* : 621—631, 1964.

Discussion

W. Larcher: With completely closed stomata we can consider the precise meaning of different expressions. "Assimilation" means CO_2 uptake and changes in actual plant substance. Photosynthesis is not identical with CO_2 uptake. Photophosphorylation, i.e. the direct binding of energy and ADP and ATP metabolism, also belongs to it. It would be better, therefore, when it can be assumed that the stomata are more or less closed, to speak of "CO_2 uptake" which we can measure, than of "photosynthesis". I repeat the discussion contribution of Mr. Ziegler at the Symposium in Innsbruck in 1961: It could be that in dry areas where in the time of greatest insolation the stomata of the plant are more or less closed, that energy is to a great extent bound by photophosphorylation and this energy is obviously not basically effective for the biosynthesis of secondary plant substances. If, however, reassimilation is possible on hydroactive closure of the stomata and probably also adequate binding of energy, then it is necessary to explain the question of when pathological changes occur in the mesophyll under the influence of drought. That is where the difference lies as to whether the plant dried out slowly or rapidly in water stress experiments. The power of adaptation should also be taken into consideration.

H. Polster: I would only add that at the moment when net assimilation and transpiration was suppressed in the dark, slight respiration still took place and that the respiration went to the compensation of net assimilation, at least to the compensation point.

W. R. Müller-Stoll: It must be taken into consideration that there may be certain differences in reaction between plants drying up under natural conditions (in the field) and desiccation experiments in pots (or with excised shoots). In the first case it is possible that an increased physiological adaptation to drought becomes apparent, dependent, of course, on the species concerned. This might cause the photosynthetic rate to remain at a relatively high level whereas the photosynthetic rate may be decreased much more by unnatural desiccation inducing resistance in a minor degree. Unfortunately, we still lack closer information on the connection between drought hardening and photosynthetic activity.

H. Polster: In our experiments we worked with whole young larches cultivated in pots in the soil. Although these cultivation conditions are obviously different from those in nature, we obtain results with these plants in situ which are quite different from those obtained with cut twigs.

M. Rychnovská: I wish to add a few words to the observation of Dr. Larcher on photosynthesis with closed stomata: It is stated in the literature that in semi-desert and desert plants the stomata remain half open and that therefore continuous uptake of CO_2 is possible. We have also made similar observations. In one species of *Festuca* we made the following field experiment: we investigated several plants under natural dry conditions and some we irrigated. Assimilation was determined as increased carbohydrate content during the day. We found that both the dry and irrigated variants had the same carbohydrate content. Irrigation, therefore did not lead to an essential increase in carbohydrate formation. This means that, in this case, stomatal movements do not appear to be of decisive importance.

H. Polster: It can certainly be assumed that half-open stomata do not limit the diffusion of CO_2 and are quite adequate for maximal assimilation under conditions of mild dryness. It will probably be otherwise with transpiration where half-open stomata probably entail restriction. The resulting favourable ratio A : T is probably the reason for carbohydrate formation not increasing in the irrigation experiment.

G. F. Makkink: Have you observed any discrepancy between the moment in which assimilation and transpiration were blocked or becoming blocked ?

H. Polster: No. Transpiration and CO_2 assimilation were always blocked at about the same time. Since there was no measurable transpiration in the dark, cuticular transpiration also evidently stopped. At the same time, as you have seen, even at the blocking of transpiration and assimilation in the dark, distinctly measurable respiration continued and was in the region of premortal damaging still just compensated by assimilation.

P. G. Jarvis: Would Dr. Polster explain the advantages obtained by using intermittent light ?

H. Polster: We work with intermittent light because we can assess the power of stomatal regulation from the amplitude of transpiration with alternating illumination and thereby the vitality of the plant under conditions of drought in the sense used by Stocker ("active vitality"). In addition we are thus able to record the entire extent of CO_2 metabolism (photosynthesis and respiration).

P. G. Jarvis: An estimate of vitality will be obtained only if the periodicity is long enough to allow stomatal movements to be evident. In the experiments of Virgin cited by Dr. Polster, the alternating periods were only some seconds long. In that case the fluctuations in transpiration were more probably the result of the changes in absorbed energy and hence in leaf temperature.

H. Polster: We work with young cut twigs of *Populus* with 3-min. alternation of light and dark, with young trees of *Larix decidua* cultivated in pots with 20-min. light intervals. The metabolism of gases of itself was not decisive for the length of the phases, but the maximal attained amplitude of assimilation, respiration and transpiration. The effect of leaf temperature was minimal and could be neglected.

THE INFLUENCE OF MICROELEMENTS ON ATP CONTENT IN PLANTS IN THE PRESENCE OF WATER DEFICIT AND UNDER THE INFLUENCE OF HIGH TEMPERATURES

V. P. BOZHENKO

V. L. Komarov Institute of Botany, U.S.S.R. Academy of Sciences,
Leningrad, U.S.S.R.

Soil and atmospheric drought is usually accompanied by high temperatures, in connection with which the plant is exposed simultaneously to the harmful effect of these three mutually related factors. Differences in the participation of the action of temperature factor, soil and atmospheric drought condition different responses in plants. It is known that there are very different protective reactions of the plant organism to unfavourable factors in the environment, including drought and high temperatures.

If we take into consideration that each cell has its own well developed autoregulation system which controls thousands of parallel chemical processes and immediately alters in response to changing conditions in the external environment, it must be quite clear how difficult it is to explain the complicated mechanisms by which plants ensure the possibility of maintaining themselves in combating unfavourable external factors. Despite this the task of plant physiology is to discover these mechanisms and to find means of actively affecting processes controlling the adaptation of plants to conditions unfavourable to growth.

In the last twenty years progress has been made in this direction. A number of scientists (Okuntsov and Levtsova 1952, Petinov and Molotkovskiy 1956, Shkolnik 1939, Shkolnik and Bozhenko 1959, Skazkin and Rozhkcva 1956) have shown that certain macro- and especially microelements can be used to increase the resistance of plants to drought and high temperatures.

A series of papers published from our laboratory (Novitskaya 1958, Shkolnik and Makarova 1957, Shkolnik, Bozhenko and Maevskaya 1960) and the papers of other scientists (Natanson 1952, Petinov and Molotkovskiy 1956, Vasileva and Startseva 1959) reported mechanisms taking part in forming the basis of the positive effect of microelements on resistance to drought and high temperatures. The manifold nature of their effect on processes determining the resistance was also pointed out. Microelements influence the synthesis and transport of carbohydrates, increase the viscosity and decrease the permeability of plasma, increase the content and hydration of hydrophil colloids (proteins and nucleoproteids), increase the bound water content and as a result of this lead to a decrease in transpiration in the hottest hours of the day. They increase the ascorbic acid content which is also correlated with drought resistance and act on the metabolism of organic acids, in this connection decreasing the ammonium content.

Our investigation of the effect of microelements on the nucleotide phosphate content of plants during drought showed that aluminium, molybdenum, and, in particular, cobalt increased the content of these substances in the over-ground organs of plants.

The work is concerned with the effect of microelements, which increase resistance to drought and high temperature, on the adenosintriphosphoric acid (ATP) content in the growing points and roots of the sunflower with normal water supply, decreased soil moisture and in the presence of high temperatures acting for short periods. These experiments were made in the years 1961—1962, being stimulated by reports in the literature (Molotkovskiy 1961, Zholkevich and Koretskaya 1960) on disturbances in energy mechanisms, associated with respiration and oxidative phosphorylation, developing under the influence of drought and high temperature, and reports on the participation of boron in energy mechanisms, i.e. its effect in increasing ATP content (Shkolnik and Maevskaya 1962). In recent years it had been assumed that other microelements also play a big role in energy metabolism (Crane 1960, Peyve 1961). It is assumed that microelements are involved in the formation of esters of inorganic phosphates, in relation to the ability of metals to change their valency, bind phosphates to phosphorylated derivatives and participate in energy balance.

Materials and Methods

Sunflower seeds (variety VNIIMK-8931) were soaked for 20 hours in a solution (0·2 g./l.) of one of the following salts: $Al(NO_3)_3$, $Co(NO_3)_2$, $ZnSO_4$, $CuSO_4$ or a solution of boric acid of the same concentration. They were then dried at room temperature and planted. The plants were grown in soil cultures in polyethylene pots under artificial illumination. Two-week-old plants were taken into the experiment, when the first pair of true leaves appeared.

Drought conditions were created by witholding irrigation. Heat exposure was carried out in a wet-chamber of Hepler ultrathermostat, where cut growing points were left for 40 min. at a temperature of 50° C.

ATP was determined in the growing points or roots of sunflower, using the method of precipitation of mercurious acetate (Mashkova and Severin 1950). The phosphorus easily hydrolysable from ATP was determined by the highly sensitive iso-butanol-benzol method (Weil-Malherbe and Green 1951).

Results and Discussion

The results of the experiments are given in Tab. 1—3. The data are the mean from 2—3 analyses.

Tab. 1 shows that all microelements investigated, with the exception of zinc, increased the ATP content of growing points at a temperature of 20° C.

Fig. 1. Sunflower plants in pots 1 and 2 were grown from seeds pretreated with cobalt, wilting plants in pots 3 and 4 are the controls.

Exposure to a temperature of 50° C for 40 min. usually led to a slight decrease in ATP content. In this case, however, all microelements (aluminium, zinc, boron, copper and cobalt) led to a greater increase in the ATP content of the growing points of plants than at 20° C.

Analysis of Tab. 2 shows the striking fact that the ATP content of the growing points was three-fold that of the roots. Pre-treatment of the seeds with boron considerably increased the ATP content of both growing points and roots under conditions of drought. As the data show, the ATP content increased in the growing points 35—65% and in the roots 30—47% after treatment of the seeds with boron.

Tab. 3 gives the results of the effect of microelements on ATP content in the presence of both decreased soil moisture and exposure of the growing points in a Hepler ultrathermostat. The table shows that, with this conjoined action of the two factors often occurring simultaneously under natural conditions, a considerable increase in ATP content was found after treatment with aluminium and boron and particularly after treatment with cobalt. Cobalt led to an increase in ATP in the growing points of 93% and in the roots of 98%.

T a b l e 1.

The effect of pre-treatment of the seeds with microelements on the ATP content in γ P per 1 g. fresh weight, in growing points of sunflower plants with normal water supply. The growing points were exposed to 20° C or 50° C for 40 min.

Experimental variant — microelement applied	Normal water supply			
	Temperature of 20° C		Temperature of 50° C	
	P content		P content	
	in γ	in %	in γ	in %
1. Water control Al(NO₃)₃	26·1 27·6	100 106	19·5 26·1	100 133
2. Water control ZnSO₄	23·1 22·8	100 98·7	19·5 26·1	100 133
3. Water control H₃BO₃	18·6 21·5	100 115	17·2 24·8	100 144
4. Water control CuSO₄	23·3 29·5	100 126	22·8 28·7	100 125
5. Water control Co(NO₃)₂	12.9 14·4	100 111	16·0 19·6	100 122

T a b l e 2.

The effect of pre-treatment of the seeds with boron on ATP content (in γ P per 1 g. fresh weight) in growing points and roots of sunflower at low soil moisture

Experimental variant	Low soil moisture, 20° C			
	Growing points		Roots	
	P content			
	in γ	in %	in γ	in %
1. Water control H₃BO₃	27·3 36·8	100 135	8·7 11·3	100 130
2. Water control H₃BO₃	20·4 33·7	100 165	7·3 10·8	100 147

It is possible that the very high ATP content in plants after treatment with cobalt was also connected with their resistance to drought being the most effective (Fig. 1).

T a b l e 3.

Effect of pre-treatment of seeds with microelements on ATP content (in γ P per 1 g. fresh weight) in growing points and roots of sunflower with combined low soil moisture and exposure to high temperature

Experimental variant — microelement applied	Low soil moisture + exposure to 50° C for 40 min.			
	Growing points		Roots	
	P content			
	in γ	in %	in γ	in %
1. Water control	20·7	100	not determined	
Al(NO$_3$)$_3$	29·5	142	not determined	
2. Water control	18·3	100	5·5	100
Co(NO$_3$)$_2$	35·5	193	10·9	198
3. Water control	22·1	100	not determined	
H$_3$BO$_3$	31·3	141	not determined	

Summary

Sunflower seeds were pretreated before sowing for 20 hours by soaking in 0·2 g./l. solutions of aluminium nitrate, cobalt nitrate, zinc sulphate, copper sulphate or boric acid. The growing points or roots of 14-day-old plants, grown under conditions of adequate water supply or drought were exposed for 40 min. to temperatures of 20° C or 50° C and ATP content was then determined. Treatment with microelements (especially with cobalt) increased the ATP content of the growing points and the roots, especially in the presence of unfavourable external conditions. The results show that one of the possible causes of increased resistance of plants to drought and high temperatures after treatment with microelements is the ability of these microelements to improve the energy balance in the plant.

References

Crane, F. L.: Role of trace elements in electron transport and oxidative phosphorylation. — Soil Sci. *85* : 78—86, 1958.

Mashkova, N. P., Severin, S. E.: [Textbook of animal physiology methods.] In Russ. — Publish. House "Sovetskaya Nauka", Moscow 1950.

Molotkovskiy, Yu. G.: [Plant metabolism characteristics in relation to resistance of plants to high temperatures.] In Russ. — Izvest. Akad. Nauk SSSR, Ser. biol. *2* : 246—249, 1961.

Natanson, N. E.: [Influence of some microelements on plasma viscosity of plants.] In Russ. — Dokl. Akad. Nauk SSSR *87* : 1067—1070, 1952.

Novitskaya, Yu. E.: [Significance of pre-sowing hardening in solutions of some microelements for drought resistance of plants.] In Russ. — Trudy BIN, Ser. IV — Eksper. Bot. *12* : 74—94, 1958.

Okuntsov, M. M., Levtsova, O. P.: [Effect of copper on water relations and drought hardening of plants.] In Russ. — Dokl. Akad. Nauk SSSR *82* : 649—651, 1952.

Petinov, N. S., Molotkovskiy, Yu. G.: [Physiological basis of resistance of some cultivated plants to high temperatures.] In Russ. — Fiziol. Rast. *3* : 516—526, 1956.

Peyve, Ya. V.: [Microelements and Enzymes.] In Russ. — Publ. House Acad. Sci. Lithuan S.S.R., Riga 1961.

Shkolnik, M. Ya.: [Effect of microelements on drought and salt resistance of plants and on chemical composition of grain.] In Russ. — Sovet. Bot. *6—7* : 218—226, 1939.

—, Bozhenko, V. P.: [Effect of microelements on drought resistance of plants.] In Russ. — *In*: Primenenie mikroelementov v s/kh i medicine [Use of Microelements in Agriculture and Medicine], pp. 151—157. Publish. House Acad. Sci. Lithuan. S.S.R., Riga 1959.

—, Maevskaya, A. N.: [Mechanism of boron action in nucleic acid biosynthesis. Effect of boron on energetic balance.] In Russ. — Dokl. Akad. Nauk SSSR *145* : 222—224, 1962.

—, Makarova, N. A.: [Effect of microelements on physiological processes determining drought resistance of plants.] In Russ. — *In*: Biologicheskie osnovy oroshaemogo zemledeliya [Biological Basis of Irrigated Agriculture], pp. 565—584. Pubish. House Acad. Sci. U.S.S.R., Moscow 1957.

—, Bozhenko, V. P., Maevskaya, A. N.: [Effect of alluminium, cobalt and molybdenum on physiological and biochemical processes determining drought resistance of plants.] In Russ. — *In*: Fiziologiya ustoychivosti rasteniy [Physiology of Plant Resistance], pp. 522. Publish. House Acad. Sci. U.S.S.R., Moscow 1960.

Skazkin, F. D., Rozhkova, V. G.: [Effect of boron on cereals in conditions of soil drought and in the critical development phase of plants.] In Russ. — Dokl. Akad. Nauk SSSR *108* : 962—964, 1956.

Vasileva, I. M., Startseva, A. V.: [Effect of drought on water relations and metabolism in red clover and their relation to mineral nutrition.] In Russ. — Izv. Kazansk. Filiala Akad. Nauk SSSR, Ser. biol. *7* : 39, 1959.

Weil-Malherbe, H., Green, R. H.: The catalytic effect of molybdate on the hydrolysis of organic phosphate bonds. — Biochem. J. *49* : 286—292, 1951.

Zholkevich, V. N., Koretskaya, G. F.: [Metabolism of roots in conditions of soil drought.] In Russ. — *In*: Fiziologiya ustoychivosti rasteniy [Physiology of Plant Resistance], p. 423. Publish. House Acad. Aci. U.S.S.R., Moscow 1960.

Discussion

W. R. Müller-Stoll: Can any information be given on the manner of uptake of microelements and their possible importance for metabolic processes in connection with the effects described by the author? The use of radioactive substances, e.g. labelled cobalt, would certainly give some indication of the efficiency of mechanisms of the uptake of microelements.

V. P. Bozhenko: Studies along the lines of those suggested by Prof. Müller-Stoll have been started by the biochemical group of our department.

Z. Šesták: I have only a remark mainly concerning the methodology of investigations on water relations, but also relevant to other problems in plant physiology. I think that it is very necessary to consider in advance the suitability of reference units in investigating experimental values and relations. It is a very complicated question and specific for each experiment, whether we should calculate the results in relation to the whole plant, the shoot, one leaf, leaf area, fresh weight, dry weight, etc. We must be particularly careful in the choice of reference units in considering problems of water relations. Thus, for example in the work of Dr. Bozhenko, data are compared on the ATP content in the growing points of plants which developed either in dry soil or in water saturated soil. These data, however, are not calculated in relation to dry weight, as we would have expected, but to the fresh weight of the analysed growing points. Changes in water content in the soil and in the water relations of the plant can strongly influence the percentual dry weight in relation to fresh weight of the tissue. I think the results obtained by using this reference unit do not express the real quantitative relations between the compared plants.

WATER STRESS AND WATER BALANCE

THURSDAY, OCTOBER 3, and
FRIDAY, OCTOBER 4

Chairmen: *W. R. Müller-Stoll* and *W. C. Visser*
Secretary: *Z. Šesták*

CHANGES IN THE TRANSPIRATION OF PLANTS IN THE COURSE OF THE DAY

M. PENKA

Department of Forest Botany and Phytocenology, Faculty of Forestry,
University of Agriculture, Brno, Czechoslovakia

For the study of water deficit in plants it is of importance to determine the changes in transpiration in the course of the day and in the course of the vegetation period. In our previous papers we have tried to solve these question of changes in the transpiration of certain irrigated and non-irrigated plants (Penka 1953a, b, c, 1956, 1963; Penka et al. 1955, 1960, 1962; Srb et al. 1960, etc.). As most of our experiments had been carried out in field conditions, we performed analogous experiments in greenhouse conditions, in order to avoid possible errors.

Material and Methods

Linum usitatissimum L., which had proved suitable in previous experiments, was used as test plant (Penka 1953a, b, 1956). Test samples of soil consisted of three kinds of natural earths, i.e.: sandy soil — P_1, clay soil — P_2, and argillaceous soil — P_3. Fine air-dried soil was prepared from all test soils and certain important pedological and hydrometeorological characteristics determined (Table 1). Special wooden vegetation containers 250 cm. $\times$ 150 cm. $\times$ 30 cm. in size were made. One of this vegetation containers was filled with fine P_1 soil, the other with fine P_2 soil, and the third with fine P_3 soil. The seeds of the test plants were sown on April 12, 1953 in rows 15 cm. apart. The vegetation containers were put in the greenhouse admitting the sun from the east and from the west. The test plants were studied up for their growth, development and transpiration. Changes in growth were determined gravimetrically. Changes in development were marked by phenological characteristics. The transpiration rate was expressed both by the rate of transpiration, and by the consumption of water for transpiration. Ivanov's gravimetric method (1918) was used for determining the intensity of transpiration. The water consumed for transpiration was calculated by means of the summation method (Penka 1953a, 1958b). On the day when measurements were carried out samples of plants were

T a b l e 1.

Certain pedological and hydropedological characteristics of soils P_1, P_2, and P_3 used. Explanatory notes: Line 1. = mechanical analysis of soils (I. argillaceous elements below 0·015 mm. in %, II. dust 0·015 to 0·048 mm. in %, III. sand dust 0·048 to 0·09 mm. in % and IV. sand 0·09 to 2·0 mm. in %); 2. = humus content in %; 3. = soil moisture expressed in % of solids; 4. = humidity in % of weight; 5. = silting point in % of soil moisture; 6. = critical water content of soil according to Sekera in % of soil moisture; 7. = retardation of growth of plant in % of soil moisture; 8. = cessation of growth of plant in % of soil moisture.

		P_1	P_2	P_3
1.	I.	15·94	41·48	62·36
	II.	6·30	18·80	22·86
	III.	5·20	11·22	5·92
	IV.	72·56	28·50	8·86
2.		0·67	3·19	2·60
3.		29·00	56·00	66·00
4.		1·01	6·53	8·61
5.		3·90	12·60	22·20
6.		5·00	16·00	21·00
7.		11·00	26·00	38·00
8.		6·00	20·30	31·80

taken at 9 a.m., 12 a.m., 2 a.m. and 5 p.m. The experiment was begun on April 12, and ended on July 10, 1953. Throughout the time of the experiment attention was devoted to changes in temperature and relative humidity of the

T a b l e 2.

Temperature and relative air humidity.

Day of measurement	Temperature in °C			Relative air humidity in %		
	Minimum	Maximum	Average	Minimum	Maximum	Average
25. IV.	17·0	29·8	22·7	75	80	77
3. V.	18·0	30·5	23.9	67	88	78
9. V.	11·3	15·2	12·5	75	80	78
16. V.	18·5	28·8	24·4	65	88	71
23. V.	22·8	30·0	26·3	58	73	67
30. V.	11·4	15·8	14·2	74	77	75
6. VI.	23·5	30·8	27·4	68	77	73
12. VI.	19·5	21·5	20·7	76	85	81
20. VI.	25·0	30·5	27·5	54	63	59
26. VI.	19·8	22.0	20·9	81	85	84
3. VII.	27·::	32·2	28·8	73	85	78
10. VII.	20·0	23·5	21·4	79	83	80

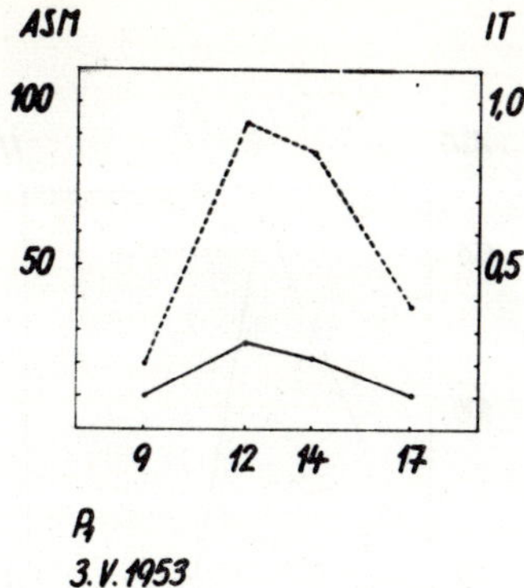

Fig. 1

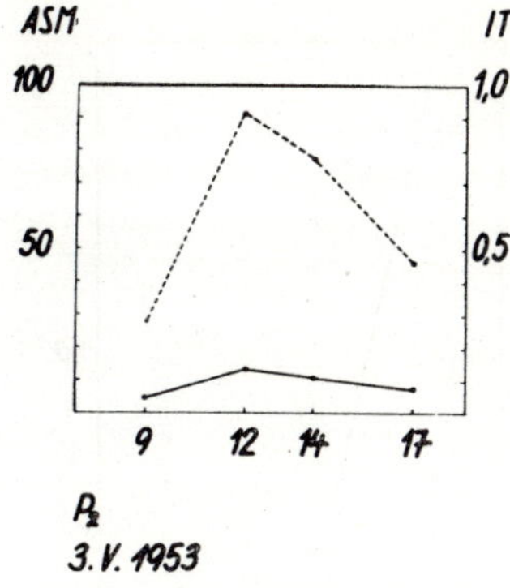

Fig. 2

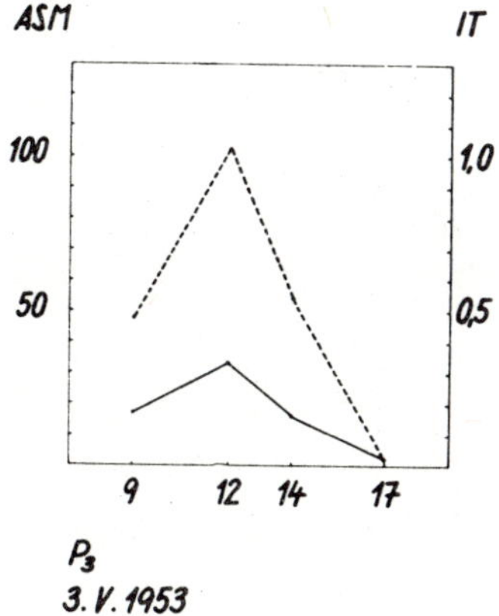

Fig. 3

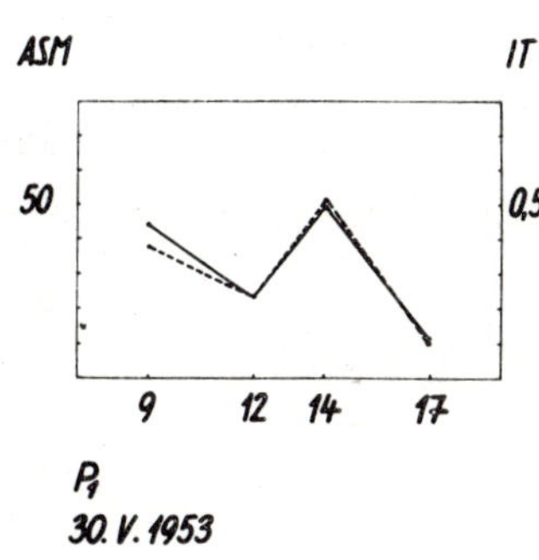

Fig. 4

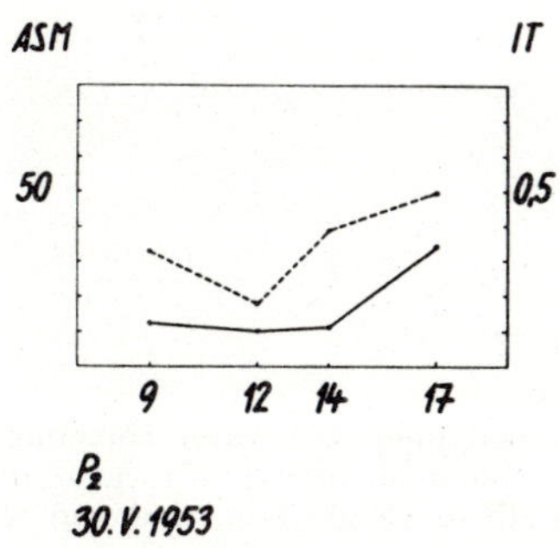

Fig. 5

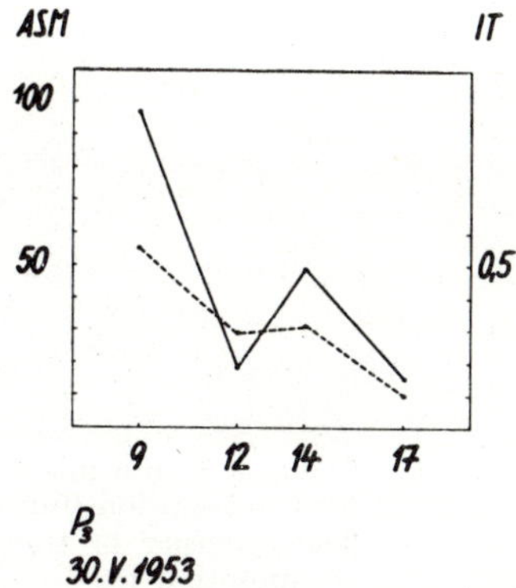

Fig. 6

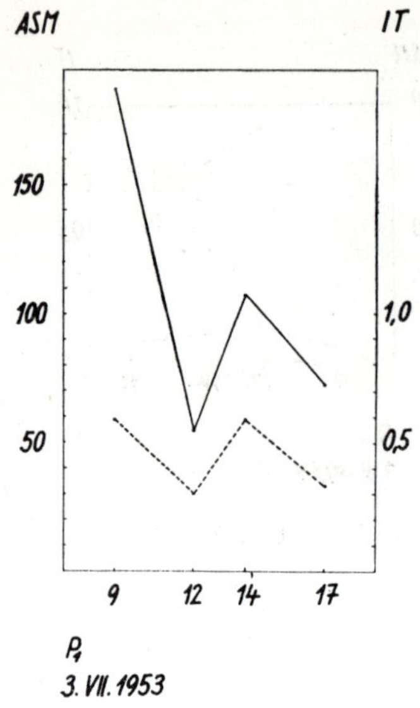

Fig. 7

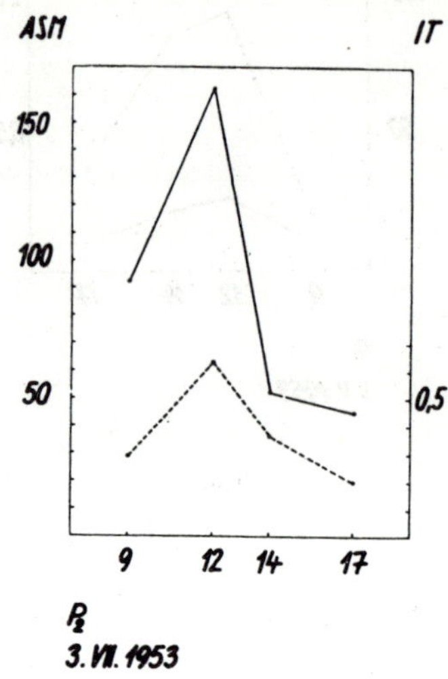

Fig. 8

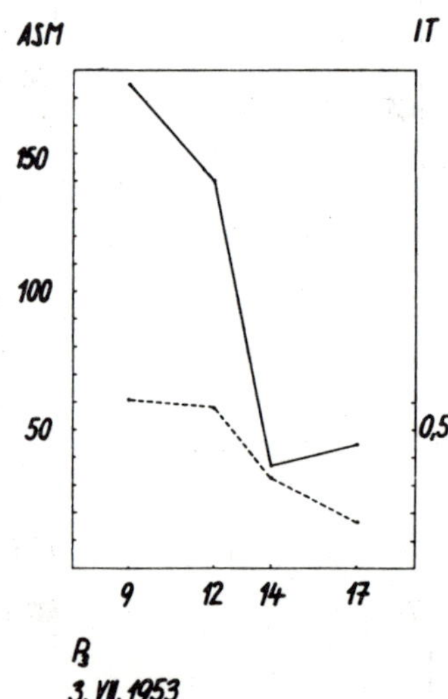

Fig. 9

Figs. 1 to 9.   Changes in the intensity of transpiration (dashed line) and water consumption for transpiration (full line) of test plants in the course of the day of measurement. The intensity of transpiration (IT) in mg./g. fresh weight/min. and the water consumption for transpiration (ASM) in mg./min. are plotted on the ordinate; the time in days is plotted on the abscissa. P_1, P_2, and P_3 stand for the respective soil on which the test plants grew up.

air (Table 2). Further details are given in the paper Penka and Nováček (1963).

Table 2 denotes the changes in temperature and relative air humidity. The mean diurnal temperature was about $22 \pm 2°$ C. A marked temperature drop occurred only on May 3 and 30. The mean diurnal relative humidity in the air was $73 \pm 2\%$.

From the viewpoint of changes in growth and development of the test plants the most balanced hydropedological characteristics were displayed by the soil P_3, though it did not have the highest humus content. Soil P_2 had the highest humus content ($3 \cdot 19\%$), then P_3 ($2 \cdot 60\%$), and finally P_1 ($0 \cdot 67\%$). For details see Tab. 1.

Results

The stage of stem extension lasted from April 16, to 27, the stage of the third pair of leaves from April 27, to May 9, and the stage of fast growth from May 9 to June 26, 1953.

The test plants reached the highest growth rates (fresh weight, total solids,

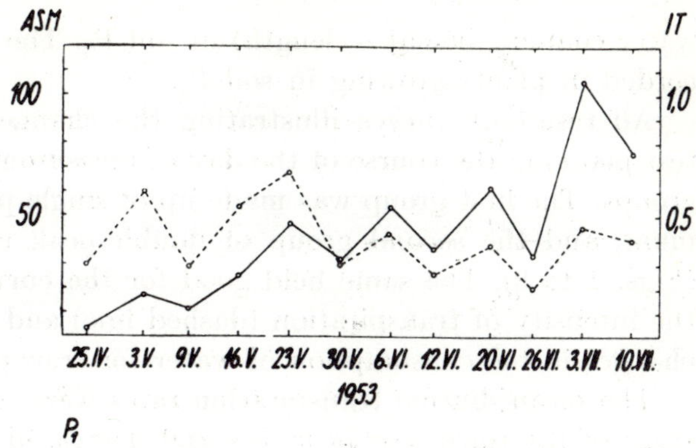

Fig. 10

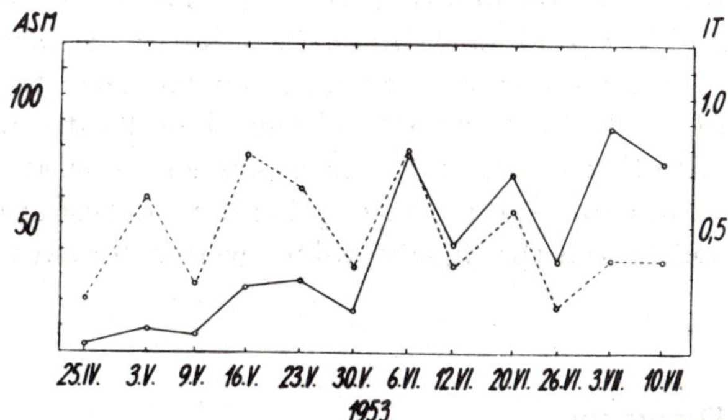

Fig. 11

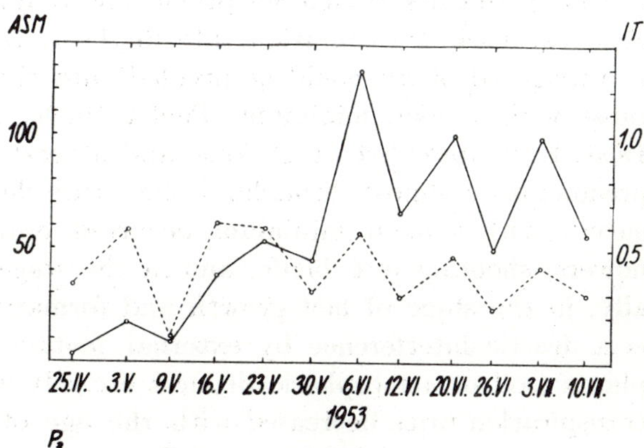

Fig. 12

Figs. 10 to 12. Changes in the mean diurnal transpiration rates (dashed line) and in the water consumption for transpiration (full line) of flax test plants from April 25 to July 10, 1953. For details see Figs. 1 to 9.

water content, height — length) in soil P_3. The lowest growth rates were recorded in plants growing in soil P_2.

All resultant curves illustrating the changes in the transpiration of the test plants in the course of the day of measurement could be divided into two groups. The first group was made up of single-peak curves with a noon maximum, and the second group of double-peak curves with a noon minimum (Figs. 1 to 9). The same held good for the curves illustrating the changes in the intensity of transpiration (dashed line) and for the curves illustrating the changes in the consumption of water for transpiration (full line).

The mean diurnal transpiration rates were rather high in the development stage of the third pair of leaves and decreased distinctly in the development stage of fast growth. On the other hand, the values for the consumption of water for transpiration were very low in the stage of the third pair of leaves and increased markedly in the subsequent stage of fast growth (Figs. 10 to 12).

From the static aspect the lowest mean transpiration rates were recorded in plants from soil P_3 and the highest rates in those from soil P_2. The consumption of water for transpiration showed an opposite trend. The lowest values for water consumption for transpiration were found in plants from soil P_2 and the highest values were recorded in plants from soil P_3.

Discussion

In experiments carried out previously we had found that curves illustrating changes in the transpiration rate in the course of the day in irrigated and non-irrigated plants could be divided into those with a noon maximum and those with a noon minimum (Penka 1953a, 1958a, b, c, 1963; Penka et al. 1955, 1960, 1962; Srb et al. 1960 and others). Curves with a noon maximum predominated almost throughout the entire development of the plants, whilst curves with a noon minimum occurred only in the stage of formation of flowers, shooting into blade, and in the stage of efflorescence, and sporadically, in the stage of fast growth and formation of stalks, especially if there is a drastic interference by external factors with the development of the plants (soil or atmospheric dryness etc.). It was also found that the diurnal transpiration rates decreased with the age of the plant, although the shape of the curves remained essentially unchanged. The resultant changes in the intensity of transpiration could be explained, in accord with the results obtained by other investigators, by changes in soil moisture (Maksimov 1916, 1917), by quantitative and qualitative changes in the water content of the plant body (Livingston 1906, Maksimov 1916, 1917, 1952; Schratz 1931, 1937; Penka 1958a) and by the transfer of free water from the nontranspiring parts of the plant to the transpiring organs (Penka 1963). All these changes then affect the water deficit in plants in the course of the day and the vegetation period (Slavík 1955; Čatský 1959, 1960, 1961).

In these experiments the above mentioned two groups of curves, illustrating the changes in the transpiration rate in the course of the day, were also obtained, although the test plants grew in greenhouse and not in field conditions. The curves illustrating the changes in the consumption of water for transpiration in the course of the day could also be divided into these two groups. Curves with a noon minimum occurred only sporadically, whilst curves with a noon maximum predominated throughout the entire duration of the experiment.

Both curves illustrating the changes in transpiration during the day show certain deviations and differences. Thus, for instance, curves with a noon maximum show a maximum in certain cases already in the morning hours, in other cases not until the afternoon hours. Curves with a noon minimum reached in certain cases only one maximum, either only in the morning hours (forenoon) or not until the afternoon (evening) hours.

If we compare the curves illustrating the changes in the intensity of transpiration and the consumption of water for transpiration, it is evident that the transpiration rates are high especially in the initial stages (shooting, third pair of leaves), whilst the values for water consumption are very low in that period. In the following stage of fast growth the transpiration rates gradually decrease, whilst the water consumption values markedly increase. These changes and differences are most clearly to be seen in Fig. 10 to 12, where the mean diurnal values of the intensity of transpiration (dashed line) and those of the consumption of water for transpiration (full line) are graphically illustrated. In these figures, "compensation periods" can be found in which the intensity of transpiration and the consumption of water are balanced and show a fundamental change. These "compensation periods" are not equally long in all plants and depend on changes in growth, falling, as a rule, into periods when the resistance of the plants to drought is very low (beginning of fast growth, formation of flowers and sap maturity — Cetl 1953, 1957; Penka 1953b). The diurnal transpiration curves from such a period are illustrated in Fig. 4 to 6.

Summary

1. Changes in the intensity of transpiration and in the consumption of water for transpiration in the course of the day in the plant *Linum usitatissimum* L. were studied in the period of shooting, the third pair of leaves and

fast growth. The test plants grew in special containers placed in a green-house (cf. Tables 1 and 2).

2. Again it was found that the resultant curves illustrating the changes in the intensity of transpiration and in the consumption of water for transpiration can be divided into two known groups, i.e. a) curves with a noon maximum, and b) curves with a noon minimum (cf. Figs. 1 to 9).

3. Curves with a noon maximum predominated almost throughout the entire duration of the experiment, whilst curves with a noon minimum occurred only sporadically, in the stage of fast growth.

4. Curves, illustrating the changes in the intensity of transpiration, showed high values especially in the stage of the third pair of leaves, whilst in the subsequent stage of fast growth a distinct decrease in these values occurred. On the other hand, the consumption of water for transpiration was very low in the period of shooting and the third pair of leaves and increased markedly in the stage of fast growth.

5. These changes are clearly to be seen in Fig. 10 to 12 where the mean diurnal transpiration rates (dashed line) and the mean diurnal values of water consumption for transpiration (full line) are plotted.

In these figures it is also possible to define the "compensation period" in which these features of transpiration are balanced and register a fundamental change.

References

Čatský, J.: The role played by growth in the determination of water deficit in plants. — Biol. Plant. *1* : 277—286, 1959.

—: Determination of water deficit in disks cut out from leaf blades. — Biol. Plant. *2* : 76—78, 1960.

—: K problematice vodního deficitu rostlin. [On the problem of water deficit in plants.] In Czech. — Thesis, Biol. Inst., Czechoslovak Acad. Sci., Praha 1961.

Cetl, I.: Návrh jednoduché metody ke zjištění odolnosti vůči suchu. [A new simple method for drought resistance determination.] In Czech. — Čs. Biologie *2* : 361—369, 1953.

—: Odolnost polních plodin vůči suchu a možnosti jejího zvýšení. [Drought resistance of field crops and possibilities of its increasing.] In Czech. — Rozpravy ČSAV, ř. MPV, *67* (8) : 1—107, 1957.

Ivanov, L. A.: [On the method of the determination of transpiration of plants in natural conditions.] In Russian. — Lesn. Zhurn. *48* : 1—7, 1918.

Livingston, B. E.: The relation of desert plants to soil moisture and to evaporation. — Carnegie Inst. of Washington Publ. *50* : 1—77, 1906.

Maksimov, N. A.: [Comparative study of transpiration in xerophytes and mesophytes.] In Russian. — Zhurn. Russk. bot. Obsh. *1* : 56—75, 1916.

—: [On the problem of diurnal course and control of transpiration in plants.] In Russian. — Trudy Tifl. bot. Sada *19* : 23—108, 1917.

—: [Selected studies on drought and winter resistance in plants. Water relations and drought resistance.] In Russian. — Publ. House Acad. Sci. U.S.S.R., 1952.

Penka, M.: Spotřeba vody polními plodinami v průběhu jejich individuálního vývoje. [Water consumption by field crops during their ontogenesis.] In Czech. — Čs. Biol. *2* : 183—190, 1953a.

—: Odolnost lnu, jarní pšenice a prosa vůči suchu ve srovnání se spotřebou vody v průběhu jejich individuálního vývoje. [Drought resistance in flax, spring wheat and millet in relation to water consumption during their ontogenesis.] In Czech. — Čs. Biol. *2* : 370—376, 1953b.

—: Hodnocení půdní vody s biologického hlediska metodou vysychacích křivek. [Soil water evaluation from biological point of view by means of dessiccation curves.] In Czech. — Folia biol. *2* : 100—111, 1956.

—: Intensita transpirace některých odrůd našich pšenic. [Transpiration rate in some wheat varieties.] In Czech. — Folia biol. *4* : 290—298, 1958a.

—: Spotřeba vody na transpiraci u některých odrůd našich pšenic. [Water consumption for transpiration in some wheat varieties.] In Czech. — Folia biol. *4* : 299—309, 1958b.

—: Content of essential oils and water regime in some officinal plants growing in Czechoslovakia. — Acta Facult. Pharmaceut. Brun. et Brat. *1* : 11—31, 1958c.

—: Transpiration rates of leaf blades of irrigated and not irrigated plants of spring wheat. — Biol. Plant. *5* : 200—210, 1963.

—, Bacíková, A.: Vodní provoz a hromadění silic u rostliny *Coriandrum sativum* L. [Water relations and essential oils in *Coriandrum sativum* L.] In Czech. — Čs. Farmacie *9* : 9—16, 1960.

—, Kocábová, A.: Příspěvek ke studiu změn obsahu silic u rostliny *Levisticum officinale* Koch. [On the changes of essential oil content in Levisticum officinale Koch.] In Czech. — Čs. Farmacie *11* : 299—233, 1962.

—, Nováček, M.: Vliv množství humusu v půdě na růst a transpiraci rostlin *Linum usitatissimum* L. [Influence of humus content in soil on growth and transpiration in flax.] In Czech.— In press, 1963.

—, Hloušková, D., Řezáč, K.: Vliv závlah na vodní provoz rostlin a na obsah silic. [Influence of irrigation on water relations in plants and their content of essential oils.]In Czech. — Fol. biol. *1* : 288—287, 1955.

Schratz, E.: Zum Vergleich der Transpiration xeromorpher und mesomorpher Pflanzen. — J. Ecol. *19* : 292—296, 1931.

—: Beiträge zur Biologie der Halophyten. IV. Die Transpiration der Strand- und Dünenpflanzen. — Jb. wiss. Bot. *84* : 593, 1937.

Slavík, B.: K dynamice vodního deficitu rostlin. [On the dynamics of water deficit in plants.] In Czech. — Preslia *27* : 124—153, 1955.

Srb, V., Klimešová, E., Klimešová, J.: Vliv závlah na růst a obsah silic u máty peprné *(Mentha piperita* Hudson*)* během ontogenese. [Influence of irrigation on growth and essential oils content in *Mentha piperita* Hudson during ontogenesis.] In Czech.— Čs. Farmacie *9* : 49—55, 1960.

Discussion

W. R. Müller-Stoll: It should be assumed that the differences in the daily course of transpiration in plants of different age and size might be explained at least in part by a different water deficit arising at different times. In pots of equal size young plants dispose of much better water source than older grown-up plants with their much higher need for water.

M. Penka: We can agree, in principle, with the opinion expressed by Prof. Müller-Stoll. However, it is necessary to add that the transpiration rate was expressed by the actual transpiration rate (mg. per g. fresh weight per minute) and also by water uptake for transpiration (in mg. per minute). In the young experimental plants the transpiration rate was very high, while water uptake for transpiration was very low. As against this, in older plants the transpiration rate was low, while the water uptake for transpiration was high. These differences in transpiration in plants of different age were found in plants cultivated in experimental pots but not in plants cultivated in field conditions. Water deficit influences transpiration of plants just as transpiration influences water deficit in plants.

THE MOISTURE CONSUMPTION OF PLANTS DESCRIBED AS A HYDROLOGICAL PHENOMENON

W. C. VISSER

Instituut voor Cultuurtechniek en Waterhuishouding, Wageningen, Netherlands

Evapotranspiration is considered to depend on three factors: the capillary movement of water through the unsaturated soil to the root surface, the transmissibility of the stomata and the evaporative capacity of the atmosphere.

It has proved to be possible to evolve a quantitative description of the flow of moisture through the soil and the plant. This flow is a function of the availability of the soil moisture and the consumptive needs of the plant depending on the potential evaporation due to radiation. In the formula, parameters of the soil, the plant and the climate describe how evapotranspiration is determined by the co-operating factors. The elaboration which leads to the formula is given elsewhere (Visser 1962, 1963a, b, c). Here we will deal particularly with the parameters depending on the properties of the plant.

The flow equation

The formula uses as starting point a description of the extraction of moisture from a cylinder of soil around each root. The moisture flow converges to the soil-root interface. The flow into and through the plant may be considered as Poisseuille flow through a number of flow sections, each with a specific length l_p, a constant conductivity k_p and cross sectional area of flow F_p. To one of these sections, representing the stomata, special consideration will be given.

The influence of the soil moisture availability V_i, the evaporative capacity of the atmosphere E_0 and the flow resistance in the stomata $\Delta s = l_s/k_s F_s$ is given by the formula:

$$g^{n-1} (E_0 - E^*)^{n-1} \left(\sum_{i=1}^{u} A_i V_i - gE^* \right) = \sum_{i=1}^{u} A_i \left(\frac{\mathrm{h}_p}{\mathrm{s} + \Delta \mathrm{s}} \right)^{n-1} \tag{1}$$

In Fig. 1 the influence of g, E_0 and V on the evapotranspiration E^* as given by formula [1] is depicted.

Meaning of the symbols

The parameters and constants appear in the formula as single values or lumped together, with the following meaning and composition:

E_0 = evaporation of open water, for which pan values or calculated values according to Penman are used;

g = a reduction factor, allowing for 1) discrepancies between the values used and the real values of E_0, 2) for differences in climatological exposition of the site and 3) for differences in flow resistance in the plant;

E^* = evapotranspiration only dependent on the soil moisture content and more specifically not influenced by the properties, accounted for by g. The real evapotranspiration E is equal to gE^*;

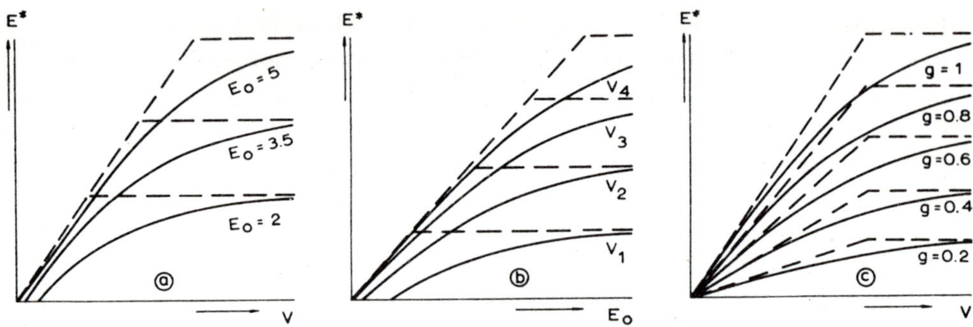

Fig. 1. Change in E_0 — Fig. a — and V — Fig. b — have the effect of changing the horizontal asymptote of the curve. Change of g — Fig. c — changes both asymptotes.

V = availability level of soil moisture content according:

$$V = \frac{1}{\psi^{n-1}} = \left(\frac{Gv^a}{(v_{\max} - v)^b} \right)^{n-1} \qquad [2]$$

G, a and b are constants of the desorption curve; $v_{\max}$ the apparent pore space and v the soil moisture content. ψ stands for the soil moisture stress at a moisture content v. Generally n has the value 2 so that $n - 1$ will be unity. The value of a is of the order of 3. The value of b may change from 0 to 10 or higher and is very variable;

A_i = availability factor for the ith layer of the profile with:

$$A = \frac{0.3^n . 4}{n - 1} \frac{k_s}{r_{\max}^n} \frac{L}{d^2 \left\{ \ln \left(\dfrac{d}{r_p} \right)^2 - 1 \right\}} \qquad [3]$$

Here L is the thickness of each layer, d the radius of the cylinder of moisture extraction around each root and r_p the mean radius of the roots. All values may vary from layer to layer;

$\dfrac{k_s}{r^n_{\max}}$ = measure for the unsaturated permeability, derived from the relation:

$$\frac{k_c}{k_s} = \left(\frac{r_{\max}}{r_c}\right)^n \tag{4}$$

This relation between the capillary conductivity k_c, the saturated conductivity k_s, the maximum pore size $r_{\max}$, the size of the largest pore r_c containing moisture and an exponent n describes the change in capillary conductivity with changing moisture content and changing size of the largest water filled pore. The value of k_c vanishes by elimination, the r_c is substituted by the values of the availability level V;

h_p = number of plants per unit area;

$s, \Delta s$ = the flow resistance in the plant according the Darcy formula for moisture flow:

$$s = \sum_{j=1}^{p-1} \frac{l_{pj}}{k_{pj} F_{pj}}$$

$$\Delta s = \frac{l_s}{k_s F_s} \tag{5}$$

In this formula l stands for the length of the section of the flow path, k for the conductivity and F for the cross sectional area of the jth section or for the sth section, the section of the stomata. For s the sum of the flow resistance is taken, with the exception of the flow resistance Δs of the stomata.

All units should be expressed in a consistent system, especially the formula [7], obtained by integration.

The influence of the plant parameters

The plant influences the magnitude of the evapotranspiration by the size and distribution of its root system, represented by d and r_p of the parameter A, by the transmissibility $k_s F_s$ of the stomata contained in the parameter Δs and by the value of g.

The stomata

The closing of the stomata is described by a zero value of the transmissibility $k_s F_s$ in formula [5] and therefore by an infinite value of Δs. This makes the right hand side of equation [1] equal to zero. This means that also the left hand side of equation [1] becomes zero. This requires alternatively:

$$\text{a)} \quad \sum_{i=1}^{u} A_i\, V_i - g\, E^* = 0$$

$$\text{b)} \quad E_0 - E^* = 0$$

$$\text{c)} \quad g = 0$$

Alternative a) and b) mean that as a consequence of the closing of the stomata the observations should shift from the curve to the two asymptotes of the curve. This shift from the curve to the asymptote, however, is not likely to depict the result of closing of the stomata correctly, for it means an increase in evapotranspiration (see Fig. 2).

The only sensible solution is that a decrease in the transmissibility $k_s F_s$ means a decrease in the value of g and because of the relation $E = gE^*$, a decrease in the real evapotranspiration E.

Now by decreasing availability value ΣAV, the stomata will close more and more and the successive evapotranspiration values will shift to curves with decreasing value of g (see Fig. 3a). Up to now, not much of such curving to the right of the lower end of the experimental curves has been observed (Fig. 3b).

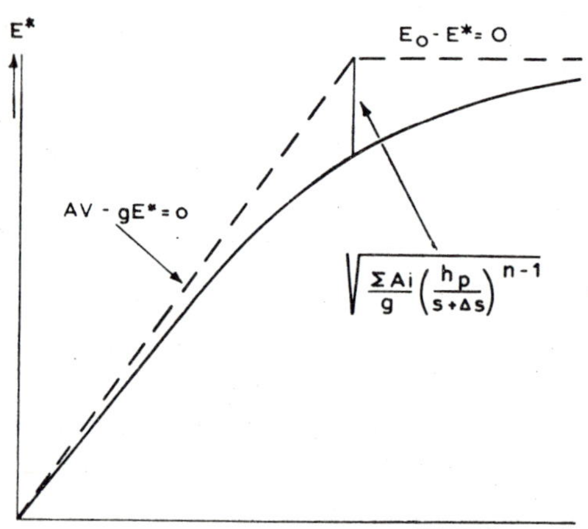

Fig. 2. An infinite value of Δs, describing closed stomata, with constant value of g, makes the right hand side of formula [1], depicted by the vertical distance between the point of intersection of the asymptotes and the curve, equal to zero. The value of g cannot be constant, for it would mean an increase in evapotranspiration. A decrease in transmissibility $k_s F_s$ can only mean a decrease in g and with $E = gE^*$ a decrease in E.

An autonomous influence of the stomata was not to be expected. The stomata regulate the loss of water to a value as nearly equal to the uptake of water by the roots as is possible. Thereby the turgor of the plant is maintained to the utmost. The explanation of the activity of the stomata cannot be found in the stomata or the plant, but in the moisture situation in the soil.

The influence of the stomata therefore does not constitute an independent variable in the formula for the evapotranspiration.

The root system

The significance of the size and distribution of the root system is described by the availability factors A_i for the successive layers of the profile.

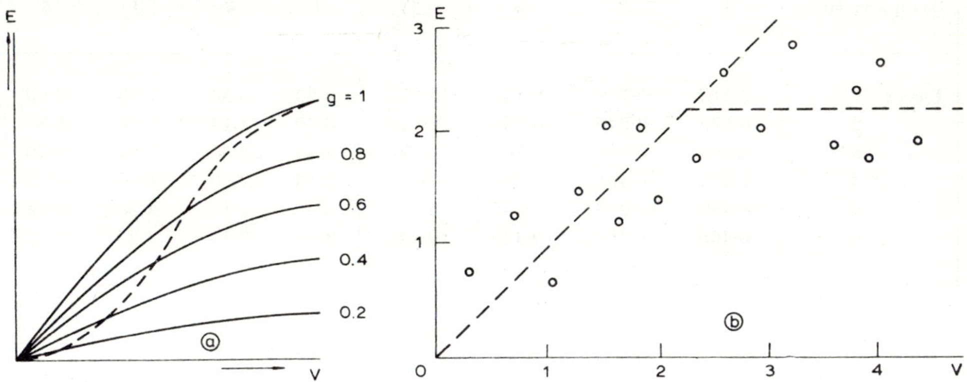

Fig. 3. If the stomata had an active influence on evapotranspiration, superimposed on the influence of the moisture content of the soil, this would mean that experimental curves would be found, shifting from a higher value of g to lower ones. The s-shaped dotted curve of Fig. 3a might be expected. Not much evidence for such influence can be obtained, however, from experimental data (see Fig. 3b).

From formulae [1] and [2], it may be deduced that the ratio A_i/A_j between the availability factors of two layers may be computed from the soil moisture contents of these layers by solving the differential equation:

$$\int_{v_1}^{v_t} \frac{1}{A_i} \left(\frac{(v_{max}-v_i)^b}{Gv_i^a} \right)^{n-1} dv_i = \int_{v_1}^{v_t} \frac{1}{A_j} \left(\frac{(v_{max}-v_j)^b}{Gv_j^a} \right)^{n-1} dv_j \qquad [6]$$

The constants G, a, b and n follow — see formulae [2] and [4] — from soil tests. We assume $a = 3$, $b = 1$ and $n = 2$, whilst G cancels out, taking the case that the soil profile is homogeneous. Integration and rearrangement leads to the formula:

$$\alpha_{ij} = \frac{A_j}{A_i} = \frac{\left(\dfrac{1}{v_{1i}} - \dfrac{1}{v_{ti}} \right)\left(\dfrac{1}{v_{1i}} + \dfrac{1}{v_{ti}} - \dfrac{2}{v_{max}} \right)}{\left(\dfrac{1}{v_{1j}} - \dfrac{1}{v_{tj}} \right)\left(\dfrac{1}{v_{1j}} + \dfrac{1}{v_{tj}} - \dfrac{2}{v_{max}} \right)} \qquad [7]$$

For other values of a, b and n the solution will differ. As i^{th} layer is taken the layer with the most reliable observations of moisture extraction. As soil

Table 1.

Date	26/5	12/6	27/6	13/7	26/5	12/6	27/6	13/7
		v_i					$1/v_i$	
Sampling time	1	2	3	4	1	2	3	4
Layer 1	0·210	0·180	0·130	0·095	4·76	5·56	7·69	10·53
2	0·270	0·240	0·190	0·170	3·70	4·17	5·26	5·88
3	0·425	0·390	0·340	0·270	2·35	2·56	2·94	3·70
4	0·470	0·455	0·400	0·320	2·13	2·20	2·50	3·13
5	0·480	0·465	0·445	0·410	2·08	2·15	2·25	2·44
6	0·465	0·460	0·455	0·445	2·15	2·17	2·20	2·25

$$\left(\frac{1}{v_t} - \frac{1}{v_1}\right) = g \qquad\qquad \left(\frac{1}{v_t} + \frac{1}{v_1} - 4\right) = h$$

	2 — 1	3 — 1	4 — 1		2 + 1	3 + 1	4 + 1
1	0·80	2·93	5·77		6·32	8·45	11·89
2	0·47	1·56	2·13		3·87	4·96	5·58
3	0·21	0·59	1·35		0·91	1·29	2·05
4	0·07	0·37	1·00		0·33	0·63	1·26
5	0·07	0·17	0·36		0·23	0·33	0·52
6	0·02	0·05	0·10		0·32	0·35	0·40

$$gh \qquad\qquad\qquad \alpha_{1j} = \frac{(gh)_j}{(gh)_1}$$

	2,1	3,1	4,1		2,1	3,1	4,1
1	5·056	24·758	65·143		1·0000	1·0000	1·0000
2	1·819	7·738	12·164		0·3598	0·3125	0·1867
3	0·191	0·761	2·768		0·0378	0·0307	0·0425
4	0·023	0·233	1·260		0·0045	0·0094	0·0193
5	0·016	0·056	0·187		0·0032	0·0023	0·0029
6	0·006	0·018	0·040		0·0012	0·0007	0·0006

sampling data two or more successive data should be taken without rain or other moisture supply in the interval.

When the values of α_{ij} are known, the value of $A_{1 \cdots u}$ are found as $A\alpha_{i,\,1 \cdots u}$ by calculating the values of $\Sigma \alpha V$ and $\Sigma \alpha$.

A follows from:

$$A = \frac{E}{\sum\limits_{j=1}^{u} \alpha_{ij} V_j - \dfrac{\sum\limits_{j=1}^{u} \alpha_{ij} P}{gE_o - E}} \simeq \frac{E}{\sum\limits_{j+1}^{u} \alpha_{ij} V_j} \qquad [8]$$

Here P stands for $(h_p/s + \Delta s)^{n-1}$ but this value is small and may be neglected so that the second term of the denominator may be omitted. Observations with small values of $gE_0 - E$ and an increasing magnitude of the second term have slight indicative value for the computation of A and may be left out.

Example

An example of calculation is given for a homogeneous soil, with a recently sown grass sod. The frequency of observation of the soil moisture content in layers of 20 cm. was once in a fortnight. The moisture content should be expressed in a per unit scale.

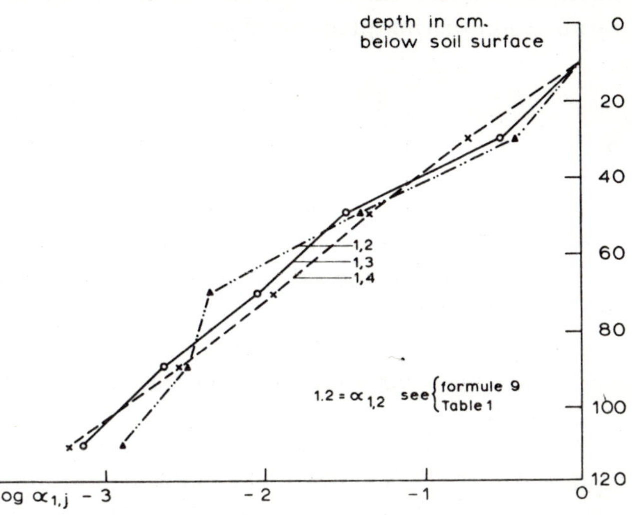

Fig. 4. The log of α_{1j} for a homogeneous soil and the three consecutive sampling data of Table 1 show a linear relation with the depth below soil surface, which proves that the moisture extraction capacity of the root system decreases according to a geometrical series.

The results for α_{1j} in Table 1 clearly show how quickly the capacity α_{ij} of the plant to extract water diminishes with increasing depth. This decrease is sufficiently close to a geometrical series with an argument of 0·225. This means that in this case the availability factor of the next deeper layer of 20 cm. is 22·5% of the factor of the higher layer (see Fig. 4).

The ratio of the α-values to each other is proportional to the number of roots n per unit area according:

$$\alpha_{1j} = \frac{n_j}{n_1} \left(\frac{\dfrac{4}{\ln \pi r^2 n_1} - 1}{\dfrac{4}{\ln \pi r^2 n_j} - 1} \right) \qquad [9]$$

as comparison with formula [3] may show. This relation enables one to compute activity constants for unit length or unit weight of the plant roots in relation to depth.

There is in Fig. 4 some scatter in the results for successive layers or sampling data. This may be due to errors in the moisture determination — here neutron probe readings — as well as to capillary movement between soil layers without the plant as mediator. It also may be due to an increase or decrease with time in the number of active roots, or to penetration of the roots to deeper layers. Work is in progress to get to a better understanding of the extractive capacity of the plant and of the errors involved in the determination of this plant property.

Summary

The way in which real evapotranspiration depends on the evaporative capacity of the atmosphere, the soil moisture content and the density and distribution of the root system, can be described by an equation based on the formula of Darcy.

The extractive capacity of the root system decreases rapidly with increasing depth. In a homogeneous soil this decrease may be described by a geometrical series.

Up to now no autonomous effect of the stomata could be shown. They apparently only regulate the evapotranspiration to be as equal to the uptake of water as possible. The explanation of the regulative activity of the stomata has to be found in the changing moisture situation in the soil and the evaporative capacity of the atmosphere. The stomatal aperture is no free variable in the process and in the given formula.

This formula enables one to elucidate the influence of environmental factors on evapotranspiration and the influence of moisture extraction by the plant on the shape of the moisture profile. It is also possible to calculate a number of evaporation parameters, which consist of soil, plant and site constants with a definite physical meaning. The plant activity constants, often not so easy to determine, may be computed by inserting the soil constants in the complex evaporation parameters.

References

Visser, W. C.: A method to determine evapotranspiration in soil monoliths. Paper read at the Colloque sur la Méthodologie de l'Ecophysiologie Végétale, Montpellier, April 1962.

—: Die Entwicklung der Zielsetzung und der Methoden bei Lysimeteruntersuchungen. Paper read at the German-Dutch Lysimeter Colloquium, Wageningen, May 1963a.

—: Moisture requirements and rate of moisture depletion of crops. Paper read at the 16th Internationale Studiedagen over Waterproblemen, Liege, May 1963, In French: Cebédeau, Nr. 235—236, 1963 b. English version. Technical Bull. I.C.W. 32, 1964.

—: Soil moisture content and evapotranspiration. Paper presented at the XIIIth General Assembly of the I.A.S.H., Berkeley, August 1963, Publ. 62 I.A.S.H. Comm. Evap., and Technical Bull. I.C.W. 31, 1963 c.

Discussion

W. Larcher: We experimental ecologists and also most experimental physiologists are not used to working with mathematical derivations and we therefore tend to avoid complicated formulae in particular. We must be grateful to Dr. Visser for the way he has taught us to read formulae. Although we cannot follow the deductions, the formula, whose correctness is confirmed in our experiments, is a very useful aid to our work. Not only do we find in it the expression of all factors which affect a given process but also the clear delimitation of the extent to which the individual factors have an effect. We need the clear thinking of theoreticians and their result, a formula, just as much as we need basic experimental research. When we have gone a little further in this direction we shall be able consciously and economically (i.e. by obtaining definite results at less outlay), to carry out extensive work and test varieties in such a way that our results can be used directly in practice.

W. C. Visser: I thank Dr. Larcher very much for emphasizing some of the points in my paper.

G. F. Makkink: Although Dr. Visser himself has already pointed out that his approach has a practical value, and Dr. Larcher has accentuated the importance of the theoretical base so clearly demonstrated in Visser's approach, I should like to stress the fact that his approach is also important because it tries to give absolute figures for the field which are needed by farmers and civil engineers. Generally, most of our work, however important in fundamental scope it may be, does not give us the possibility of calculating absolute figures for the field under ordinary circumstances.

P. Strebeyko: I don't understand your calculations in all details. It's rather difficult. Therefore I'd like to ask you to explain if there are elements of Poiseuille's law? What about the water viscosity?

W. C. Visser: The Poiseuille equation, combined with the equation for the desorption curve, leads to an expression for the unsaturated permeability. The unsaturated permeability, inserted in the Darcy equation gives us the flux. Integration gets rid of the parameters for unsaturation. Only the upper integration limit shows up. The Poiseuille law, therefore, is behind the saturated conductivity k_s and the maximal pore radius r_{max}. Variation in viscosity will be expressed as variation in k_s.

S. J. P. K. Bezuidenhout: What in your opinion is the most variable factor, measured under field conditions that causes deviations from the ideal curve derived from your formula?

W. C. Visser: The formula will prove to be correct if applied under conditions for which it is evolved. This means that longer time intervals have to be used during which the constants are considered to remain constant. The formula does not describe short time variations. Used under the right conditions deviations will mainly follow from the variations in the soil moisture content at small distances.

P. G. Jarvis: Dr. Visser's formula takes into account the water content of the soil and its influence on the water potential and capillary conductivity of the soil. As far as I can see it does not take into account the osmotic potential of the soil solution. This is, of course, small compared with the water potential of the atmosphere and will make very little difference to the overall flux when the supply of water is plentiful. However, it contributes to the control of the base level of water potential and turgidity in the plant and hence has an influence on growth and metabolism. A lower soil osmotic potential would be expected to result in a lower water potential for the same flux and hence a lower turgidity and, when water is in short supply, a more rapid approach to stomatal apertures limiting the overall flux.

W. C. Visser: The formula states that a certain flux is the result of a soil moisture status, evaporative capacity of the atmosphere and a conductivity for moisture flow of the plant. The description of these three properties is a second problem. The poisture potential in the soil may be measured in several ways, dependent on the definition one gives of the problem of water transport. The osmotic potential may be used if required. Here the hydrological potential is used because the determination of the desorption curve with osmotic potentials is time consuming and the investigation directed to explain the rate of water loss, requires a shift from tensions to moisture content. The use of water potential, therefore, is not a basic feature of the method. It seems questionable, however, whether the osmotic potential would yield better results, taking into account that the salts from the soil solution will have their influence on the osmotic potential of the cell sap and in a gradient the larger part of the salt influence will cancel out.

We expect that the plant will be able to change its turgidity without changing the gradient of the moisture flux and expect that active or passive change of the osmotic value in the plant need not be of importance for transpiration. The transpiration in that case should not influence growth or metabolism functionally but should only be more or less correlated to it.

WATER RELATIONS OF SOME SPECIES OF WORMWOOD IN KAZAKHSTAN

V. M. SVESHNIKOVA

V. L. Komarov Institute of Botany, Academy of Sciences U.S.S.R., Leningrad, U.S.S.R.

One of the most widespread forms of plant stands in arid and semi-arid zones are half-shrubs. Among these plants which occur in the deserts and semi-deserts of Central Asia and Kazakhstan, a special place is occupied by wormwood. Representatives of this genus are strikingly well adapted to existence under conditions of severe drought. It could, therefore, be assumed that the leading place of wormwood in the phytocenosis, developing in arid environments, is conditioned to a certain extent by the character of its water relations.

A study of the water relations of some species of wormwood was made in the subzone of the arid steppes which are very extensive in central Kazakhstan. The following species of wormwood were chosen for investigation: *Artemisia frigida*, *A. lercheana*, *A. pauciflora*, *A. sublessingiana* and *A. nitrosa*, which are edificators of basic zonal plant communities.

In order to understand the character of adaptation of wormwood to arid conditions it was considered to be necessary to investigate the diurnal and seasonal changes in the basic elements of water balance and the amplitude of the variations in the basic indicators of water relations.

The results of experiments made under field conditions (from May 15, to September 15), permitted both the determination of the characteristic content of the different forms of water in leaves, of osmotic pressure, transpiration rate, etc. and the size of the variations in these indicators during the vegetation period. Thus, for example, the curve of the water content of the leaves of *Artemisia lercheana*, *A. sublessingiana* and *A. nitrosa* gradually fell from the beginning to the end of vegetation. However, the greatest water deficit in the leaves occurred in *A. lercheana* at the end of June, in *A. sublessingiana* at the end of July and in *A. nitrosa* only half way through August.

The curve of water content in *A. frigida* showed an equilibrium with small variations. The curve of *A. pauciflora* was typical in the unchanging level of leaf hydration with a sudden break in June. This was connected with a sudden change in water relations in the soil where these plants were growing.

An analysis of variations in maximal and minimal hydration showed the greatest differences in *Artemisia nitrosa* (34·8%) and *A. lercheana* (32·6%), i.e. in wormwoods that are less protected by soil moisture which also points to the great lability of its water balance. On the other hand, the amplitude in *A. frigida* (11%) and *A. pauciflora* (16·7%), occurring in better conditions of water supply, was insignificant.

The study of different forms of water by means of the Gusev's method in 50 % sucrose solution in the leaves showed that in full vegetation *A. sublessingiana* had the greatest amount of bound water (approx. 70%), then *A. lercheana* (approx. 55%) and *A. frigida* had much less (40%). From half way through August it showed a clear tendency to rise. At the end of the vegetation period maximum values of bound water content were registered, amounting to 90% in *A. lercheana*, and 70—76% in *A. frigida* and *A. sublessingiana*.

It is interesting that the bound water content and also the other indicators of water relations varied most in *A. lercheana* (21—90%), then in *A. frigida* (11—71%) and and only slightly in *A. sublessingiana* (49—76%).

The amount of free water in the leaves in the vegetation period amounted to 30—35% in *A. frigida*, to 20—25% in *A. lercheana* and 17—20% in *A. sublessingiana*. In the critical time, at the end of the dry period the free water content was decreased to 4·0—5·8% in *A. lercheana* and to 15·3—33·3% in *A. frigida*.

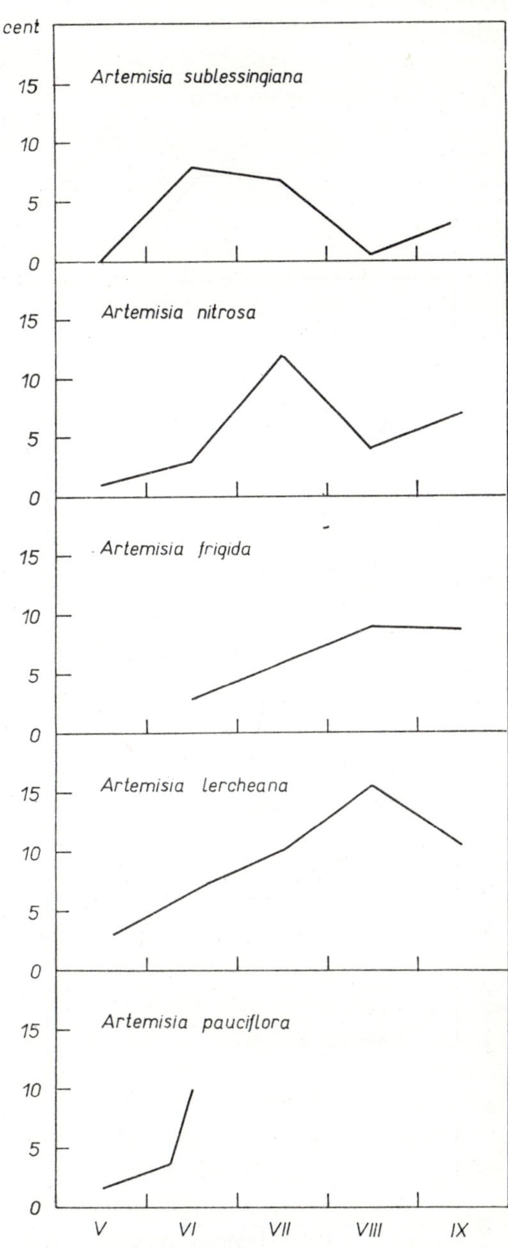

Fig. 1. Changes in water content in leaves during vegetative season (in % fresh weight).

The mean values of water deficit (Fig. 2), determined by the method of Krasnoselskaya-Maksimova (1917) varied between 2 and 16% in wormwood. The maximal value was found in *A. lercheana* (16%), growing in soil with a deficient water supply. The water deficit in the leaves of *A. sublessingiana*, growing in soil with a large water supply, was the smallest (7%). *Artemisia pauciflora* ended vegetation at a deficit of 10%.

The wilting of leaves was first noted in *A. sublessingiana* (after 1—2 hours in the shade), in *A. frigida* and *A. pauciflora* after $^1/_2$—2 hours. Wilting in *A. lercheana* and *A. nitrosa* started a little later (after $1^1/_2$—2 hours) and at a leaf hydration of 45—25%. This means that *A. lercheana* and *A. nitrosa* had the greatest resistance ot water loss in the leaves and *A. sublessingiana* and *A. frigida* had less.

In investigating osmotic pressure as an indicators clearly reflecting provision of the plant with water, the following results were obtained (Fig. 3):

The leaves of *A. lercheana* had the highest concentration of cell sap, increasing in parallel with increasing drought; the cell sap concentration increased less in *A. nitrosa* and *A. sublessingiana*. Mean osmotic pressure values were lower in *A. frigida*.

A comparison of maximal and minimal values of osmotic pressure showed that the variation was great in *A. lercheana* and small in *A. nitrosa*. The lowest osmotic pressure was found in *A.*

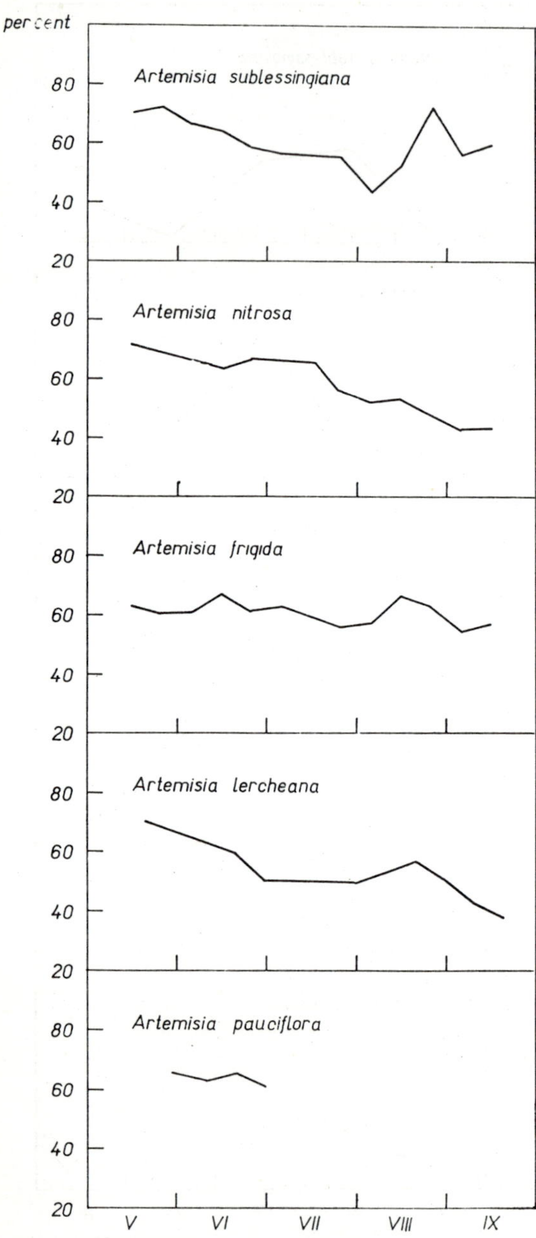

Fig. 2. Water deficit in leaves (in % fresh weight).

sublessingiana (18 atm.) and in
A. pauciflora (20 atm.). The
highest concentration was found
in *A. lercheana* (61·2 atm.) and
A. nitrosa (55·2 atm.).

The curve of water loss by
transpiration (Fig. 4) had
little variation during the vegetative period in *A. frigida*
and *A. sublessingiana* — 420—
750 mg./g./hr. in the first species and 600—1,250 mg./g./hr.
in the second. The seasonal
transpiration curves in *A. lercheana* and *A. nitrosa* showed
considerable variations, transpiration rate changed from
500—2,050 mg. in *A. lercheana*
and from 960—2,200 mg./g/hr.
in *A. nitrosa*. The highest transpiration rates were recorded
in *A. pauciflora*. Maximal transpiration rates attained in *A. frigida* were 670—1,960 mg./g./hr.,
then in *A. sublessingiana* 860—
1,960 mg./g./hr. In *A. lercheana* the value was 670—2,420
mg./g./hr., in *A. nitrosa* 100 to
3,180 mg./g./hr. and in *A. pauciflora* 1,250—3,200 mg./g./hr.

In connection with water
loss by transpiration we were
interested in the ratio between
the green and woody tran-
spiring parts and also in the
ratio of each of these parts to
the total weight of overground
organs in all species of wormwood investigated.

It was shown that the leaf
weight made up 17—40% of
the weight of the total sub-

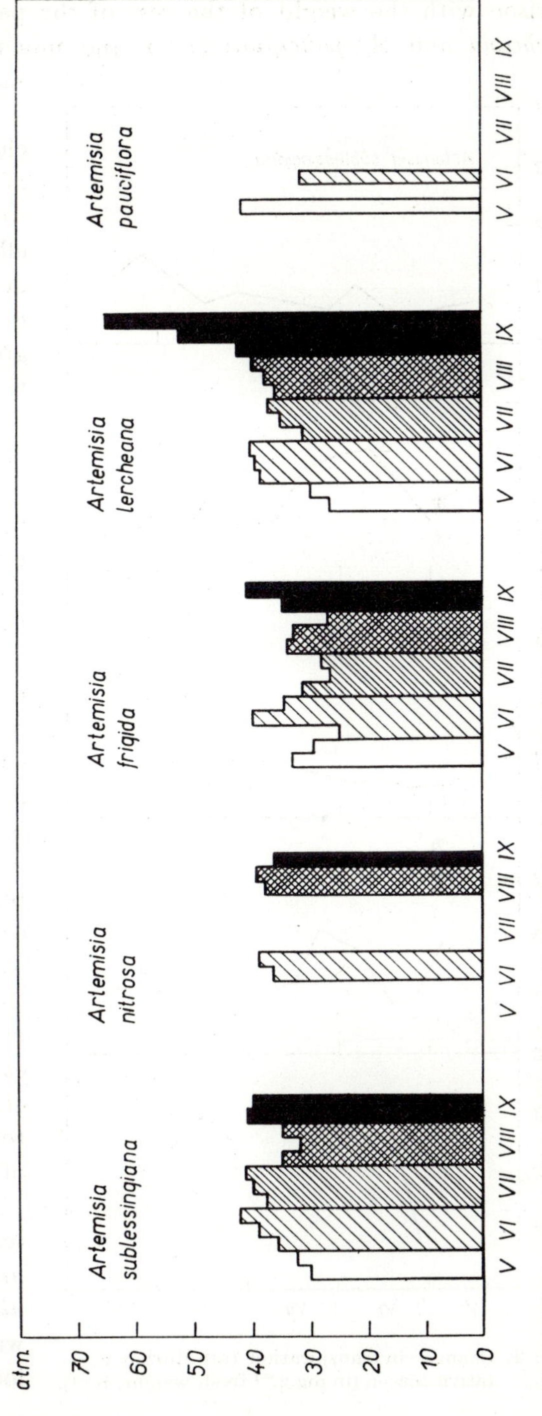

Fig. 3. Mean values of osmotic pressure in plant leaves (in atmospheres in decade).

stance and the weight of the woody parts 83—60%. The leaf weight in comparison with the weight of the rest of the parts was relatively small in *A. lercheana* and *A. pauciflora* (1 : 5) and much greater in *A. frigida* and *A. sublessingiana* (1 : 1·7).

The results obtained showed a close agreement in the reactions of the given species of wormwood to water supply and the effect of high temperatures in the desert-steppe zone. The agreement is displayed in the rapidly progressive water loss, in high bound water values, in cell sap concentration and in the transpiration rate.

In wormwoods with a high water content in the leaves, wilting started earlier and in species with a small water content in the leaves, it started much later.

A detailed analysis of the water relations in the species of wormwood investigated made it possible to explain species differences in the reconstruction of water balance during the vegetative season. In *A. lercheana* it is displayed sharply in variations in osmotic pressure, total and bound water and also in the minimal free water content in the leaves; water loss in the leaves started earlier than in the other species. *A. lercheana* was more resistant to wilting than the other species of wormwood.

The character of the water relations in *A. lercheana* and *A. nitrosa* was similar, but in *A. nitrosa* the leaves were richer in water and less resistant to water loss in wilting. On the other hand

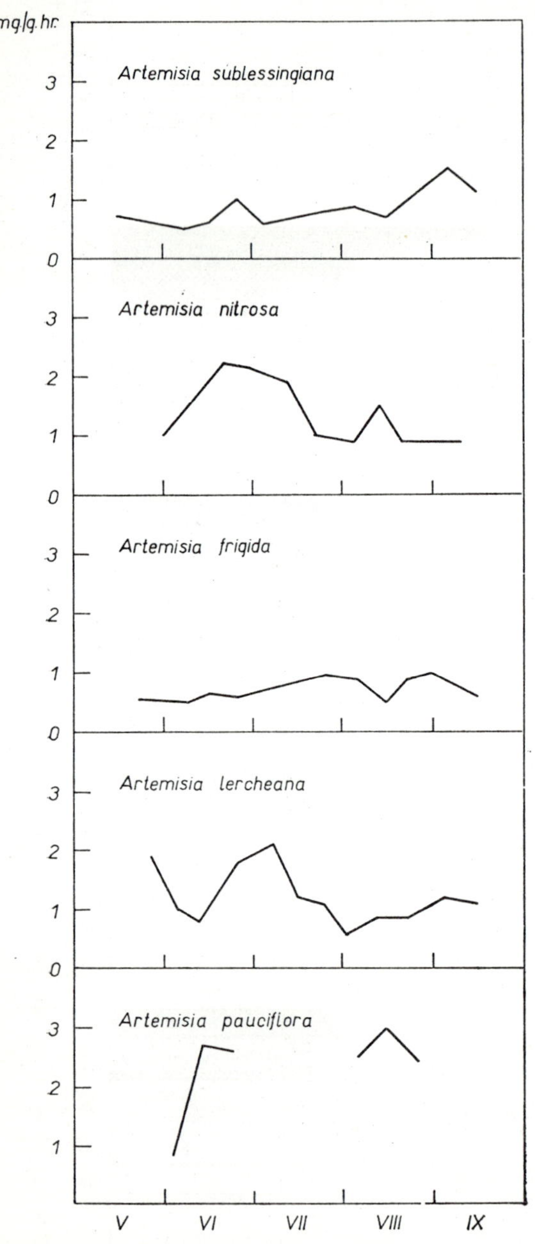

Fig. 4. Changes in transpiration rate during vegetative season (in mg.g.⁻¹ fresh weight. h.⁻¹).

the changes and size of the different indicators of water relations in *A. frigida* differed sharply from those of the other species. Water content showed almost no change during vegetation (total and bound water), the amount of free water was greater. In *A. sublessingiana*, irrespective of similar relations to *A. frigida*, there was a sharp and earlier onset of variations in water content in the leaves and greater water loss from the leaves and also a smaller free water content.

Comparison of the values of basic indicators of water relations in *A. pauciflora* shows some special properties: the highest transpiration rate and the water content of the leaves is almost unchanged during the vegetative season, further there is a rapid loss of water reserves and a low osmotic pressure. In *A. pauciflora* all these indices did not change with the onset of a period of drought as observed in the other species.

Thus *A. lercheana* and *A. frigida* can very clearly be distinguished on the basis of the indicators of water relations during the vegetative period. The character of the water relations in *A. nitrosa* is similar to that in *A. lercheana*. The water balance in *A. frigida* and *A. sublessingiana* are very similar and quite different from the water balance in *A. pauciflora*.

In our opinion, based on the power of rapidly reorganizing water balance and the ability to retain water on wilting, *Artemisia lercheana* and *A. nitrosa* are the most resistant species of wormwood, capable of surviving a very long dry summer without loss of leaves. *A. frigida* and *A. sublessingiana* can be placed among species less resistant to insufficient soil water according to the character of the water relations. Finally, *A. pauciflora* differs in having a stable water balance as a result of which the plant looses its leaves with the onset of a dry period.

Summary

An investigation was made of the characteristics of the water relations of the following types of wormwood which are edificators of the main zonal plant communities in dry conditions in Kazakhstan; *Artemisia frigida*, *A. lercheana*, *A. pauciflora*, *A. sublessingiana* and *A. nitrosa*.

Since the character of adaptation of these species to existence in arid and temi-arid conditions is partly conditioned by their water relations, observaiions were made under field conditions of the diurnal and seasonal changes sn the basic indicators of water balance: content of different forms of water in the tissues, osmotic pressure, transpiration rate, water deficit and resistance to wilting.

Considerable agreement in the reactions of the above species were found to inadequate water supply and the effect of high temperatures in desert-steppe zones and definite species differences were found in the reorganization of water balance in the individual species during the vegetation season.

References

Krasnoselskaya—Maksimova, T. A.. [Daily variations in the water content of leaves.] In Russ. — Trudy Tifl. bot. Sada *19* : 1—22, 1917.

Discussion

M. Rychnovská: 1. How did you obtain the curve of water content of the leaves in the vegetation season ? 2. How was the curve depicting the concentration of cell sap obtained ? 3. How was the transpiration curve obtained ?

V. M. Sveshnikova: 1. The curve of the seasonal pattern of the water content of leaves was obtained by measuring this index in the day (every two hours) during the growth and development of the plant (May—September). 2. The cell sap concentration was measured cryoscopically using the method of Walter. Samples were taken 5 times a day every eight days during the vegetative season. 3. Transpiration rate was measured by the method of Ivanov-Huber. Measurements were made every hour from 6—20 hours each 8th day. The curve of the seasonal pattern is derived from mean daily values, each point is one from the day on which transpiration was measured.

CONTRIBUTION TO THE STUDY OF VARIETAL DIFFERENCES IN WATER RELATIONS IN SPRING BARLEY BY QUANTITATIVE ANALYSIS OF TRANSPIRATION CURVES

M. ZEMÁNEK

Cereals Research Institute, Kroměříž, Czechoslovakia

The physiological characterization of productive and less productive varieties of agricultural plants has great theoretical and practical significance. In relation to indices of water economy, the solution of this problem is unusually difficult since there is a dynamic relationship between changes in the indices of the water relations of plants and changes in factors of the external environment and in the ontogenesis of plants. Some work (Cetl 1957, Penka 1958 and others) has shown that specific differences in indices of water relations do not exist between all varieties but that in some varieties they do exist. Certain characteristics, for example, cuticular and stomatal transpiration rate are to a certain degree species and varietally specific. Little work has been done on the specific indices of the water regime of productive and non-productive varieties of plants and similarly, little has been done in solving the question of the suitability of known methods for such a study.

Material and Methods

Two varieties of *Hordeum sativum* ssp. *distichum* — Valtický and Selekční hanácký — were used for the experiments. Under optimal conditions of water supply the variety Valtický gives higher grain yields than the variety Selekční hanácký. The two varieties do not differ basically in the level of total dry weight production. Under dry conditions Selekční hanácký is able to produce a higher total dry weight and a higher grain content (up to 15%). The experimental plants were cultivated in vegetation pots and water to 60% water capacity throughout the vegetation period.

On the evening before the experiment the pots with the plants were taken into the laboratory and weighed. At 8 a.m. the leaves were cut off from the plants and placed in the basket of an Arland balance. Then, as the average sample they were weighed at 10-minute intervals in the first hour of wilting and at 20-minute intervals in the second and third hours of wilting. Weighing was carried out in 4 repetitions for each variant. The weighing of leaves was carried out in the phases of tillering, earing and grain formation, twice at each stage. When carrying out the weighing the temperature of the laboratory was 22—25° C, the relative air humidity was 60—70% and the illumination 600 lux. At the same time water deficit was determined by the method of Stocker (1929). Transpiration curves were plotted and evaluated by the method of Slavík (1958).

Results and Discussion

The greatest differences in the course of the transpiration-loss curves between the variety Valtický and Selekční were found in the tillering phase and are shown in Fig. 1. The slope of the first linear phase corresponds to total transpiration rate and the slope of the second linear phase to cuticular transpiration (Table 1). In the phase of earing the varietal differences were smaller but the characteristic slope of the second linear phase remained specific for each variety. In the phase of swelling no basic

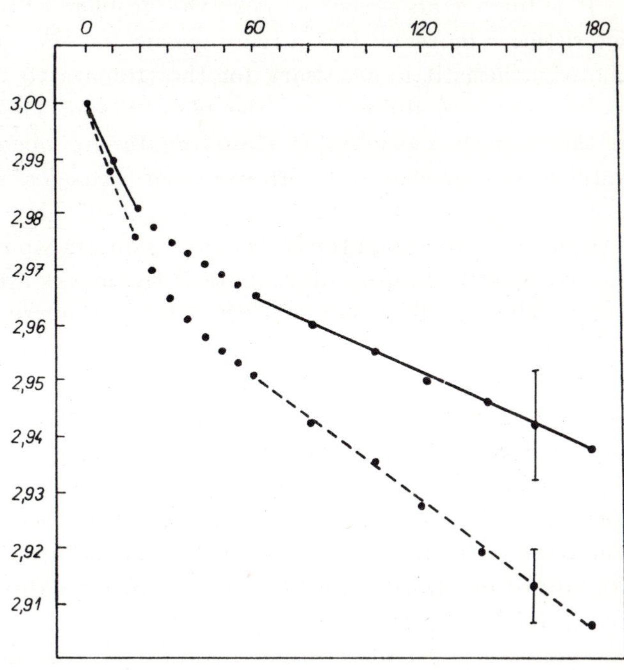

Fig. 1. Transpiration-loss curves in var. Valtický (- - - - -) and var. Selekční hanácký (————). Ordinate: logarithm of leaf water weight calculated to the original weight $G_0 = 1000$; abscissa: time in minutes. Vertical lines represent the confidence interval.

Table 1.

Cuticular and stomatal transpiration rates in two barley varieties in the tillering phase

Variety	Total transpiration rate in mg. $H_2O.g.^{-1}$ H_2O . min.$^{-1}$	Cuticular transpiration rate in mg. $H_2O.g.^{-1}$ H_2O . min.$^{-1}$	Stomatal transpiration rate in mg. $H_2O.g.^{-1}$ H_2O . min.$^{-1}$
Valtický	3·340	0·898	2·442
Selekční hanácký	2·650	0·513	2·137

differences were found between the two varieties. The water deficit of the leaves at the beginning of the experiment was the same in each variety and reached 1—3%.

It is further necessary to solve the problem of which of the observed characteristics is more or less advantageous for the variety. For the process of photosynthesis it is necessary for the stomata to be open as long as possible so that CO_2 can diffuse into the plant tissue. This condition is best fulfilled in the variety Valtický. If the stomata are open, however, higher rate of water loss is reached and with excess of water loss over uptake, a water deficit is created leading to hydroactive closing of the stomata and thereby to the restriction of photosynthesis. Thus a high transpiration rate will not lead to the hydroactive closing of stomata if the leaves are adequately supplied with water. The condition can be fulfilled on supplying the plant with sufficient water and with a higher relative air humidity. Similarly the rate of water loss in the second linear phase contributes to the creation of water deficit and thereby to limitation of photosynthesis and other physiological processes. This effect will be inversely proportional to the water supply of the plant and the relative air humidity. Therefore, a low rate of water loss in this phase of the loss-curve will be of advantage for varieties cultivated under dry conditions, while for varieties cultivated under optimal conditions of water supply it appears to be an index of little significance. It can therefore be assumed that low water loss in the second linear phase would be a suitable characteristic for varieties cultivated under conditions of inadequate water supply. This fact was already pointed out by several workers, e.g. Cetl (1953), and the problems have been analysed by Slavík (1955) and others. The differences in the dry weight production of barley species can be explained by the studied characteristics of the water regime of the plant in addition to other factors. This problem, however, requires more detailed elaboration, particularly in relation to the causes of different water regimes in plants.

Summary

1. The rate of water loss from cut wilting leaves of spring barley changes in dependence of time and is due to water loss from the stomata and from the entire surface of the leaves. It was confirmed that the method of transpiration curves can be used to study varietal differences in spring barley.

2. Significant differences were found in the course of the closing phase of transpiration-loss curves and also in the slope of the second linear phase of the curves in two varieties of spring barley, Valtický and Selekční hanácký.

3. It was found that the slope of the second linear phase of the transpira-

tion-loss curve was specific for a given variety, i.e. in the variety Selekční hanácký it was less and in the variety Valtický it was more, both in the phase of sprouting and earing. In the phase of swelling of the grain no significant differences were found.

4. The differences found in the transpiration-loss curves and the production properties of the varieties show that it is possible to assume a connection between the physiological characteristics of the water relations and the yield of plants. This must, however, be studied in more detail.

References

Cetl, I.: Návrh jednoduché metody ke zjištění odolnosti rostlin vůči suchu. [Simple determination of drought resistance of plants.] — Českoslov. Biol. *2* : 361—369, 1953.

—: Odolnost polních plodin vůči suchu a možnost jejího zvýšení. [Drought resistance of field crops and possibility of its increasing.] — Rozpravy ČSAV *67* (8) : 1—108, 1957.

Penka, M.: Spotřeba vody na transpiraci u některých odrůd našich pšenic. [Water consumption for transpiration in some varieties of wheat.] — Českoslov. Biol. *7* : 98—109, 1958.

Slavík, B.: K dynamice vodního deficitu rostlin. [On the dynamics of plant water deficit.] — Preslia *27* : 124—153, 1955.

—: Grafické stanovení intensity průduchové a kutikulární složky transpirace rostlin. [Graphical determination of the intensity of stomatal and cuticular transpiration in plants.] — Českoslov. Biol. *7* : 347—352, 1958.

Stocker, O.: Das Wasserdefizit von Gefässpflanzen in verschiedenen Klimazonen. — Planta *7* : 382—387, 1929.

Discussion

W. R. Müller-Stoll: How does agricultural practice assess the two varieties of barley compared in these experiments (suitability and yield) and how do the observations from practice
agree with the results obtained?

M. Zemánek: In agricultural practice the variety Valtický is more used because, as compared with Selekční hanácký, it is less subject to lodging and gives higher yields under the usual
conditions of cultivation. Further, Selekční hanácký dries out at the time of ripening. It is
probable that a series of factors come into play at this time. However, they have not been
investigated. Our pot experiments showed that if Valtický is cultivated under dry conditions
it gives a higher grain yield.

G. F. Makkink: I am not sure that it is enough to refer the loss of water to the maximum
water content in the top. I believe that it must be referred to the maximum water content of
the whole plant even though the root has often a limited mass as compared with the top. The
threshold value for stomata closing represents a certain percentage of water in the whole plant
which in field conditions represents an absolute quantity. What is your opinion?

M. Zemánek: We measured transpiration rate in average samples of cut overground parts
and referred it only to the water content of these parts and not to the water content of the
roots, which is, without doubt, important.

H. Polster: You worked with samples which represent the leaf average. Would it not have
been better to work with cut whole plants? During the weight the plants were illuminated by
only about 600 lux. Did you not then only measure cuticular transpiration? From the steep
fall in the curves that you showed us it is evident that drying and the clossing movements of
the stomata were accentuated by decrease in illumination.

M. Zemánek: If whole plants are taken the varieties differ in stalk weight and the measurements are affected by further factors which must be taken into consideration. Better results
are obtained if we only work with the leaves. It is probable that the varietal differences found
are the result of the reaction of the variety to the experimental conditions, particularly to
light. This is certainly a factor which we must pay more attention to. It will be necessary to
study the transpiration curves under different conditions of illumination, temperature and
relative air humidity.

G. Hygen: Were the plants cultivated at 600 lux or only weighed?

M. Zemánek: The plants were cultivated in pots with normal illumination in a glasshouse and the evening before the day of measuring they were brought into the laboratory and weighed further at 600 lux.

W. Larcher: Are you only able to cultivate plants in a glasshouse? Have you no possibility of cultivating them outside ?

M. Zemánek: Plants can also be cultivated in fields. With regard to the slope of the second linear phase we got the same results whether the plants were cultivated in fields or in pots. If, however, we wish to record the first linear phase and the closing phase it is more complicated to transport samples from the field to the laboratory.

A REPORT ON THE PROBLEM OF THE WATER ECONOMY OF OAT PLANTS AND OF REGULATING TRANSPIRATION BY TREATMENT WITH VARYING AMOUNTS OF NITROGEN ·

RUTH ZWICKER

Laboratory of Water Relations in Crop Plants, Adolf Zade Institute for Agriculture, Karl Marx University, Leipzig, German Democratic Republic

The water economy of plants, and specially transpiration, have been the subject of numerous investigations. The results obtained by the use of nitrogen and its effect on transpiration are not uniform. It has been found that nitrogen both increases and decreases transpiration (Biebl 1955). The results obtained by studying the influence of the seasonal rhythm of oat plants treated with different concentrations of nutrient solutions of potassium (Arland and Zwicker 1959) were also of interest in relation to nitrogen. In view of the conflicting reports concerning the effect of nitrogen on transpiration a number of experiments were conducted in an attempt to determine the response and, if possible, elucidate the mechanism by which transpiration is regulated.

Material and Methods

Oats "Svalöfs Goldregen III" were cultivated in pots by the use of quartz sand with different concentrations of nutrient solutions of nitrogen under conditions of seasonal rhythm. Each pot was filled with 6,000 g. of quartz sand. The composition of the nutrients was as follows: 1·92 g. KH_2PO_4, 0·54 g. KCl, 2·97 g. $CaCl_2$, 1·00 g. $MgSO_4$, 0·50 g. NaCl. NH_4NO_3 was applied in varying amounts, viz. 0·25, 0·50, 1·00, 2·00 and 3·00 g. N. The trace elements were added by use of A-Z-solution after Hoagland. On top of that 1,000 g. of quartz sand without fertilizer was added to obtain equal germination of the seeds. During the growing season the pots (with 30 plants) were placed on carriages in the open air and only at night or on rainy days the carriages were drawn into the greenhouse. Demineralized water was used to keep the quartz sand at a saturation of 60 per cent of its water capacity.

The measurements of transpiration were made from shoots at different leaf stages, beginning at the one-leaf stage and continuing until flowering, using the "Anwelk"-method originally devised by Arland (1929, 1952, 1956, 1959). The investigations were carried out under open-air conditions and also in a room where a balance was put up.

The concentration of the cell sap of leaves or shoots was determined by use of the cryoscopic method (Walter 1931). The samples were taken at the end of a dark period (7 a.m.) as well as under open-air conditions on a fine day at 2 p.m.

Total-N and protein-N were determined by use of the Kjeldahl-method using the device of Mothes and Engelbrecht (1952). The determination of glucose was made by use of the method developed by Hagedorn-Jensen (Pringsheim and Leibowitz 1932), and that of potassium by use of a flame photometer.

Results and Discussion

Six-leaf plants tested on May 11, 1959 (Fig. 1) were kept in the open air from the last watering (on the previous day at 5 p.m.) till the beginning of the measurements at 7 a.m. so that they were exposed to the normal seasonal cycle of darkness and light. It was found that under conditions of low light

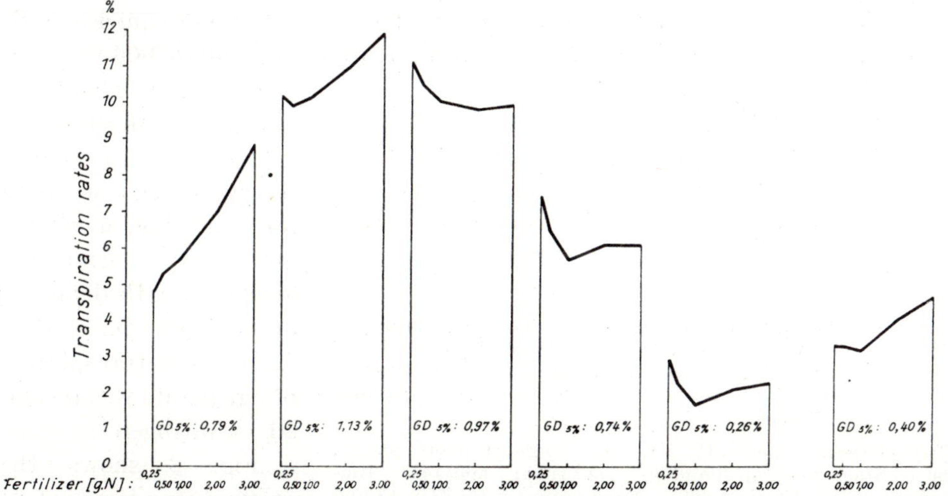

Fig. 1. Transpiration rate of oat plants grown in sand culture with different concentrations of nutrient solutions with nitrogen. Measurements under open-air conditions. Mean values of 3 sets of 10 plants.

Day of measurement:			May 11, 1959			May 12, 1959
Age of plants (days):			45			46
Stage (leaves):			6			6
Day time:	7—8 a.m.	11—12 a.m.	2—3 p.m.	7—8 p.m.	10—11 p.m.	5—6 a.m.
Sand temperature (°C):	13	24	26	20	13	9
Air temperature (°C):	14	26	28	19	14	9
Humidity of air (%):	75	50	43	55	72	100

intensity and also of low temperature, between 7 and 8 a.m., the transpiration was increased by nitrogen. At 8 a.m. the plants obtained sunlight. The increasing of the light intensity and of the temperature, i.e. the decrease of humidity of the air, were followed by an increase of transpiration so that the results of measurements carried out between 11 and 12 a.m. showed a higher loss of water than in the early morning. In the following period (2 to 3 p.m.) there were only a few changes in the environmental factors. While variants 1 and 2 increased, variants 3, 4 and 5 regulated transpiration, so that now the transpiration rate was reduced to a relative minimum by plants treated with amounts of 2·00 g. N. In the late afternoon (5 p.m.) the pots were again supplied with water. The results of measurements obtained at 7 and 10 p.m. showed a decreased transpiration in lowered light intensities with a relative minimum in plants treated with amounts of 1·00 g. N. Next morning with the increase of solar radiation the transpiration rates of all plants rose, but more quickly in the case of plants treated with 1·00 to 3·00 g. N. The transpiration of oat plants fertilized with 0·25 and 0·50 g. N increased more slowly, so that as on the previous day, a relative minimum was established in plants with deficiency of nitrogen. On dark or rainy days transpiration of oat plants was increased by nitrogen.

Fig. 2 shows that at flowering-time, plants fertilized with optimal amounts of nitrogen

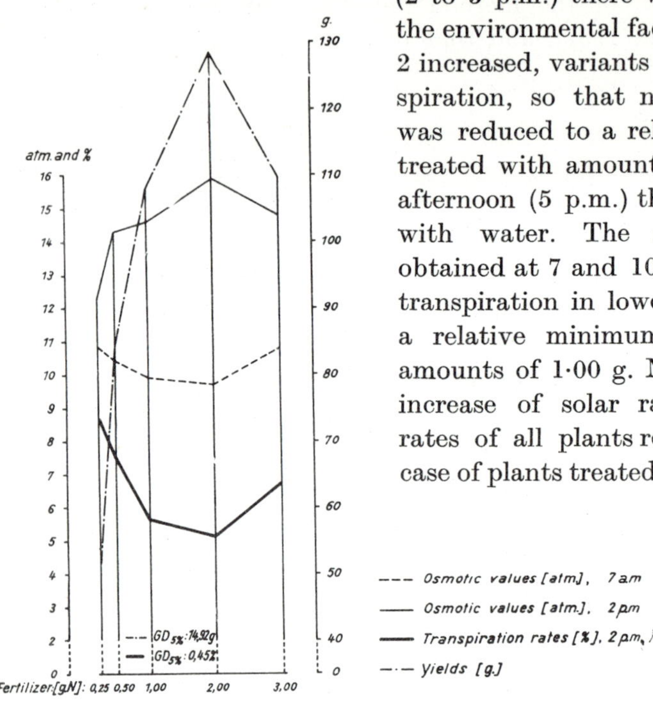

Fig. 2. Osmotic values of the cell sap of leaves and transpiration rates of flowering oat plants grown in sand culture with different concentrations of nutrient solutions of nitrogen. Measurements under open-air conditions. Mean values.

Day of measurement: 17. VI. 1961 — Age of plants (days): 69 — Stage (leaves): 8 and 7 — Day time: 7 and 14 — Illumination (lx): 2 and 64,000 — Sand temperature (°C): 17 and 27 — Air temperature (°C): 18 and 29 — Humidity of air (%): 78 and 43.

(2·00 g. N) effected the best growth accompanied by low osmotic values at the end of the dark period (7 a.m.). They gave the highest yields. Furthermore, Fig. 2 also shows that the regulated transpiration at 2 p.m. in the flowering stage during a fine day in plants treated with optimal amounts of nitrogen is accompanied by a marked increase of the osmotic value. An analysis of the cell sap was made, and the results obtained showed (Table 1) that the increased concentration of the cell sap of plants optimally fertilized arises not only from a loss of water but also from an absolute increase of glucose. The analysis made of plants with a deficiency of nitrogen showed either no change in the glucose rate or a slight reduction. In the diurnal rhythm during fine days the difference between the maximum and the minimum hydrature in optimally fertilized plants was larger than in those with

T a b l e 1.

Analysis of the cell sap of oat plants (June 17, 1961, flowering stage) treated with different amounts of nitrogen (Mean values of 2 sets of 30 plants: 4th, 5th, 6th, 7th and 8th leaves).

Fertilizer g. N	Cell sap ml.	Nitrogen		Potassium		Glucose	
		g.	%	g.	%	g.	%
7 a.m.							
0·25	8	0·0008	0·010	0·089	1·11	0·102	1·28
0·50	17	0·0024	0·014	0·217	1·28	0·191	1·12
1·00	36	0·0040	0·011	0·342	0·95	0·437	1·21
2·00	42	0·0020	0·005	0·260	0·62	0·485	1·15
3·00	28	0·0019	0·007	0·177	0·63	0·281	1·00
2 p.m.							
0·25	4·5	—	—	0·060	1·33	0·064	1·42
0·50	10·0	0·0008	0·01	0·113	1·13	0·153	1·53
1·00	25·0	0·0047	0·02	0·232	0·93	0·388	1·55
2·00	34·0	0·0088	0·03	0·244	0·72	0·797	2·34
3·00	23·0	0·0062	0·03	0·184	0·80	0·337	1·47

a deficiency of nitrogen. This indicates that optimally fertilized plants have an "elasticity" of their own when the weather is fine, while plants with a deficiency under the same environmental factors are "inflexible". These did not much increase the concentration of the cell sap but did increase the loss of water and therefore transpiration. The extent and the regulation of transpiration may be taken as a measure of the ability of the plant to react economically to changes in environmental factors, and specially to solar radiation.

Up to the two-leaf stage, the fresh weight and the area of the leaves were controlled by the concentration of nutrient solutions. Growth of both shoots and roots was limited by high concentration. A large growth of roots and shoots through low concentration of nutrient solution (deficiency of nitrogen) in the one- and two-leaf stage was followed by a high content of water and by a low osmotic value of the cell sap in the shoots. In proportion to the high uptake of water in plants with deficiency of nitrogen in the first leaves, a high uptake of potassium was also observed. The synthesis of protein in the one- and two-leaf stage was more complete by plants with a deficiency of nitrogen than by those optimally fertilized (Zwicker 1963).

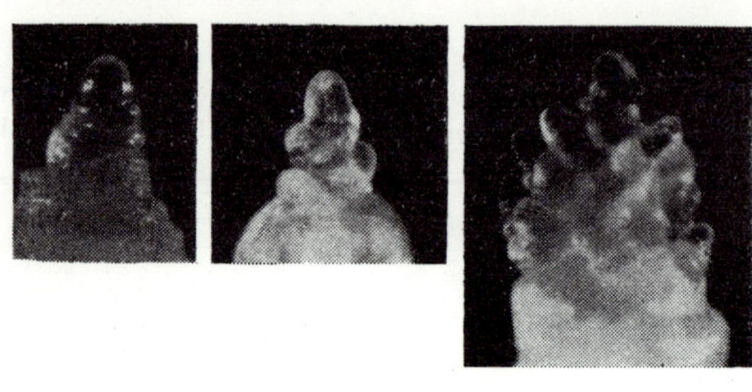

Fig. 3   Fig. 4   Fig. 5

Fig. 3. Apical point of oat plants in the p h o t o - s t a g e.

Figs. 4 and 5. Apical point of oat plants in the s p e c t r o - s t a g e.

The apical point of oat plants in the two-leaf stage reached the photo-stage (Fig. 3) (Kuperman 1955). Neither blue nor red radiation influenced the transpiration rates specially. Only the concentration of the substrate or the concentration of the cell sap of leaves influenced the extent of the transpiration: The higher the concentration — the lower the transpiration.

In comparison with the first and second leaf, the fresh weight and the extent of the third leaf of plants with deficiency of nitrogen were limited and the largest growth was established in plants fertilized with amounts of 0·50 g. N. In the four-leaf stage the fresh weight and the area of the youngest leaf were also reduced in plants treated with amounts of 0·50 g. N, and now plants fertilized with amounts of 1·00 g. N showed the best growth of all. Compared with the one-leaf stage the synthesis of protein by plants treated with 1·00 g. N increased. With the increased protein synthesis, the water content of the plants also increased (Zwicker 1963). In plants with a deficiency of nitrogen in the three- and four-leaf stage nitrogen began to wander from the oldest leaves into the youngest. In comparison with the first leaf, the content of potassium was reduced in the upper leaves of plants with a deficiency of nitrogen. Considerable differences in the content of potassium were found among the first four leaves of plants treated with a deficiency of nitrogen (Zwicker 1963). Similar differences were found in the osmotic values of the first four leaves of these plants, while both the content of potassium and the osmotic values in leaves of optimally fertilized plants were uniform. It is

T a b l e 2.

Analysis of the cell sap of oat plants (May 17, 1961, spectro-stage) treated with different amounts
of nitrogen (Mean values of 2 sets of 60 plants).

Fertilizer g. N	Cell sap ml.	Refracto-meter value %	pH	Nitrogen		Potassium		Glucose	
				g.	%	g.	%	g.	%
0·25	6·50	9·4	—	0·003	0·046	0·106	1·63	0·092	1·42
0·50	11·00	8·0	—	0·007	0·064	0·159	1·45	0·148	1·36
1·00	27·25	7·5	5·25	0·009	0·033	0·204	0·75	0·307	1·13
2·00	32·00	5·6	5·75	0·017	0·053	0·218	0·68	0·320	1·00
3·00	29·75	4·5	5·95	0·027	0·091	0·217	0·73	0·180	0·61

supposed that the biochemical behaviour of the leaves influences the physiological reactions of the w h o l e plant.

In the three- and four-leaf stage the spectro-stage (Kuperman 1955) was
reached (Figs. 4, 5).

Experiments treated with artificial blue light (by use of "TL" 55 lamps
from the Philips Laboratory) under laboratory conditions caused a relative
minimum of transpiration by o p t i m a l l y fertilized plants (2·00 g. N),
just like open-air conditions at midday or in the early afternoon (2 to 3 p.m.)

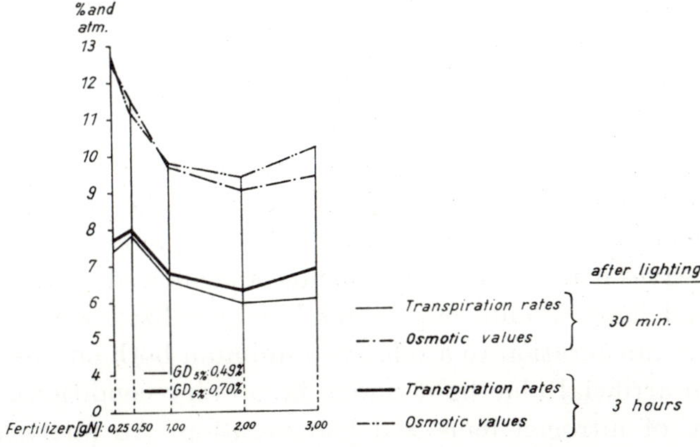

Fig. 6. Transpiration rate and hydrature of oat plants in the spectro-stage grown in sand culture
with different concentrations of nutrient solutions of nitrogen. Measurements under
laboratory conditions by use of blue light (lamps: "TL" 55 Philips Laboratory). Mean
values of 3 gels of 20 plants.

Day of measurement: 17. V. 1961 — Age of plants (days): 32 — Stage (leaves): 4 —
Treatment before measurement: dark, 17° C, 86 % humidity — Treatment during
measurement: Lamps: "TL" 55 (day light with much blue light) after lighting 30 min.
or after lighting 3 hours — Illumination (lx): 6,000 — Sand temperature (°C): 18 and
23 — Air temperature (°C): 20 and 26 — Humidity (%): 70 and 62

on fine days (Fig. 6). Reproducible results by the use of blue light under laboratory conditions have been obtained for the rhythm of long-day-cycles (15-hours-day) at which the plants have been kept at temperatures of 20 to 25°C in the light period and at 18 to 20°C in the dark period. Corresponding with the relative minimum of transpiration in optimally fertilized plants a relative minimum of hydrature was also observed. The analysis obtained from the cell sap of these plants (Table 2) showed, in the same way as under open-air conditions (Table 1), a high uptake of water and also an absolute increase of glucose in optimally fertilized plants. In long-day-cycles the rapidity of metabolism between roots and shoots (Kursanow 1959, Mothes 1956) led quickly to the transport of nitrogen from the oldest leaves of plants with deficiency of nitrogen. These yellowed and faded. In summer-time, growth and development of oat plants with deficiency of nitrogen were considerably retarded, while optimally fertilized plants seemed not to be disturbed. These results obtained from young plants are of ecological importance in so far as they show the need for nitrogen in summer-time. The effect of blue light on transpiration of oat plants from the spectro-stage until flowering in fine days corresponds with the thesis: "The lower the transpiration — the higher the yields" (Arland 1960). Blue light proved to be a factor in regulating physiological processes (Kleschnin 1960), at which first transpiration and then its relation to hydrature, growth, development and yield have been studied.

Summary

Hydrature and transpiration of oat plants treated in sand culture with different amounts of nitrogen were studied from the one-leaf stage until flowering. In comparison with the two-leaf stage (photo-stage), plants with three and four leaves reached the spectro-stage and showed a marked sensitivity for blue and red light. Plants o p t i m a l l y fertilized with nitrogen were able to regulate transpiration to a relative minimum both on fine days at midday and also in artificial blue light under laboratory conditions, while plants with deficiency of nitrogen increased transpiration. An analysis of the cell sap showed that the increased concentration in o p t i m a l l y fertilized plants on fine days arises not only from a water loss but also from an increase in glucose content. At flower-time, watering in the late afternoon led during the night to a relative minimum of hydrature in plants with nitrogen deficiency, and to a relative maximum in o p t i m a l l y fertilized plants. These brought forth the highest yields. Blue light proved to be a regulator of physiological processes.

References

Arland, A.: Das Problem des Wasserhaushaltes bei landwirtschaftlichen Kulturpflanzen in kritisch-experimenteller Betrachtung. — Wiss. Arch. f. Landwirtsch. (Abt. A Pflanzenbau) *1* : 1—160, 1929.

—: Die Transpirationsintensität der Pflanzen als Grundlage bei der Ermittlung optimaler acker- und pflanzenbaulicher Kulturmassnahmen. — Abh. Sächsischen Akad. d. Wiss. Leipzig, Mathem-naturwiss. Kl. *44*, H. 2, Berlin 1952.

—: Ein Beitrag zur Anwelkmethode. — Sitzungsber. d. Dtsch. Akad. d. Landwirtschaftswiss. Berlin *5*, H. 6, 1956.

—: „Tyrannei der Erde". Ein Problem des modernen Landbaues. — Abh. d. Sächs. Akad. d. Wiss. Leipzig, Mathem.-naturwiss. Kl. *46*, H. 3, Berlin 1959.

—, Zwicker, R. (Ref.): Anwelktranspiration und Hydratur verschieden mit Kalium ernährter Haferpflanzen unter besonderer Berücksichtigung des Jahresrhythmus. — Ztschr. Acker- u. Pflanzenbau *108* : 449—472, 1959.

Biebl, R.: Der Einfluss der Mineralstoffe auf die Transpiration. — *In*: Ruhland, W.: Hdb. Pflanzenphysiol. *IV* : 382—426, Berlin-Göttingen-Heidelberg 1958.

Kleschnin, A. F.: Die Pflanze und das Licht. — Berlin 1960.

Kuperman, F. M.: Biologichesky kontrol na sluzhbu urozhayu. [Biological analysis used for yield control.] — Moscow 1960.

Kursanow, A. L.: Wechselbeziehungen der physiologischen Prozesse in der Pflanze. — Sowjetwissenschaft, naturwiss. Beitr.: 899—920, 1961.

Mothes, K.: Stoffliche Beziehungen zwischen Wurzel und Spross. — Ang. Bot. *30* : 12 5—128, 1956.

—, Engelbrecht, L.: Über Allantoinsäure und Allantoin. I. Ihre Rolle als Wanderform des Stickstoffs und ihre Beziehungen zum Eiweissstoffwechsel des Ahorns. — Flora *139* : 586—616, 1952.

Pringsheim, H., Leibowitz, J.: Einfache Kohlehydrate. — *In*: Klein, G.: Hdb. d. Pflanzenanalyse II. Spezielle Analyse, I. Teil: 774—821, Wien 1932.

Walter, H.: Die Hydratur der Pflanze und ihre physiologisch-ökologische Bedeutung. — Jena 1931.

Zwicker, R.: Untersuchungen zur Arlandschen Anwelkmethode unter besonderer Berücksichtigung der Stickstoffernährung bei Hafer. — Habilitationsschrift Leipzig 1963.

Discussion

J. Úlehla: Is it not possible to count with reactivity of the roots to light in the experiments of Dr. Zwicker along the lines of the results of Prof. Kopetz?

R. Zwicker: In reply to that I would wish to say that we only dealt with the reaction of the shoot to light within the limits of the given data. In further research we investigated questions of root respiration and the relationship between root respiration and transpiration in plants with different nitrogen nutrition and illuminated with different blue light. The results are published as the concluding report of the research.

ON THE PROBLEM OF CUVETTE-CLIMATE
(PRELIMINARY COMMUNICATION)

G. LERCH

Institute of Botany, Pedagogical University, Potsdam, German Democratic Republic

Ecological investigations on gas exchange of plants usually demand that the test objects be enclosed in cuvettes for the period of experiment. Those containers made of glass or other transparent material work as heat traps, since in their interior considerably high temperatures will arise even at fairly normal insolation. This is because the natural air exchange between the plant surface and the surrounding atmosphere is interrupted by the walls of the cuvette which prevent the heat arising within the container from radiating out. As a result of this situation rates of respiration and transpiration increase very rapidly to an abnormally high level, and in consequence of it movement of stomata, photosynthesis and the entire water relations of the enclosed plant are changed in a quite unnatural manner. Experimental data obtained under such conditions are practically useless.

Many solutions have been tried to solve this problem, and in the laboratories the so-called cuvette-climate has been checked now in a satisfactory manner.

It is still difficult, however, in field investigations, when the test plants must be kept within the cuvettes for several hours or even days, to avoid super-temperatures. One possibility is to move a strong air current through the cuvette maintaining the natural air exchange as well as possible. High air speed, however, bears two sources of complication: Firstly the danger of mechanical injury to the measuring instrument. The small test-cuvettes in the normal Infrared-Recorders do not usually allow an air speed of more than 100 l./hr. Secondly the exactness and sensitivity of the measuring equipment decreases when the speed of the test air current is much increased. We have, therefore, a contradiction here: On the one hand there is the need to keep the air flow at a moderate speed for measuring purposes; on the other hand the air current must be strong enough to cool the interior of the cuvette. We tried to solve this problem by the following method (see Fig. 1):

We are working with two different air currents induced by separate sources of energy. For ten minutes atmospheric air streams through the cuvette at a speed of 700—1,500 l./hr. cooling its interior. This air current is moved by a strong vacuum cleaner. After that membrane pumps suck a second air current at a speed 60—90 l/hr. for two minutes through the cuvette and through the two infrared-recorders (URAS) measuring net-assimilation and transpiration. The speed of air currents may be regulated according to size of cuvettes and external conditions.

Since we are using cuvettes with a volume of 300—1,000 cm.^{3}, the actual air speed within the cuvettes never exceeds 0·5 m./sec., so that stomata will not be injured by the cooling stream.

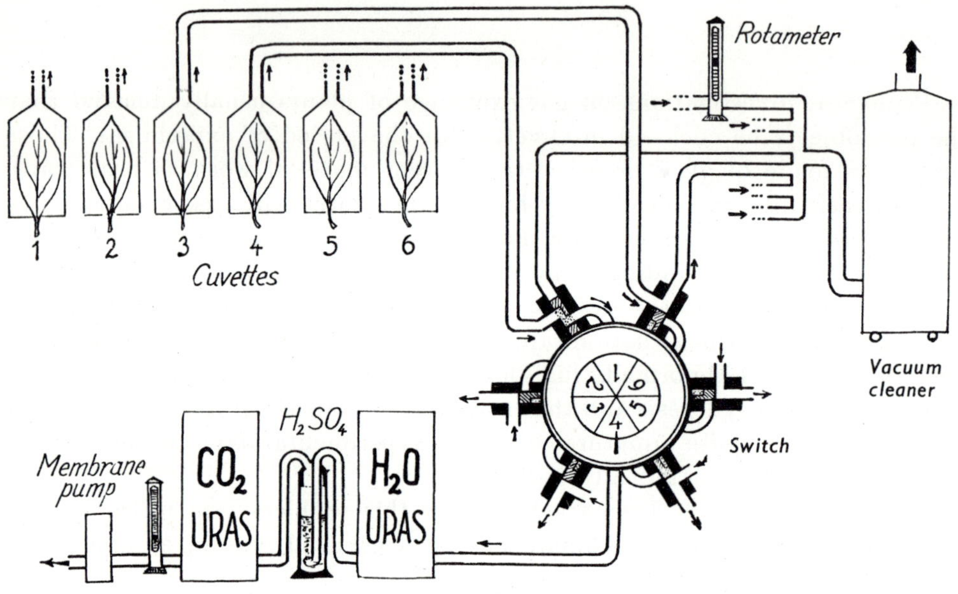

Fig. 1

As we are working with 6 checkpoints we have developed a special switching apparatus which automatically connects each of the 6 cuvettes one after another with the measuring equipment at intervals of 2 minutes, meanwhile the other 5 cuvettes are cooled by the stronger air flow.

We started working in this way during this summer and were able to avoid super-temperatures to a satisfactory degree. Even under strong and direct isolation with air temperatures above 30° C temperatures within the cuvettes did not exceed those of the external atmosphere by more than 1 to 2° C. More experience will have to be gathered, but we hope that this method will be a useful way of carrying out investigations in the field.

Discussion

H. Polster: Is it not possible that stomatal movements occur with the alternation of flow rate of 700—1,500 l./hr. and 70—120 l/hr. and that these affect the investigation?

G. Lerch: The rate of air flow is relatively small and always remains below 0·5 m./sec. There was no wind and the leaves in the cuvette did not move so that it is not necessary to assume that the stomata were affected.

P. G. Jarvis: We think it necessary to have two separately controlled air movements in our cuvettes, but they operate simultaneously. We consider it necessary to be able to vary the rate of flow of air leaving the cuvette for analysis, which is, of course, measured in units such as litres per hour, without altering the conditions at the leaf. Thus we have an internal circulation of air in the cuvette so that the boundary layer resistance to gas exchange is defined and reproducible. This air circulation should be measured in units such as meters per second. By changing the air flow through the cuvette just prior to measuring gas exchange you are altering completely the conditions at the leaf and I wonder if you get steady rates of exchange for measurement. Secondly, I would have thought that the high rates of air flow necessary to give adequate cooling would disturb gas exchange. Tranquillini, I think, finds that air flow rates of more than 4 m./sec. are necessary to give adequate cooling and such rates seem rather unnatural for most plant communities.

G. Lerch: In natural conditions the air speed is not constant and it must be only sufficient to prevent overheating. This year we confirmed that under our experimental conditions transpiration and photosynthesis rates were quite normal. We found no increase above the usual level. The transpiration rates thus measured were compared with values obtained by weighing cut leaf parts and we found very good agreement.

Discussion

GENERAL DISCUSSION

FRIDAY, OCTOBER 4

Chairman: *W. C. Visser*
Secretary: *Z. Šesták*

B. Slavík: Water relations like other physiological processes are controlled by plant qualities which have an ontogenetic character, i.e. they change regularly during individual development. This is valid not only for the ontogenesis of the whole plant but also for the ontogenesis of the individual organs and tissues. For example, with ageing there are changes in the sensitivity of the tissue to water stress, in the hydroreactivity of the stomata, in the relative rate of cuticular transpiration, in the absorption capacity of roots, etc.

It is even further true that during ontogenesis the causal relations and correlations between individual components of water relations are subject to change. For example, young leaves are preferentially supplied with water, leaves of the medium insertion level are more hydrotable than young and old leaves. And finally, it is without doubt that there is a change in quantitative relations between water relations and other processes, e.g. photosynthesis, mitotic and extension growth, respiration, etc. We have been told here of many important facts about these relations. I am of the opinion that the investigation of ontogenetic changes in these relationships and finding an explanation of their cause could provide a key to controlling plant productivity. I assume that into all our schemes of water flow and water potential distribution in the body of plants and its effect on physiological activity it will be necessary to include the ontogenetic time factor.

G. Meinl: In my contribution to discussion I would like to suggest some very interesting theoretical considerations on water stress and water relations for further investigation, from the viewpoint of agricultural research:

1. What methodological requirements, or improvements are necessary to detect the different specific behaviour of closely related varieties and forms in so far as concerns the properties we have been considering here? We can no longer be satisfied with the statistical significance of the differences obtained since even in applied research analytical investigation of the causes is important.

2. It is necessary to determine what interrelations exist between the individual organs of plants with regard to water relations.

3. What dependence or relationship is there between water supply and rhythmic processes (diurnal growth, mitotic frequencies, organ movement, etc.)?

We regard the question of the necessary transpiration rate as of cardinal importance. This has, hitherto, never been satisfactorily settled. The rate of dry matter production is clearly dependent on water supply but its relation to transpiration rate remains, however, despite this, relatively unexplained.

P. G. Jarvis: I think many of us are interested in how different species or ecotypes differ from one another in their response to soil water shortage.

We may regard the plant and drying soil as a system, composed of a series of resistances.

through which water moves from soil to atmosphere along a gradient of decreasing water potential. As plant physiologists we are interested in the water potentials which develop at various points in this pathway, especially in the plant. These do not enter directly into Dr. Visser's considerations. The water potential in which we are probably most interested is that which can and does develop in the leaves of plants to maintain a given flux when soil moisture becomes limiting.

As soil water content decreases, both the soil water potential and capillary conductivity decrease. To maintain a flux, dictated by the climatic conditions at the leaf, the water potential in the root must decrease and this results in a reduced leaf water potential. Following Gardner (1960), the leaf water potential is dependent as follows:

$$\frac{dq}{dt} = \frac{(\psi_{root} - \psi_{soil})}{R_{soil}} = \frac{(\psi_{leaf} - \psi_{root})}{R_{plant}}$$

$$\psi_{leaf} = \psi_{soil} + \frac{dq}{dt}(R_{soil} + R_{plant}) \qquad R_{soil} = \frac{I}{k.L.A}$$

$\dfrac{dq}{dt}$ = water flux, L = effective root length per unit volume of soil, ψ = water potential, R = resistance to water movement, A = constant, k = capillary conductivity of the soil at ψ_{soil}

We may summarize the ways in which plants may differ in their response, as follows:

1. Differences in root density. A larger density of roots results in lesser importance of the resistance in the dry soil to water movement to the roots. The pathway is shortened.

2. Differences in the resistance in the plant. This regulates the potential drop through the plant and thus determines how much the water potential in the leaf will be lower than that at the root surface.

3. Differences in the leaf in the relation between leaf water potential and turgidity. This varies widely. In plants such as *Lupinus* or tomato a large amount of water is lost for only comparatively small decreases in leaf water potential; for *Pinus* and *Picea*, loss of the same amount of water is associated with a much lower leaf water potential.

4. Differences in the relation between turgidity and stomatal aperture. This seems to vary greatly with the age and condition of the plant. Stomatal aperture, of course, determines gas exchange.

5. Differences in the direct sensitivity of metabolism to reduction in turgidity and water potential. For example, effect on photosynthesis, as shown by Dr. Slavík, or on respiration, not controlled by stomatal aperture.

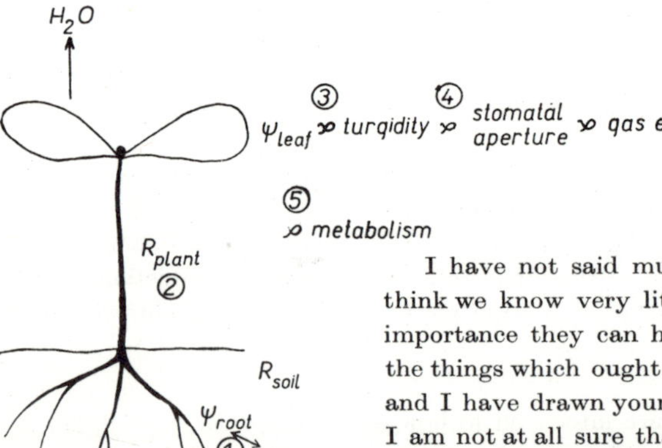

I have not said much about these five points because I think we know very little about how they can vary or what importance they can have. We think that these are some of the things which ought to be studied in plant water relations and I have drawn your attention to them in this way because I am not at all sure that many of the contributions we have heard are contributing to an understanding of these important sources of difference.

W. Larcher: The research concerning water stress in plants could become more efficient by better coordination of the different approaches in the study of this highly important ecophysiological problem.

1. Coordination of the information:

It is becoming increasingly difficult to survey the large amount of widespread literature, especially since the experimental ecological research is being done more and more by agricultural or forestry research stations, which frequently publish their results in special bulletins or even circulars, issued by these stations. We need a review journal (like the Excerpta Botanica) reporting short abstracts (by the authors themselves) covering the whole of experimental botany.

2. Coordination in methods and planning:

Even in the design of a research programme we should try to build up as many connecting links as possible with research work already published or in progress at other institutes. Such connecting links may be: Use of the same or similar plants, analogous research programmes, identical experimental techniques, etc. At least an exact description of the kind and the pretreatment of plants used in the experiments should be given. This should be as specific as possible, giving dimension parameters, i.e. such relationships as surface area to dry weight etc. Also all important particulars of the methods applied ought to be included. The person reading the paper should not have any doubts, if he would like to repeat the same experiment or if he likes to extend it to other species of plants. A worker in basic research who developes new methods of general importance should contrive to make these methods more widely applicable for applied research such as in agriculture and forestry (e.g. point out how the method may be adapted for ready use when comparing different species, varieties or proveniences) or at least he should call attention to the possibility of using these methods in applied experimental botany.

3. Coordination in evaluation:

Evaluation should consider, from as many viewpoints as possible, the requirements of different disciplines which might be interested in the observations and data presented. In basic physiological research, for example, there is a trend in recent years to express the results in absolute physical measuring units. In ecophysiology and applied botany, however, we prefer other measuring units, which are perhaps less exact but more expressive. In climatology — a science, able to furnish plant ecology and applied botany with many important data — some observations are again expressed in units, different from those used in experimental botany. The interrelations between the different disciplines dealing with plant water stress could become much closer, and often valuable results would be appreciated and evaluated to a further extent, if all the publications presented the necessary conversion factors (or at least approximative factors) from one system of measuring units to others, e.g. radiation energy units to illumination units, gas exchange values from leaf surface units to weight units, threshold values expressed also in climatological units etc. To reach this goal, considerable work must be done in some cases to establish exact relationships or equivalency (e.g. DPD — relative turgidity — water saturation deficit of tissues — "Hydratur" etc.).

Finally, the investigator should always clearly indicate the limits to applicability of the results. In general he will be able to evaluate these limitations better than the reader of the publication. In some cases the data obtained may be used quantitatively, but in other cases the results may only serve to establish trends or correlations between different processes (e.g. between water stress and growth, dry matter production etc.). It is especially important to communicate to which degree the presented results may be extended in later evaluations, calculations or applications (e.g. calculation of community water consumption using plant transpiration values, estimation of the behaviour of the whole plant from results of experiments using twigs).

We should try to evaluate our results in such a way that anyone, will be able to make maximum use of them, even if he persues a different line of research.

N. A. Gusev: During the Symposium we have listened to a whole series of interesting reports, clearing up questions of the origin of water deficit, changes in osmotic pressure, transpiration, water potential, etc.

It seems to me, however, that it would be necessary to pay more attention to processes which take place in living protoplasm, i.e. the hydration and swelling of colloids. This means paying increasing attention to colloid-chemistry and biochemistry in our research work. It is necessary to elaborate new methods and to work with new methods in which physics can be of very great assistance. In addition, it is necessary to make a closer study of the laws of thermodynamics, entropy and of chemical potentials, activity, etc. Such a study would permit the better investigation of the behaviour of water in the system: soil — plant — atmosphere.

I hope that such symposia will be organized more often since mutual contact of scientists from different countries can greatly enrich our knowledge.

W. R. Müller-Stoll: In summarizing the impressions gained during the Symposium, I should like to draw your attention to some general points of view which seem to be important for future work on water relations in plants.

The first point concerns the role of scientific logic in the interpretation of experimental results. Correlations between observed facts, found by way of purely statistical methods, in general do not permit the drawing of conclusions upon actually existing connections. For this purpose very thorough investigations will have to be made in order to clear up, step by step, all links of a causal chain.

The second problem concerns the evaluation of experimental results obtained under different conditions. Knowledge acquired by physiological experiments in general is based upon keeping all factors constant except one. Experimental ecology, however, has to deal with a complex constellation of factors each freely variable, the analysis of which has to be undertaken. A finding made by physiological experiment, therefore, does not justify the assumption that it is of the same importance under ecological conditions. We have to be very cautious in transferring physiological experiment to ecological relations and the reverse. This fact also renders the use of physiological results for practical needs a difficult problem and requires as extensive research as possible on the topic concerned.

In the third place, the connections between water relations and the basic metabolic phenomena in plants should be pointed out. We are sure that there are very close, but still widely unknown correlations. At present greater attention is being paid to the relations between water economy and photosynthetic activity (productivity). In addition, there are a lot of other problems to be investigated concerning, in particular, energy relations, e.g. reactions of the protoplasmatic system, respiratory pathways, protein metabolism, nucleic acids, nucleotides and enzymes on changes in the water situation. Research on causal connections between water relations and the different components of energy metabolism might be an important task for the future.

S. J. P. K. Bezuidenhout: I have listened with much interest to all the papers and discussions during this symposium, but I was unable to gather a complete picture on the whole problem of water stress in plants. In my view, the aspect neglected most was sufficient data showing the recovery of plants, rewatered after different periods of water stress. A simultaneous study of photosynthesis, respiration and transpiration may provide valuable data on the ability of plants to recover from water stress.

As both photosynthesis and transpiration are intimately related physiologically and since both are linked by the mechanism of the stomata, much more attention should be paid to the closing response of stomata during increased water deficit and re-opening after rewatering of the plants.

P. G. Jarvis: Of course the five points I have mentioned would be expected to vary with the age and condition of the plant. Consideration of them could also be used in comparing plants of different ages and conditions. Ideally, we should like to know how these properties varied both between and within ecotypes or species with age and conditions of the environment. This would be a lot of work.

R. Zwicker: I would like to point out that in studying water relations it is important to bear in mind relationships among transpiration, gutation and bleeding. We are working in Leipzig with questions of mineral nutrition and also with these excretion fluids, their amount and composition.

DISCUSSION ON TERMINOLOGY AND LINGUISTIC PROBLEMS

WEDNESDAY, OCTOBER 2

Chairman: *P. E. Weatherley*
Secretary: *Z. Šesták*

P. Strebeyko: The word "stress" has been used for many years to express a special state of the living body in the case of illness, indeed it is commonly used in medicine and animal physiology. This state is characterised by certain activities of the organism and some of these such as the speed of blood circulation or the activity of the brain provide useful indices of "stress".

Can we talk about water stress in plants? I am not so sure. Yet it ought to be possible because the plant is a living organism too and lack of water seems to constitute a pathological state. On the other hand can we talk about "moisture stress" in the soil? This is doubtful because, from the physiological point of view, it would refer to quite a different phenomenon.

If we agree that the "water stress" can exist in a plant, we must find some indices for it as in animal pathology. What indices can we find?

1. I suppose the first one could be respiration. It concerns the production of energy and is connected with active water intake.

2. The second index of the water stress could be the circulation of cytoplasm in the cell. This phenomenon is connected with respiration and expresses a state of sensitivity and excitation.

3. The third index could be stomatal movement. This phenomenon is complicated in its mechanism but it is closely connected with the water economy of the plant.

I put these three indices forward for discussion. They are based on a physiological point of view and give a physiological meaning for the term "stress".

B. Slavík: May I ask the English participants of our sessions to be so kind as to explain the real meaning of the word "stress", for I feel that this is rather a linguistic question than a terminological one?

P. E. Weatherley: I agree with Prof. Strebeyko that the word s t r e s s in animal physiology and medicine has been used to mean a certain physiological state of the organism, and that we plant physiologists seem to use it in a wider and perhaps more confused way. Physicists, however, know exactly what they mean by s t r e s s and it might be useful to examine their definition and to see how far it can be applied to our water stress problems.

In physics a s t r e s s is a system of forces applied to a body. The change in dimensions of the body resulting from the stress is called a s t r a i n. Thus if a wire is fixed at its upper end and loaded with weights at its lower end, the load is the stress, the resulting increase in length of the wire is the strain.

The strain may take a variety of forms according to the shape of the body and the mode of application of the forces. Thus the strain may be a change in volume if the body is uniformly compressed in all directions, a s h e a r if the forces are tangential, a t o r s i o n if the body is subjected to twisting forces, or a b e n d i n g in the case of a cantilever.

It seems to me that we have here a key to the definition of the term s t r e s s as applied to plants, but I think it must involve us in the use of the term s t r a i n too. By analogy with the purely physical systems just mentioned, water s t r e s s is the system of forces acting on the plant, whereas the response of the plant to these forces we should call the water s t r a i n.

On this definition the water stress would be measured in terms of lowering or depression of water potential. But in the context of the soil : plant : atmosphere system it is evident that the water stress is as heterogenous as the system itself. Thus the water stress in the soil surrounding the roots is usually of quite different magnitude from that of the atmosphere surrounding the leaves and these external stresses are bridged by the plant which develops what might be termed an internal stress connecting the two external stresses. In fact the soil : plant : : atmosphere system presents us with a catena of stresses. The water stress is usually lowest in the soil but there may be a gradient of increasing stress as we approach the roots. Inside the plant the stress increases as we pass from the roots to the xylem (hydrostatic tension) and again increases in the mesophyll. In the gas phase there is the steepest gradient of stress and the greatest value is reached in the atmosphere at large. Thus we cannot refer to t h e water stress in the system but must specify that part of the catena on which we are focussing attention.

The distinction between internal and external stress may be a useful one. The internal stress is, in a sense, a response to the external stress, but not necessarily a passive response. As I said the plant bridges two phases of the environment which are usually in very different states of water stress. The plant may control the position it takes up between these two, and furthermore the time relations of its responses is of interest to us as physiologists and ecologists. Again the internal stress may not simply have the pattern of a simple gradient from root to leaf, it is possible that there may be regions of the cell or parts of plant cut off from the main stream by low permeability barriers and these regions may respond much more slowly to changes in the external stress (e.g. cell vacuoles may respond more slowly than cell walls).

The external stress of the environment causes then a complex spatio-temporal pattern of internal stress in the plant. This internal stress is not the strain corresponding to the physical deformation of the body, the phenomena analogous with strain are the physiological responses following the development of the internal stress. And just as the physical strain may take many forms — so also may the physiological strain. It is here that the three indices: respiration, cytoplasmic circulation and stomatal aperture mentioned by Prof. Strebeyko come in. Others of varying importance could be added. Changes in cell volume, in cell permeability, in photosynthesis, in the girth of a tree trunk and in growth may all be manifestations of water strain as are the development of water deficits, the occurrence of wilting or even the death of the plant. It is these "physiological deformations" which constitute water s t r a i n in plants. Clearly the term w a t e r s t r a i n covers a wide range of phenomena, some of them measurable, others perhaps qualitative only. Some are of greater biological significance than others, indeed it may be that the water strain can only be assessed by studying several or even many of the strain phenomena. But since growth is the result of so many integrated processes its retardation may perhaps be the most significant measure of strain.

For physical systems the ratio of stress : strain is found to be a constant, e.g. the modulus of elasticity which defines an important property of the material under consideration. Would such a modulus have any significance in the physiological realm? It is clear that the water stress is so heterogeneous and the strain so complex that a precise ratio would be impossible to arrive at. But it would seem that there are two distinct possibilities: first the relationship between external stress (Se) and internal stress (Si). The ratio Se/Si [or (Se/Si—1)] might give an assessment of drought avoidance in the sense of the extent to which Si is "protected" against high values of Se. It would be a measure of the plant's homeostatic performance with respect to water stress. This would correspond with Stocker's 'constitutional' drought resistance. Si could be measured as the W.P.D. of the tissues, but evidently Se would need further definition: it could refer only to the environment as a whole if soil and atmosphere were contrived to be at the same W.P.D., otherwise both would have to be defined and possibly one kept at zero (soil or atmosphere at saturation). Also the necessary times to attain equilibria or steady states would have to be known.

The second possibility is the relationship between the internal stress and the strain it produces. With certain restrictions this is perhaps easier to measure. Thus for leaf tissue the water stress (DPD) : water deficit (relative turgidity) relationship (Weatherley and Slatyer, Nature, *179* : 1085, 1957) might be regarded in this light. Such a relationship can hardly be expressed as a single modulus, but the pattern of the relationship expresses the way in which the tissue suffers a "physiological deformation" in response to a given internal stress. This is similar to Stocker's p r o t o p l a s m i c drought resistance and Maximov' t r u e drought resistance.

Since the external stress, internal stress and strain are causally related:

$$Se \;\rightarrow\; Si \;\rightarrow\; strain$$

a third possibility is the relationship between the first and last i.e. Se : strain. This in a sense is the most practical relationship in that it is often more readily measurable and ultimately is of the greatest ecological significance, although it is of course a compound of the two single-step relationships and the ultimate elucidation of each separately is no doubt desirable. Here as in the study of all these stress : strain relationships the time aspect is of importance. The slowness of the development of a physiological strain in response to an external stress may be of importance; an excised *Pelargonium* leaf may remain exposed to the atmosphere at room temperature for days or even weeks if well illuminated, without seeming to suffer any irreversible strain. No doubt the stomata close and Si remains small so that this is in fact a case of stress avoidance, or more precisely internal stress only responding very slowly to an external stress.

Finally, it is evident that the strain is itself a complex of linked physiological processes whose interrelationships it may be desirable to investigate. A possible hypothetical example might be as follows:

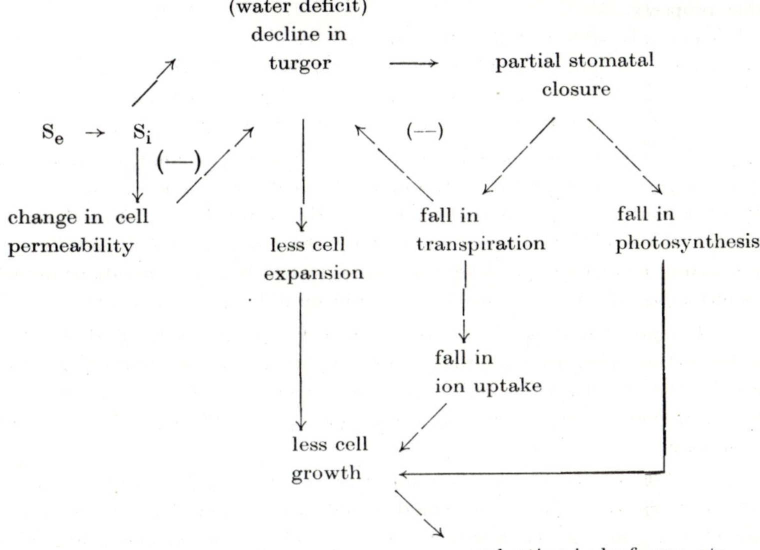

A study of the interrelationships between say water deficit and photosynthesis or transpiration and growth, however important they may be, are not in the category of stress : strain investigations and I suggest it is preferable to keep them separate.

To sum up: The precise definition of the term s t r e s s as used in physics can be applied in the plant realm but it necessarily follows that we must introduce the additional term s t r a i n. The water s t r e s s in any part of the environment : plant system is the water potential depression in that part. The corresponding strain is the physiological change (or physiological deformation) which results from the stress. It is this term s t r a i n which I suggest might be used for "stress" in Prof. Strebeyko's sense. The stress : strain modulus must in most cases be replaced by a more complex relationship often involving a time factor.

P. G. Jarvis: Professor Strebeyko's first two suggestions involve using respiration intensity as an index of water stress in the plant. We have found (see our contribution) that root tips of different species have different respiration intensities at the same water potential. Are we to say that those plants with the same respiration intensity suffer the same stress, or that respiration of the plants responds differently to the same stress (in this case, water potential)?

W. C. Visser: The problem of how to express the moisture status in the soil is not only a terminological one, but physical-physiological facts are also of importance. Stress as a physical expression hints at one single factor, but the relevant part of the water status for any problem will most probably be of a multifactorial nature. This follows directly for any flow function, where the velocity, V, is equal to the transmissibility kF times the moisture gradient $d\psi/dl$. Now V is equal to the change of quantity with time of $dQ/dt = kF\, d\psi/dl$. This means that any flow function relates quantity to gradient and the water status, therefore, has to be expressed in these studies with both quantity and stress.

A further point of importance is that the flow depends on a gradient — a difference of two stresses — where the level of the pressure is unimportant. For many problems — one of them the level of the yielding capacity of the plants as Mr. Neumann showed — is not related to the gradient but to the average stress. Here it is of importance to state how this stress should be defined, matrix stress, osmotic stress, etc. This shows that the stress as such has to be expressed in more detail, which emphasizes again that the moisture status has to be expressed as a multi-value property.

It seems advisable to follow the terminology of the physicists even if the subjects on which to determine the physical properties are of quite a different nature. The biologist had better accept a somewhat irrelevant existing nomenclature than devise names of his own and sever to a certain extent the link between fundamental disciplines.

P. G. Jarvis: I think it is clear from Dr. Visser's comments that the plant is essentially a physiological pathway for water in a physical environment. The environmental stress to which the plant is subject is solely physical. Hence we should define the physiological response in terms of the physical stress imposed by the environment. In addition, the physical basis gives a convenient reference level for studies in which several plants or processes are compared. If a physiological base were used this would be different in each case.

P. Strebeyko: I understand the point of view represented by Prof. Weatherley. I agree with Dr. Jarvis that there are some difficulties with respiration as an index. It should be emphasized, however, that the term "stress" should not mean quite different phenomena in plant physiology, in animal physiology and in human physiology. Physiological terminology must be held in common.

G. F. Makkink: In this Symposium on water stress which has been so well organized, there exists another stress which Dr. Slavík could not prevent. I mean the language stress. Where a stress exists there is also a deficit. Here the deficit is in understanding. I feel that this deficit in our symposium is quite considerable. How can we get rid of it? First, we must know its cause. I think it is due to an accidental fact, namely, that the foreign languages used are those which we have learnt at school. Which these were was, in fact, the choice of our grandfathers who drew up the curricula of the schools by law (at least in the Netherlands 100 years ago and I think in many other countries too). In their time our grandfathers were right: their space was that of the nation and its neighbours, their time seems to have been unlimited. We, however, are living in a world which is one and is already even adding interplanetary space, and we are always confronted with a lack of time. Is it not absurd to tolerate being the victims of our grandfathers in international communication? The language stress has two components: language diversity and language difficulty. We must reduce both. It is possible to learn at least one language common to all schools, but it is impossible to make national tongues easy, omitting, for instance, genders, cases and aspects from Russian or 30 vowels from English. Therefore, we must seriously consider the normalized language Esperanto which is, in my experience, very suitable for all our purposes.

How is it to get on the curricula of the schools? This cannot only be the work of the man in the street. It is the duty of people of university level whose insistent requests will finally be heard by our governments. We must use our scientific organizations for this.

Concluding Speech

by *Georg Hygen*

Since this is our last formal session, I may take the liberty of trying to express some of the feelings which I believe we all have at the conclusion of this Symposium.

One thing which struck me very strongly right from the beginning, was the feeling that we belong together. The very essence of science is unity and cooperation.

We are all aware of the embarrassing facts which decrease the permeability of national border membranes these days, and thereby hamper mutual contact. The more strongly do we feel obliged to Dr. Slavík and his extremely efficient and lovable staff of assisting angels of both sexes for all they have done to make this Symposium such a success.

The site of our conference has been particularly well suited to impress upon us the feeling of unitedness. For here in this magnificient city of Praha the very buildings spell out the alphabet of our common European culture.

It is a pity that the common language of this old cultural community has not been preserved also. Instead of resorting to Latin, which put all people on the same footing because it was native to none of them, we now have to cope with a much more difficult situation.

We who do not belong to the Slav language group are extremely grateful to our friends here for the very sincere and successful effort you have made to overcome the linguistic barrier.

May I also extend special thanks to our colleagues from the Soviet Union for coming here in spite of this barrier. It is true that we did not understand much of your talks in Russian, but all the same we have appreciated meeting you and working together on a common task.

If I may judge from my own limited experience, it seems to be much easier for Russians to learn Western languages than it is for us to learn Russian. (I have tried myself, but I had to give in because of lack of time, and I know many collegues who are in the same situation.)

I could not say whether this marked difference in linguistic capability may be due a more complicated construction of the Russian language, or to a more efficient structure of the linguistic centers in Russian brains. However this may be, it would be unrealistic not to recognize the factual situation. This means that the main effort to remove the linguistic barrier which still exists would have to be made on your side, although this does not by any means relieve us from the obligation to do what we can on our side.

I sincerely believe that future historians may find reason to blame us as scientists for working calmly in our laboratories and letting the politicians make such a mess of the whole world as they have made in our 20th century.

Being biologists I think we are bound to feel the principal unity of all humanity more strongly and deeply than people of most other professions.

This is both a privilege and an obligation. However small and insignificant our group may seem, we form a tiny rivulet among innumerable others who may eventually joint to form a spring-flood of goodwill sweeping all barriers away.

But by now I have certainly moved too far beyound the surface of the water in which we rightly belong during this Symposium. It may be high time to dive back into our safe little pool, where we do not risk to suffer drought damage from political storms.

From the layman's viewpoint the elucidation of the water relations of plants would probably seem to be one of the simplest problems any scientist could possibly work with. Indeed, one should think that no other process in the living world might be more smoothly controlled and registered than the simple uptake and loss of water from higher plants.

However, we all know from bitter experience that this task is very far from being simple.

When we cast a glance backwards through history, as Professor Němec induced us to do in his engaging opening speech, we find that in spite of the enormous amount of so-called facts which has been accumulated since the times of Sachs and Pfeffer, many of the most controversial problems of those remote days still remain open.

In this connection I may ask you, Dr. Slavík, to bring our warmest thanks and regards to Professor Němec, who himself represents a living personification of the truly great tradition in Czechoslovakian botany which runs from the days when Sachs himself worked here in Praha, right up to our present Symposium. Please tell Professor Němec that we were very pleased and honoured to see him among us.

I must also ask you to bring our sincere thanks to the Academy of Sciences and to the Director of the Institute of Experimental Botany, both for the active part they have taken in making this Symposium possible, and also for the very nice reception they gave us on Monday.

This Symposium has, I think, brought us a little step forward. Perhaps we have not solved so many of the open problems, but we have at least realized more fully the extension and depth of our ignorance, and this may in itself form a firmer base for further advance. Upon this background, I fully agree with Dr. Bezuidenhout, who suggested to me the other day that we ought to come together again in a not too distant future. May I be allowed, Mr. Chairman, to take a rapid vote on this suggestion. —

No-one seems to be against, so Dr. Bezuidenhout's suggestion is unanimously approved. Good. But I do not think that we are in a position today to discuss how this idea could possibly be put into practice, or where a future Symposium could most conveniently be held. With your permission, I may therefore take upon myself the task of acting as a sort of interim secretary, with the purpose of keeping contact with a key man from each of the countries represented here, and also from a number of others, in order to investigate the possibility of a future Symposium.

In my opinion, 1966 would be the most suitable year for such an event, after the giant congress in Edinburgh next summer has been well digested.

But let us leave the future for the moment and return to the present. And then I must end where I began, by expressing our sincere gratitude and admiration for the arrangement of this conference. I think this arrangement may best be characterized by a single small word which comprises everything: P e r f e c t. — Just perfect.

There are many names I would like to mention specially in this connection, but I dare not do it for fear of omitting the many helpers who have mainly worked behind the screen. However, one name stands out in our minds above all, the name of Slavík.

You may perhaps have considered Dr. Slavík and his staff as a team of master conjurers because they have been able to remove every obstacle for the conference and every stone from our individual paths. But I assure you that such a result cannot be achieved by tricks. Those of us who have some little experience in a similar line, are fully able to appreciate the great amount of effective planning and hard work which has been going on here for our benefit.

But work is not enough. The success of a conference like this one depends above all on the rather undefinable factor which we call the atmosphere, and this elusive substance cannot be produced to order, for it emanates only from warm human hearts.

You said yesterday, Dr. Slavík, that you found it difficult to express what you had to say in a foreign language. For my part, I should find it difficult to express what I would have liked to say on this occasion in any language. (Perhaps that is why I have used such a long time in trying to do it.)

But in the end let me again resort to a few small and simple words which we direct personally to every one of your helpers, and above all to yourself: Děkuji za všechno! Thanks for all.

Conclusion

by *Bohdan Slavík*

Our last working session will shortly come to an end. This mainly implies the ending of the mutual discussions of our common problems. This, of course, does not mean that we shall end all our discussions in reality, since during the excursion in the course of the next few days we shall certainly use every opportunity, every suitable and unsuitable moment and place, to return to our beloved questions and problems.

Nevertheless, I feel that this is a suitable occasion to thank most sincerely all who have taken part in this Symposium. I do this for three reasons: First, that you came here to take part in this Symposium. We are very glad that we have been able to welcome you. Secondly, I wish to thank you for your enthusiastic participation in all our discussions and for bringing with you papers with very interesting and stimulating results and dealing with important problems. Thirdly, that you have contributed to creating such a pleasant atmosphere of cooperation and mutual understanding of which Professor Hygen spoke just before me.

I thank you once again from my whole heart and feel that we have not only established scientific contact but that there is a real friendship between us. We believe that this friendship will continue and that it will result in the not too distant future in a Second Symposium on Water Stress.

Good-bye.

AUTHOR INDEX

Page numbers in *italics* refer to the bibliography.